CLINICAL ISSUES IN ANATOMY

FREDERIC H. MARTINI, PH.D.
&
KATHLEEN WELCH, M.D.

with

WILLIAM C. OBER, M.D.
ART COORDINATOR AND ILLUSTRATOR

CLAIRE W. GARRISON, R.N.
ILLUSTRATOR

HUMAN ANATOMY, FIFTH EDITION
MARTINI / TIMMONS / TALLITSCH

PEARSON

Benjamin
Cummings

San Francisco   Boston   New York
Cape Town   Hong Kong   London   Madrid   Mexico City
Montreal   Munich   Paris   Singapore   Sydney   Tokyo   Toronto

Editorial Director:	*Daryl Fax*
Executive Editor:	*Leslie Berriman*
Project Editor:	*Nicole George O'Brien*
Editorial Assistant:	*Blythe Robbins*
Project Management, Design, Composition:	*Elm Street Publishing Services, Inc.*
Managing Editor:	*Deborah Cogan*
Production Supervisor:	*Jane Brundage*
Manufacturing Buyer:	*Stacey Weinberger*
Executive Marketing Manager:	*Lauren Harp*
Text Printer:	*R.R. Donnelly, Willard*

ISBN 0-8053-7218-0

PEARSON

Benjamin Cummings

www.aw-bc.com

1 2 3 4 5 6 7 8-DOW-09 08 07 06 05

CONTENTS

PREFACE

This *Clinical Issues in Anatomy* supplement is designed to demonstrate the importance of anatomy in clinical practice. The enclosed sections include discussions of disorders that impact or alter anatomical features and diagnostic procedures whose interpretation relies on an understanding of normal human anatomy. By placing these discussions in a separate supplement, rather than incorporating them into the textbook, these topics can be considered in greater depth, in a format that facilitates its use as a portable reference.

The organization of the *Clinical Issues (CI)* supplement corresponds to that of the Fifth Edition of *Human Anatomy*, by Martini, Timmons, and Tallitsch. Clinical topics related to each chapter are grouped within the corresponding chapter in the CI supplement. Within the CI, when a topic is cross-linked to specific sections of the textbook, the related page number is indicated as *HA, p.000*. On that page in the textbook, this discussion is indicated by a green "**CI:**" followed by the title of the CI section. The CI also contains a series of radiological scans referenced in figure captions throughout the text using the same indicator.

Few instructors will cover all the material in the *Clinical Issues* supplement. Because courses differ in their emphases and students differ in their interests and backgrounds, the goal in designing the supplement has been to provide maximum flexibility of use. The diversity of applied topics offers instructors many opportunities to integrate the treatment of normal anatomy, clinical anatomy and pathology, and other health-related topics. Readings from the CI that are not covered in class can be assigned, recommended, or used by individual students for reference. Experience indicates that each student will read those selections that deal with disorders affecting friends or family members, address topics of current interest and concern, or include information relevant to a chosen career path.

ACKNOWLEDGMENTS

We would like to thank Bill Ober, M.D., and Claire Garrison, R.N., who prepared the illustrations for this project. We also express our thanks to Executive Editor Leslie Berriman, Project Editor Nicole George-O'Brien, and to Elm Street Publishing's Heather Johnson and Eric Arima.

Frederic Martini, Ph.D.
martini@maui.net

Kathleen Welch, M.D.
kwelch@maui.net

DISEASE, PATHOLOGY, AND DIAGNOSIS   *HA p. 4*

The formal name for the study of disease is **pathology**. Different diseases typically produce similar signs and symptoms. For example, a person whose lips are paler than normal and who complains of a lack of energy and breathlessness might have (1) respiratory problems that prevent normal oxygen transfer to the blood (as in *emphysema*), (2) cardiovascular problems that interfere with normal blood circulation to all parts of the body (heart failure); or (3) an inability to transport adequate amounts of oxygen in the blood, due to blood loss or problems with blood formation. In such cases, doctors must ask questions and collect information to determine the source of the problem. The patient's history and physical exam may be enough for diagnosis in many cases, but laboratory testing and imaging studies such as x-rays are often needed.

A **diagnosis** is a decision about the nature of an illness. The diagnostic process is often a process of elimination, in which several potential causes are evaluated and the most likely one is selected. This brings us to a key concept: *All diagnostic procedures presuppose an understanding of the normal structure and function of the human body.*

■ THE DIAGNOSIS OF DISEASE
HA p. 4, 322

Homeostasis is the maintenance of a relatively constant internal environment suitable for the survival of body cells and tissues. A failure to maintain homeostatic conditions constitutes **disease**. The disease process may initially affect a specific tissue, an organ, or an organ system, but it will ultimately lead to changes in the function or structure of cells throughout the body. Some diseases can be overcome by the body's defenses. Others require intervention and assistance. For example, when trauma has occurred and there is severe bleeding or damage to internal organs, surgical intervention may be necessary to restore homeostasis and prevent fatal complications.

A person experiencing serious symptoms usually seeks professional help and thereby becomes a patient. The clinician, whether a nurse, a physician, or an emergency medical technician, must determine the need for medical care on the basis of observation and assessment of the patient's symptoms and signs. This is the process of diagnosis: the identification of a pathological process by its characteristic symptoms and signs.

▢ SYMPTOMS AND SIGNS   *HA p. 14*

An accurate diagnosis, or the identification of the disease, is accomplished through the observation and evaluation of symptoms and signs.

A **symptom** is the patient's perception of a change in normal body function. Examples of symptoms include nausea, fatigue, and pain. Symptoms are difficult to measure, and a clinician must ask appropriate questions, such as the following:

CHAPTER 1
AN INTRODUCTION TO CLINICAL ANATOMY

"When did you first notice this symptom?"

"What does it feel like?"

"Does it come and go, or does it always feel the same?"

"Does anything make it feel better or worse?"

The answers provide information about the location, duration, sensations, recurrence, and triggering mechanisms of the symptoms important to the patient.

A **sign** is a physical manifestation of a disease, and this is one reason why clinical anatomy is an important discipline. Unlike symptoms, signs can be measured and observed through sight, hearing, or touch. The yellow color of the skin caused by liver dysfunction and a detectable breast lump are signs of disease. A sign that results from a change in the structure of tissue or cells is called a **lesion**. We shall consider lesions of the skin in detail in a later section dealing with the integumentary system.

The physical examination is a basic and vital part of the diagnostic process. Common techniques used in physical examination are *inspection* (viewing), *palpation* (touching), *percussion* (tapping and listening), and *auscultation* (listening):

- **Inspection** is careful observation. A general inspection involves examining body proportions, posture, and patterns of movements. Local inspection is the examination of sites or regions of suspected disease. Of the four components of the physical exam, inspection is often the most important, because it provides a large amount of useful information. Many diagnostic conclusions can be made on the basis of inspection alone; most skin conditions, for example, are identified in this way. A number of endocrine problems and inherited metabolic disorders can produce changes in body proportions. Many neurologic disorders affect speech and movement in distinctive ways.

- **Palpation** is the clinician's use of the hands and fingers to feel the patient's body. This procedure provides information about skin texture and temperature, the presence and texture of abnormal tissue masses, the pattern of the pulse, and the location of tender spots. Once again, the procedure relies on an understanding of normal anatomy. In one spot, a small, soft, lumpy mass is a salivary gland; in another location, it could be a tumor. A tender spot is important in diagnosis only if the observer knows what organs lie beneath it.

- **Percussion** is tapping with the fingers or hand to obtain information about the densities of underlying tissues. For example, when tapped, the chest normally produces a hollow sound, because the lungs are filled with air. That sound changes in pneumonia, when the lungs contain large amounts of fluid. To get the clearest chest percussions, the fingers must be placed in the right spots.

- **Auscultation** (aws-kul-TĀ-shun; *auscultare*, to listen) is listening to body sounds, typically with a stethoscope. This technique is particularly useful for checking the condition of the lungs during breathing. The wheezing sound heard in people with asthma is caused by a constriction of the airways, and pneumonia produces a gurgling sound, indicating that fluid has accumulated in the lungs. Auscultation is also important in diagnosing heart conditions. Many cardiac problems affect the sound of the heartbeat or produce abnormal swirling sounds during blood flow.

Every examination also includes measurements of certain vital body functions, such as the body temperature, weight, blood pressure, respiratory rate, and heart (pulse) rate. The results, called **vital signs**, are recorded on the patient's chart. Vital signs can vary over a normal range that differs according to the age, sex, and general condition of the individual. Table 1 indicates the representative ranges of vital signs in infants, children, and adults.

The medical history and physical examination may not provide enough information to permit a precise diagnosis. Diagnostic procedures can then be used to focus on abnormalities revealed by the history and physical examination. For example, if the chief complaint is knee pain after a fall, and the examination reveals swelling and localized, acute pain on palpation, the **preliminary diagnosis** may be a torn cartilage. An x-ray, MRI scan, or both may be performed to determine more precisely the extent of the injury and to ensure that there are no other problems, such as broken bones or torn ligaments. With the information the diagnostic procedure provides, the **final diagnosis** can be made with reasonable confidence. Diagnostic procedures extend, rather than replace, the physical examination.

Two general categories of diagnostic procedures are performed:

1. **Tests performed on the individual.** Information about representative tests of this type that provide anatomical information is summarized in Table 2. These procedures allow the clinician to visualize internal structures (endoscopy; x-rays; scanning procedures such as CT, MRI, and radionucleotide scans; ultrasonography; mammography). There are also a variety of functional tests that monitor physiological processes or assess the patient's homeostatic responses in other ways.

2. **Tests performed in a clinical laboratory on tissue samples, body fluids, or other materials collected from the patient.** Because these are outside the scope of clinical anatomy, we will not explore these in detail.

Many of the diagnostic procedures and disorders noted in Table 2 will be unfamiliar to you now. The main purpose here is to give you an overview; you can refer to Table 2 as needed throughout the course.

TABLE 1	**Normal Range of Values for Resting Individuals by Age Group**		
Vital Sign	Infant (3 months)	Child (10 years)	Adult
Blood pressure (mm Hg)	90/50	90–125/60	95/60 to 140/90
Respiratory rate (per minute)	30–50	18–30	8–18
Pulse rate (per minute)	70–170	70–110	50–95

TABLE 2   Representative Diagnostic Tests, Their Principles, and Their Uses

Procedure	Principle	Examples of Uses
Endoscopy	Insertion of fiber-optic tubing into a body opening or through a small incision (laparoscopy and arthroscopy); permits visualization of a body cavity or the interior of an organ; allows direct visualization and biopsy of structures and detection of abnormalities of surrounding soft tissue	*Bronchoscopy:* bronchi and lungs *Laparoscopy:* abdominopelvic organs *Cystoscopy:* urinary bladder *Esophagoscopy:* esophagus *Gastroscopy:* stomach *Colonoscopy:* colon *Arthroscopy:* joint cavity
Standard X-rays	A beam of x- passes through the body and then strikes a photographic plate; radiodense tissues block x-ray penetration, leaving unexposed (white) areas on the film negative	Limb bones: to detect fracture, tumor, growth patterns Chest: to detect tumors, pneumonia, atelectasis, tuberculosis Skull: to detect fractures, sinusitis, metastatic tumors Mammogram: x-rays of each breast taken at different angles for early detection of breast cancer and other masses, such as cysts
Contrast X-rays	X-rays taken after infusion or ingestion of radiodense solutions (Scan 10c)	Barium swallow (upper GI): series of x-rays after the ingestion of barium, to detect abnormalities of esophagus, stomach, and duodenum Barium enema: series of x-rays after barium enema, to detect abnormalities of colon IV pyelography: series of x-rays after intravenous injection of radiopaque dye filtered by kidneys; reveals abnormalities of kidneys, ureters, and bladder; allows assessment of renal function
Digital subtraction angiography	Produces strikingly clear images of blood vessel distribution by computer analysis of images taken before and after dye infusion	Analysis of blood flow to the heart, kidneys, and brain to detect blockages and restricted circulation
Computerized tomography (CT or CAT)	Produces cross-sectional images of body area viewed; together, all sections can produce a three-dimensional image for detailed examination. (Scan 9)	CT scans of the head, abdominal region (liver, pancreas, kidney), chest, and spine, to assess organ size and position, to determine progression of a disease, and to detect abnormal masses
Spiral CT scans	Produce three-dimensional images by computer reconstruction of CT data (Scans 10a,b and 10d)	Often a research tool, but of clinical use at large regional hospitals and universities
Nuclear scans	Radioisotope ingested or injected into the body becomes concentrated in the organ to be viewed; gamma radiation camera records image on film. Area should appear uniformly shaded; dark or light areas suggest hyperactivity or hypoactivity of the organ	Bone scan: to detect tumors, infections, and degenerative diseases Scans of the brain, heart, thyroid, liver, spleen, and kidney, to assess organ function and the extent of many diseases
Radioactive iodine uptake test (RAI)	Radioactive labeled iodine compound is given orally; thyroid scans are taken to determine percentage uptake of radioiodine by thyroid gland	Aids in the determination of hyperthyroidism and hypothyroidism and in detection of thyroid nodules
Positron emission tomography (PET)	Radioisotopes are given by injection or inhalation; gamma detectors absorb energy and transmit information to computers to generate cross-sectional images	Used to measure metabolic activity of heart and brain and to analyze blood flow through organs Primarily a research tool; rapid functional MRI more widely used in clinical settings
Magnetic resonance imaging (MRI)	A magnetic field is produced to align hydrogen protons and is then exposed to radio waves that cause the aligned atoms to absorb energy. The energy is later emitted and captured to produce an image (Scans 1–8)	Gives excellent contrast of normal and abnormal tissue; reveals extent of tumors, demyelination and other brain and spinal cord abnormalities, obstructions or aneurysms in arteries, and ligaments and cartilages at joints
Ultrasonography	A transducer contacting the skin or other body surface sends out sound waves and then picks up the echoes	Avoids x-ray exposure, used to view soft tissues not shielded by bone throughout the body. Used in obstetrics to detect ectopic pregnancy, determine size of fetus, and check fetal rate of growth; upper abdominal ultrasound detects gallstones, visceral abnormalities, and measures kidneys

TABLE 2	**Representative Diagnostic Tests, Their Principles, and Their Uses** *(Continued)*	
Echocardiography	Ultrasonography of the heart	Used to assess the structure and function of the heart and heart valves
Electrocardiography (ECG)	Graphed record of the electrical activity of the heart, using electrodes on the skin surface	Useful in detection of arrhythmias, such as premature ventricular contractions (PVCs) and fibrillation, and to assess damage after a heart attack
Electroencephalography (EEG)	Graphed record of electrical activity in the brain through the use of electrodes on the surface of the scalp	Analysis of brain wave frequency and amplitude aids in the diagnosis of tumors and seizure disorders
Electromyography (EMG)	Graphed record of electrical activity resulting from skeletal muscle contraction, using electrodes inserted into the muscles	Determination of neural or muscular origin of muscle disorder; aids in the diagnosis of muscular dystrophy, pressure on spinal nerves, and peripheral neuropathies
Pulmonary function tests	Measurement of lung volumes and capacities by a spirometer or other device	Aids in the differentiation between obstructive and restrictive lung diseases; used to test for and monitor asthma
Cytology	Removal of cells for laboratory analysis	Detects precancerous cells or infections; most often used to assess mucosal cells of cervix (Pap smear)

■ THE PURPOSE OF DIAGNOSIS   *HA p. 21*

Two hundred years ago, a physician would arrive at a diagnosis and consider the job virtually done. Once the diagnosis was made, the patient and family would know what to expect. In effect, the physician was more of an oracle than a healer. Wounds could be closed and limbs amputated, but few effective treatment options were available. Less obvious diagnosis than trauma often reflected the culture and beliefs of the era. Curses and bewitching vied with "unbalanced humours" as explanations of disease. Therapy was often a combination of bleeding (often performed by barbers rather than by surgeons), dietary changes, and herbal medicines (often laxatives). Strong laxatives might have helped in cases of intestinal parasites, but the combination of bleeding and laxatives was potentially dangerous because it reduced both blood volume and blood pressure.

Fortunately, a vast array of treatment options guided by a rational, accurate diagnosis are available today. A modern physician addressing a new problem presented by a patient follows the *SOAP* protocol:

*S is for **subjective.*** The clinician obtains subjective information from the patient and the medical history.

*O is for **objective.*** The clinician performs the physical examination and obtains objective information about the physical condition of the patient. The examination may include the use of diagnostic procedures.

*A is for **assessment.*** The clinician arrives at a diagnosis and, if necessary, reviews the literature on the condition. A preliminary conclusion as to the **prognosis** (probable outcome) is made.

*P is for **plan.*** A treatment plan is designed. This can be very simple (lose weight, exercise, and take two aspirin) or highly complex (referral for radiation, chemotherapy, or surgery). If the treatment is complex, one or more treatment options are usually reviewed with the patient and, in many cases, the patient's family. Treatment begins only after informed decisions are made.

The SOAP protocol is both simple to remember and remarkably effective.

The primary goal of an introductory anatomy course is to provide you with the foundation for other, more specialized courses. In the unit of this manual that deals with body systems, you will be introduced to clinical conditions that demonstrate the relationships between normal and pathological anatomy. The goal is to acquaint you with the mechanics of the process involved. This knowledge will not enable you to make accurate clinical diagnoses; situations in the real world are much more complicated and variable than the examples provided here. Making an accurate clinical diagnosis is generally a complex process that demands a far greater level of experience and training than this course can provide.

DISORDERS AND DIAGNOSIS AT THE CELLULAR LEVEL

In this section we will introduce the concept of pathogens, foreign cells and cell products that can cause disease. We will also consider the ways these pathogens, as well as normal and abnormal body cells, can be examined in detail.

THE NATURE OF PATHOGENS   *HA p. 27*

The cellular organization of a "typical" cell described in Chapter 2 of the text is that of a *eukaryotic* cell (ū-kar-ē-OT-ik; *eu*, true + *karyon*, nucleus). The defining characteristic of eukaryotic cells is the presence of a nucleus. All eukaryotic cells have similar membranes, organelles, and methods of cell division. All multicellular animals, plants, and fungi (plus many single-celled organisms) are composed of eukaryotic cells.

The eukaryotic plan of organization is not the only one in the living world, however. Some organisms do not consist of eukaryotic cells. These organisms are of great interest to us, because they include many of the **pathogens** that are recognized causes of human diseases. Representative pathogens are introduced in Figure 1●.

▣ BACTERIA

Prokaryotic cells do not have nuclei or other membranous organelles. Nor do they have a cytoskeleton, and their cell membranes are typically surrounded by a semirigid cell wall made of carbohydrate and protein.

Bacteria are probably the most familiar prokaryotic cells. They are generally less than 2 μm in diameter. Many bacteria are quite harmless, and many more—including some that live within our bodies—are beneficial to us in a variety of ways. Other bacteria are dangerous pathogens that, given the opportunity, will destroy body tissues. These bacteria are dangerous because they absorb nutrients and release enzymes that damage cells and tissues. A few pathogenic bacteria also release toxic chemicals. Bacterial infections are responsible for many serious diseases, as indicated in Table 3. We consider these and other bacterial infections in various chapters of the text and in other sections of the *Clinical Issues* supplement.

Figure 1a● shows the structure of a representative bacterium. Figure 2● shows the three basic shapes of bacteria: round, rodlike, and spiral. A round bacterium is called a **coccus** (KOK-us; plural, *cocci*, KOK-sē). A rodlike bacterium is a **bacillus** (ba-SIL-us; plural, *bacilli*, ba-SIL-ē). Shapes of spiral bacteria vary, and so do their names. A **vibrio** (VIB-rē-ō) is comma shaped; a **spirillum** (spi-RIL-um; plural, *spirilla*) is rigid and wavy; and a **spirochete** (SPĪ-rō-kēt) is shaped like a corkscrew.

Some cocci and bacilli form groupings of cells. The Latin names used to describe these groupings are also used to identify specific bacteria. For instance, pairs of cocci are called *diplococci* (diplo-, double). *Streptococci* and *streptobacilli* form twisted chains of cells (strepto-, twisted), and *staphylococci* look like a bunch of grapes (staphylo-, grapelike).

CHAPTER 2
THE CELL

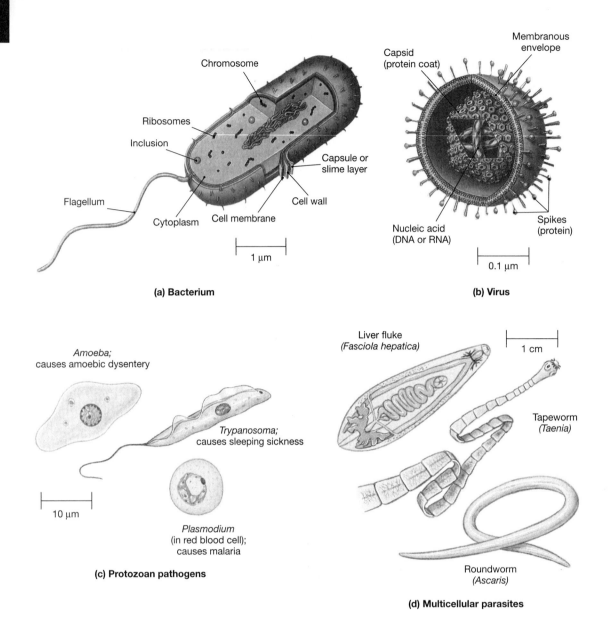

(a) Bacterium

(b) Virus

(c) Protozoan pathogens

(d) Multicellular parasites

● **FIGURE 1**

Representative Pathogens. (a) A bacterium, with prokaryotic characteristics indicated. Compare with Figure 2.3 (*HA p. 28*), which shows a representative eukaryotic cell. **(b)** A typical virus. Each virus has an inner chamber containing nucleic acid, surrounded by a protein capsid or an inner capsid and an outer membranous envelope. The herpes viruses are enveloped DNA viruses; they cause chicken pox, shingles, and herpes. **(c)** Protozoan pathogens. Protozoa are eukaryotic single-celled organisms, common in soil and water. **(d)** Multicellular parasites. Several groups of organisms are human pathogens, and many have complex life cycles.

Coccus   Vibrio   Bacillus   Spirillum   Spirochete

● **FIGURE 2**

Common Bacterial Shapes

TABLE 3	Examples of Bacterial Diseases and the Primary Organ Systems Affected	
Organism	**Disease**	**Affected Organ System**
Bacilli		
Bacillus anthracis	Anthrax	Integumentary and respiratory systems
Mycobacterium tuberculosis	Tuberculosis	Respiratory system
Corynebacterium diphtheriae	Diphtheria	Respiratory system
Cocci		
Staphylococcus aureus	Various skin infections	Integumentary system
Streptococcus pyogenes	Pharyngitis (strep throat)	Respiratory system
Neisseria gonorrhoeae	Gonorrhea	Reproductive system
Vibrios		
Vibrio cholerae	Cholera	Digestive system
Spirochetes		
Treponema pallidum	Syphilis	Reproductive and nervous systems
Borrelia burgdorferi	Lyme disease	Skeletal system (joints)
Rickettsias		
Rickettsia prowazekii	Epidemic typhus fever	Cardiovascular and integumentary systems
Coxiella burnetii	Q fever	Respiratory system
Chlamydias		
Chlamydia trachomatis	Trachoma (eye infections)	Integumentary system
	PID (pelvic inflammatory disease)	Reproductive system

■ VIRUSES

Another type of pathogen conforms neither to the prokaryotic nor to the eukaryotic organizational plan. These tiny pathogens, called **viruses**, are not cellular. In fact, when free in the environment, they do not show any of the characteristics of living organisms. They are classified as **infectious agents**—factors that cause infection—because they can enter cells (either prokaryotic or eukaryotic) and replicate themselves.

Viruses consist of a core of nucleic acid (DNA or RNA) surrounded by a protein coat called a *capsid*. (Some varieties have an *envelope*, a membranous outer covering, as well.) The structure of a representative virus is shown in Figures 1b and 3●. Important viral diseases include influenza (flu), yellow fever, some leukemias, AIDS, hepatitis, polio, measles, mumps, rabies, herpes, and the common cold (Table 4).

■ PRIONS

Recently, it was determined that certain rare and previously mysterious conditions making up the *transmissible spongiform encephalopathic (TSE)* disease are caused by a novel class of infectious agents called prions. **Prions** [from *protein infectious "ions"* (particles)] are unique among agents of transmissible disease, because they contain no nucleic acids (either DNA or RNA). Rather, they appear to be abnormal three-dimensional forms of the ordinarily harmless protein

PrPc found in cells. Apparently, an abnormally folded protein can serve as a template for converting normal proteins to the pathogenic form. When present in large quantities, these proteins cause degenerative cellular gaps in brain tissue, which takes on a microscopic "spongy" appearance. Partially metabolized fragments of abnormal prion proteins may form microscopic deposits called amyloid plaques in the brain as well.

The same clinical TSE disease can have a genetic, a sporadic, or an infectious origin. Rare genetic variations in the PrPc protein cause inheritable forms of TSE. Sporadic cases may come from spontaneous change of the PrPc protein to the abnormal shape. In addition, the disease can be transmitted from unrecognized infected individuals to recipients of corneal transplants or pituitary hormone extracts. Some are also known to have contracted TSE diseases from exposure to contaminated medical instruments or by eating affected tissues. While TSE diseases are not transmitted by normal household interactions, they can be acquired by contact with the abnormal proteins.

The first recognized human prion disease was *kuru*, a deadly disease affecting members of a society in New Guinea that practiced ritual cannibalism. The prions were passed from person to person when uninfected individuals ate infected brains. The infection, which could lead to death within a year, caused half of all childhood and adult deaths in the affected part of New Guinea. Other known prion diseases include

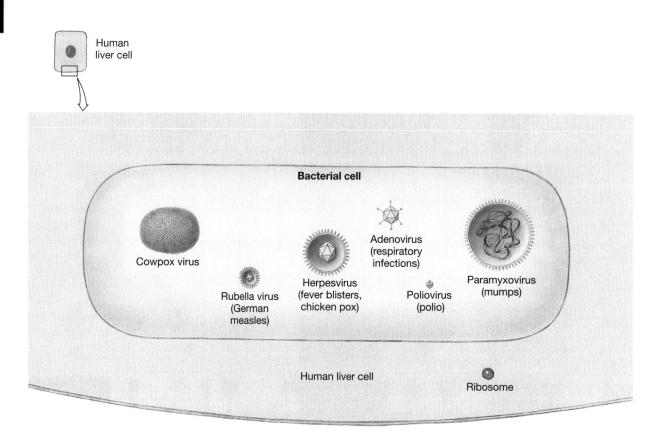

● **FIGURE 3**
Viruses. A variety of viruses, shown with a typical bacterial cell, a human liver cell, and a ribosome for scale.

TABLE 4	Examples of Viral Diseases and the Primary Organ Systems Affected		
Nucleic Acid	**Virus**	**Disease**	**Affected Organ System**
RNA	Influenza A, B, C	Flu	Respiratory system
	Paramyxovirus	Mumps	Digestive and reproductive systems
	Hepatitis A, C	Infectious hepatitis	Digestive system (liver)
	Rhinovirus	Common cold	Respiratory system
	Human immunodeficiency virus (HIV)	AIDS	Lymphatic system
DNA	Herpesvirus		
	Herpes simplex 1	Cold sore/fever blister	Integumentary system
	Herpes simplex 2	Genital herpes	Reproductive system
	Varicella-zoster	Chicken pox	Integumentary system
	Varicella-zoster	Shingles	Nervous system
	Hepatitis B	Hepatitis	Digestive system (liver)
	Epstein–Barr	Mononucleosis	Respiratory and lymphatic systems

inherited and sporadic *Creutzfeldt–Jakob disease* (which usually affects older people) and *fatal familial insomnia.* Prion infection also occurs in domesticated animals. In sheep, the condition is called *scrapie;* in cows, it is called *bovine spongiform encephalopathy (BSE).* Infected cows ultimately develop an assortment of strange neurological symptoms (such as pawing at the ground and exhibiting difficulty in walking), giving the condition the common name "mad-cow disease."

In 1995, European researchers reported a puzzling variant of Creutzfeldt–Jakob disease among teenagers and young adults. A number of fatal cases in England showed brain changes similar to those of BSE, leading investigators to attribute the outbreak to the consumption of meat products from prion-infected cows. This discovery led to a temporary ban on British beef from the European community, the slaughter and destruction of infected and potentially infected cows, and a change in livestock feeding practices throughout the world. Presumably, many cows became infected by eating feed containing beef by-products and bone meal contaminated with prions from the neural tissue of infected animals. The use of such feed has now been banned, and greater care is taken when butchering to prevent contact of brain and spinal neural tissue with meat intended for consumption. In 2003 and 2004 cows in Canada and the United States were found to have BSE. Public health measures included quarantine and destruction of herds containing the affected cows and more widespread testing of meat products.

■ UNICELLULAR AND MULTICELLULAR PARASITES

Bacteria and viruses are the best-known human pathogens, but some pathogens are eukaryotic. Examples of the most important types are shown in Figure 1c,d●. **Protozoa** (Figure 4●) are unicellular eukaryotes that are abundant in soil and

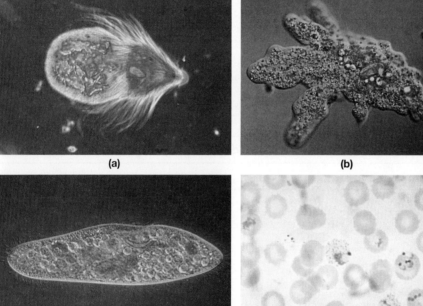

(a)

(b)

(c)

(d)

● **FIGURE 4**
Representative Protozoa.
(a) *Trichonympha,* a flagellate from a termite gut. (b) *Amoeba proteus,* a free-living form found in ponds. (LM × 310) (c) *Paramecium caudatum,* a free-living ciliate. (d) *Plasmodium vivax,* the parasite that causes malaria, stained within human blood cells. (LM × 125)

TABLE 5   Examples of Protozoan Diseases and the Primary Organ Systems Affected

Type of Protozoa	Name (Genus)	Disease	Affected Organ System
Flagellates	*Trypanosoma*	African sleeping sickness	Cardiovascular system
	Leishmania	Leishmaniasis	Lymphatic and integumentary systems
	Giardia	Giardiasis	Digestive system
	Trichomonas	Trichomoniasis	Reproductive system
Amoeboids	Entamoeba	Amoebic dysentery	Digestive system
Ciliates	*Balantidium*	Dysentery	Digestive system
Sporozoans	*Plasmodium*	Malaria	Various systems
	Toxoplasma	Toxoplasmosis	Various systems

water. They are responsible for a variety of serious human diseases, including *amoebic dysentery* and *malaria* (Table 5). Protozoa include (1) *flagellates,* which use flagella for propulsion; (2) *amoeboids,* among which are mobile, amoeba-like forms that engulf their prey; (3) *ciliates,* which are covered with cilia; and (4) *sporozoans*—parasitic forms with complex life cycles. **Fungi** (singular, *fungus*) are eukaryotic organisms that absorb organic materials from the remains of dead cells. Mushrooms are familiar examples of very large fungi. In a *mycosis,* or fungal infection, a microscopic fungus spreads through living tissues, killing cells and absorbing nutrients. Several relatively common skin conditions (including *athlete's foot*) and a few more serious diseases (such as *histoplasmosis*) are the result of fungal infections (Table 6).

Larger multicellular organisms, generally referred to as *parasites,* can also invade the human body and cause diseases. The multiplication of these larger parasitic organisms in or on the body is called an **infestation**. Diseases caused by multicellular parasites are listed in Table 7. **Helminths** are parasitic worms that can live within the body. They include **flatworms,** such as the *flukes* and *tapeworms,* and **roundworms,** or *nematodes.* These organisms, which range in size from microscopic flukes to tapeworms a meter or more in length, typically cause weakness and discomfort, but do not *by themselves* kill their host. However, complications resulting from the parasitic infection, such as malnutrition, chronic bleeding, and

TABLE 6 Examples of Fungal Diseases and the Primary Organ Systems Affected

Organism (Genus)	Disease	Affected Organ System
Aspergillus	Aspergillosis ("farmer's lung disease")	Respiratory system
Blastomyces	Blastomycosis	Integumentary system
Histoplasma	Histoplasmosis	Respiratory system
Epidermophyton, Microsporum, and *Trichophyton*	Ringworm	Integumentary system
	Tinea capitis (scalp)	
	Tinea corporis (body)	
	Tinea cruris (groin)	
	Tinea unguium (nails)	
Candida	Candidiasis	Integumentary system
Coccidioides	Coccidioidomycosis ("San Joaquin valley fever")	Respiratory system

TABLE 7 Examples of Diseases Caused by Multicellular Parasites and the Primary Organ Systems Affected

Group	Organism	Disease or Condition	Affected Organ System
Helminths			
Roundworms	*Ascaris*	Intestinal infestation	Digestive system
	Enterobius	Pinworm infestation	Digestive system
Flatworms	*Wuchereria*	Elephantiasis	Lymphatic system
Flukes	*Fasciola, Clonorchis* (liver flukes)	Fascioliasis	Digestive system
	Schistosoma (blood fluke)	Schistosomiasis	Cardiovascular, digestive, urinary systems
Tapeworms	*Taenia*	Tapeworm infestation	Digestive system
Arthropods			
Arachnids (eight legs)	Mites	Vectors of bacterial and rickettsial diseases	Various systems
	Ticks	Vectors of bacterial and rickettsial diseases	Various systems
	Spiders, scorpions	Inflammation from bites	Integumentary system
Insects (six legs)	Lice	Vectors of bacterial and rickettsial diseases	Various systems
	Human lice	Pediculosis	Integumentary system
	Mosquitoes	Vectors of bacterial and rickettsial diseases	Various systems
	Flies	Passive carriers of bacterial diseases	Various systems
	Wasps, bees	Inflammation from stings	Various systems

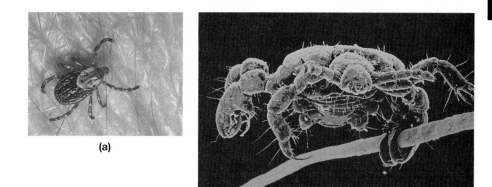

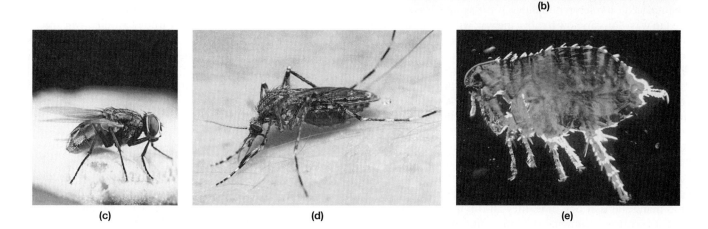

● **FIGURE 5**

Representative Disease-Carrying Arthropods. (a) *Dermacentor andersoni,* a wood tick. **(b)** *Phthirus pubis,* a crab louse, holding onto a human pubic hair. (SEM × 55) **(c)** *Musca domestica,* the housefly, which can transport microbes on its body. (× 3) **(d)** The *Aedes* mosquito, a vector for dengue fever. **(e)**

secondary infections by bacterial or viral pathogens, can ultimately prove fatal.

Arthropods (Figure 5●) make up the largest and most diverse group of animals on Earth. The major arthropods that affect humans are the *arachnids,* including scorpions, spiders, mites, and ticks, and the *insects,* such as mosquitoes, flies, lice, fleas, and bedbugs.

■ METHODS OF MICROANATOMY

HA p. 27

Over the last 50 years, our technological gadgetry has improved remarkably, enabling us to view the insides and outsides of cells in new ways. Sophisticated equipment has permitted the detailed analysis of physiological processes within cells. The basic problems facing cytologists stem from the considerable difference in size between the investigator and the object of interest. Cytologists (cell biologists) and histologists (biologists who study tissues) measure intracellular structures in *micrometers* (μm), also known as **microns**. Although the range of cell sizes is considerable, an "average cell" is a cube roughly 10 μm 10 μm × 10 μm. To fill a cubic millimeter, we would need a million cells. Because the human eye cannot recognize details smaller than about 0.1 mm, cytologists rely on special equipment that magnifies cells and their contents.

■ LIGHT MICROSCOPY

Historically, most information has been provided by **light microscopy**, a method in which a beam of light is passed through the object to be viewed. A light microscope can magnify cellular structures about 1000 times and can show details as fine as 0.25 μm. A camera can be attached to the microscope and used to produce a photograph called a **light micrograph (LM)**. Unfortunately, you cannot simply pick up a cell, slap it onto a microscope slide, and take a photograph. Because individual cells are so small, you must work with large numbers of them. Most tissues have a three-dimensional structure, and small pieces of tissue can be removed for examination. The component cells are prevented from decomposing by first exposing the tissue sample to a poison that will stop metabolic operations, but will not alter cellular structures.

Even then, you still cannot look at the tissue sample through a light microscope, because a cube only 2 mm (0.078 in.) on a side will contain several million cells. You must slice the sample into thin sections. Living cells are relatively thick, and cellular contents are not transparent. Light can pass through the section only if the slices are thinner than the individual cells. Making a section that slender poses interesting technical problems. Most tissues are not very sturdy, so an attempt to slice a fresh piece will destroy the sample. (To appreciate the problem, try to slice a marshmallow into thin sections.) Thus, before you can make sections, you must embed the tissue sample in something that will make it more stable, such as wax, plastic, or epoxy. These materials will not interact with water molecules, so your sample must first be dehydrated (typically by immersion in 30 percent, 70 percent, 95 percent, and, finally, 100 percent alcohol). If you are embedding the sample in wax, the wax must be hot enough to melt; if you are using plastic or epoxy, the hardening process generates heat on its own.

After embedding the sample, you can section the block with a machine called a *microtome,* which uses a metal, glass, or diamond knife. For viewing by light microscopy, a typical section is about 5 μm (0.002 in.) thick. The thin sections are then placed on microscope slides. If the sample was embedded in wax, you can now remove the wax with a solvent, such as xylene. But you are not done yet: In thin sections, the cell contents are almost transparent; you cannot yet distinguish intracellular details by using an ordinary light microscope. You must first add color to the internal structures by treating the slides with special dyes called *stains.* Some stains are dissolved in water and others in alcohol. Not all types of cells pick up a given stain to the same degree—if they pick it up at all; nor do all types of cellular organelles. For example, in a sample scraped from the inside of the cheek, one stain might dye only certain types of bacteria; in a semen sample, another stain might dye only the flagella of the sperm. If you try too many stains at one time, they all run together, and you must start over. Following staining, you can put cover slips over the sections (generally after you have dehydrated them again) and can see what your labors have accomplished.

Any single section can show you only a part of a cell or tissue. To reconstruct the tissue structure, you must look at a series of sections made one after the other. After examining dozens or hundreds of sections, you can understand the structure of the cells and the organization of your tissue sample—or can you? Your reconstruction has left you with an understanding of what these cells look like after they have (1) died an unnatural death; (2) been dehydrated; (3) been impregnated with wax or plastic; (4) been sliced into thin sections; (5) been rehydrated, dehydrated, and stained with various chemicals; and (6) been viewed with the limitations of your equipment. A good cytologist or histologist is extremely careful, cautious, and self-critical and realizes that much of the laboratory preparation is an art as well as a science.

■ ELECTRON MICROSCOPY

More elaborate procedures can allow for the examination of finer details. In **electron microscopy**, a beam of electrons is passed through or reflected off the surface of a suitably prepared object. In **transmission electron microscopy**, the electrons pass through an ultrathin section. Once through the section, they strike a photographic plate and produce an image known as a **transmission electron micrograph (TEM)**. Transmission electron microscopy can magnify structures up to approximately 500,000 times, revealing details less than a nanometer in size. For instance, with a transmission electron microscope, you can visualize large organic molecules. In **scanning electron microscopy**, a beam of electrons reflects

off the surface of an object such as a cell, a broken portion of a cell, or an extracellular structure. (The surfaces are specially coated to enhance reflectivity.) After bouncing off the surface, the electrons strike a photographic plate, producing an image known as a **scanning electron micrograph (SEM)**. Scanning electron microscopy can magnify structures up to only about 50,000 times, but the technique provides a three-dimensional perspective on cellular anatomy that cannot be obtained by other methods.

This level of detail poses problems of its own. At the level of the light microscope, if you were to slice a large cell as you would slice a loaf of bread, you might produce 10 sections from the one cell. You could review the entire series under a light microscope in a few minutes. If you sliced the same cell for examination under an electron microscope, you would have 1000 sections, *each* of which could take several hours to inspect! Figure 6a,b● compares SEM and TEM views of cells that line the intestinal tract, and Figure 6c● shows a diagrammatic representation of the intact cell.

■ CAUSES OF CANCER   *HA p. 45*

Relatively few types of cancer are inherited; only 20 hereditary types have been identified to date, and together they account for less than 1 percent of cancer cases. By definition, inherited cancer involves genes provided by the sperm or oocyte at fertilization; as a result, these genes are in every cell of the individual's body. Such people have a much higher risk of developing a specific cancer than the general population. However, not *everyone* with these genes gets cancer, and this indicates that other genes and/or environmental factors must act as a "trigger." For the general population, the interaction of genetic and environmental factors causes most cancers.

■ GENETIC FACTORS

Two related genetic factors are involved in the development of cancer: hereditary predisposition and oncogene activation.

An individual born with genes that increase the likelihood of cancer is said to have a *hereditary predisposition* for the disease. Such a person may never develop cancer, but his or her chances are higher than average. The inherited genes generally affect the abilities of tissues to metabolize toxins, control mitosis and growth, perform repairs after injury, or identify and destroy abnormal tissue cells. As a result, body cells become sensitive to local or environmental factors that would have less effect on cells from people lacking these genes. Roughly 15 percent of cancers "run in families" and reflect a hereditary disposition for cancer of a specific type.

The majority of cancers result from somatic-cell mutations that modify genes involved in cell growth, differentiation, or mitosis. As a result, an ordinary cell is converted into a cancer cell. The modified genes are called **oncogenes** (ON-kō-jēnz); the normal genes are called **proto-oncogenes**. *Oncogene activation* occurs by the alteration of normal somatic genes. Because these mutations do not affect reproductive cells, the cancers caused by active oncogenes are not inherited.

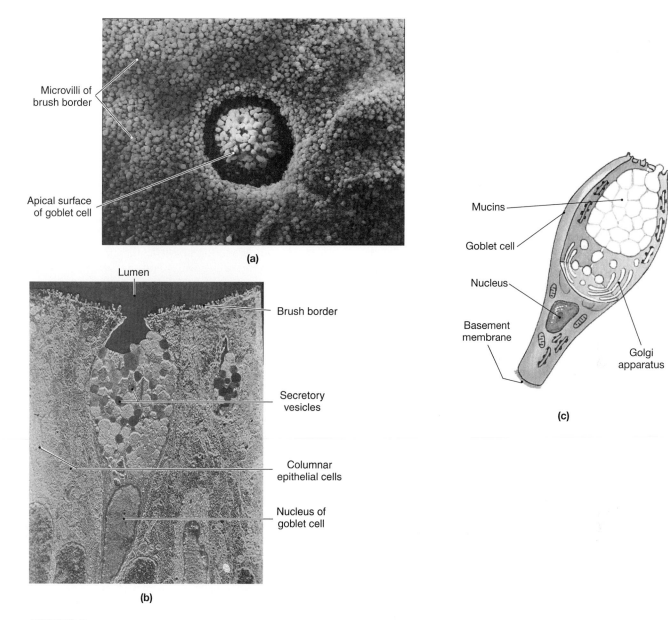

● FIGURE 6
A Comparison of Histological Techniques. (a) Cell surfaces can be seen with a scanning electron microscope. **(b)** Cells similar to those in part (a), but viewed with a transmission electron microscope. **(c)** A composite drawing that summarizes the information provided by both scanning and transmission electron microscopy.

■ ENVIRONMENTAL FACTORS

Many cancers can be directly or indirectly attributed to environmental factors called **carcinogens** (kar-SIN-ō-jenz). Carcinogens stimulate the conversion of a normal cell to a cancer cell. Some carcinogens are *mutagens* (MŪ-ta-jenz)—that is, they damage DNA strands and may cause chromosomal breakage. All forms of high-energy radiation, including cosmic rays, x-rays, UV rays, and radioisotopes, are mutagens that have carcinogenic effects.

The environment contains many chemical carcinogens. Plants manufacture poisons that protect them from insects and other predators, and although their carcinogenic activities are often relatively weak, many common spices, vegetables, and beverages contain compounds that are carcinogens if consumed in large quantities. Animal tissues may also store or concentrate toxins, and hazardous compounds of many kinds can be swallowed in contaminated food. A variety of laboratory and industrial chemicals, such as coal tar derivatives and synthetic pesticides, have been shown to be carcinogenic. From studies that compared cancer incidence in twins with other data, it has been estimated that 70–80 percent of all cancers are the result of chemical or environmental factors, and 40 percent are due to a single source of carcinogens: cigarette smoke.

Specific carcinogens will affect only those cells capable of responding to that particular physical or chemical stimulus. The responses vary because differentiation produces cell types with specific sensitivities. For example, benzene can produce a cancer of the blood, cigarette smoke a lung cancer, and vinyl chloride a liver cancer. Very few stimuli can produce cancers throughout the body. Radiation is a notable exception. In general, cells undergoing mitosis are most likely to be vulnerable to chemical or radiational carcinogens. As a result, cancer rates are highest in epithelial tissues such as the skin and the lining of the intestines, where stem cell divisions occur rapidly, and lowest in nervous and muscle tissues, where divisions do not normally occur. Cancer statistics and treatment are considered in a later section (p. 16).

NORMAL AND ABNORMAL TISSUES

Tissue structure changes over a lifetime both through development and aging and by adapting to changing conditions and stresses. In this section we will consider examples of varied functions among normal and abnormal tissues. We will also take a closer look at the nature of cancer, a disease that will be discussed repeatedly in later sections.

■ INFANTS AND BROWN FAT   *HA p. 68*

At birth, an infant's temperature-regulating mechanisms are not fully functional. Infants also lose heat quickly because they have a relatively large surface area (where heat is lost) and a relatively small volume (where heat is generated). Newborns must be dried promptly and kept bundled up; for those born prematurely, a thermally regulated incubator is required. Infants' body temperatures are also less stable than those of adults. Their metabolic rates decline while they sleep, then rise after they awake.

Although infants cannot shiver, they do have a mechanism for raising body temperature rapidly. The adipose tissue between the shoulder blades, around the neck, and possibly elsewhere in the upper body is histologically and functionally different from most of the adipose tissue in adults. The tissue is highly vascularized, and the individual adipocytes contain numerous mitochondria. Together these characteristics give the tissue a deep, rich color responsible for the name **brown fat**. The individual adipocytes are innervated by sympathetic autonomic fibers. When these nerves are stimulated, lipolysis accelerates in the adipocytes. The cells do not capture the energy released through fatty acid catabolism, and it radiates into the surrounding tissues as heat. This heat quickly warms the blood that passes through the surrounding network of vessels, and it is then distributed throughout the body. In this way, an infant can accelerate metabolic heat generation by 100 percent very quickly.

With increasing age and size, body temperature becomes more stable, so the importance of this thermoregulatory mechanism declines. Adults have little if any brown fat; with increased body size, skeletal muscle mass, and insulation, shivering is significantly more effective in elevating body temperature.

■ LIPOSUCTION   *HA p. 68*

One much-publicized method of battling obesity is the process of liposuction. **Liposuction** is a surgical procedure for the removal of unwanted adipose tissue. Adipose tissue is flexible but not as elastic as areolar tissue, and it tears relatively easily. In liposuction, a small incision is made through the skin and a tube is inserted into the underlying adipose tissue. Suction is then applied. Because adipose tissue tears easily, chunks of tissue containing adipocytes, other cells, fibers, and ground substance can be vacuumed away. Estimates of the number of liposuctions, among the most common cosmetic surgeries performed today, with an estimated 372,831 procedures performed in 2002.

CHAPTER 3
THE TISSUE LEVEL OF ORGANIZATION

This practice has received a lot of news coverage, and many advertisements praise the technique as easy, safe, and effective. In fact, it is not always easy, and it can be dangerous and have limited effectiveness. The density of adipose tissue varies from place to place in the body and from individual to individual, and it is not always easy to suck through a tube. Blood vessels are stretched and torn, and extensive bleeding can occur. An anesthetic must be used to control pain, and anesthesia always poses risks; heart attacks, pulmonary embolism, and fluid balance problems can develop, with fatal results. The death rate for this procedure is 1 in 5000. Finally, adipose tissue can repair itself, and adipocyte populations recover over time. The only way to ensure that fat lost through liposuction will not return is to adopt a lifestyle that includes a proper diet and adequate exercise. Over time, such a lifestyle can produce the same weight loss, *without liposuction,* eliminating the surgical expense and risk.

■ CARTILAGES AND KNEE INJURIES

HA p. 72

The knee is an extremely complex joint that contains both hyaline cartilage and fibrocartilage. The hyaline cartilage caps bony surfaces, while pads of fibrocartilage within the joint prevent bone-to-bone contact when movements are under way. Many sports injuries involve tearing of the fibrocartilage pads or supporting ligaments; the loss of support and cushioning places more strain on the hyaline cartilages within joints and leads to further joint damage. Articular cartilages not only are avascular, but also lack a perichondrium. As a result, they heal even more slowly than other cartilages. Surgery usually produces only a temporary or incomplete repair. For this reason, most competitive sports have rules designed to reduce the number of knee injuries. For example, in football "clipping" is outlawed because it produces stresses that can tear the fibrocartilages and the supporting ligaments at the knee.

Recent advances in tissue culture have enabled researchers to grow fibrocartilage in the laboratory. Chondrocytes removed from the knees of injured patients are cultured in an artificial framework of collagen fibers. Eventually, they produce masses of fibrocartilage that can be inserted into the damaged joints. Over time, the pads change shape and grow, restoring normal joint function. This labor-intensive technique has been used to treat severe joint injuries, particularly in athletes.

■ PROBLEMS WITH SEROUS MEMBRANES *HA p. 73*

Several clinical conditions, including infection and chronic inflammation, can cause the abnormal buildup of fluid in a body cavity. Other conditions can reduce the amount of lubrication, causing friction between opposing layers of serous membranes. This can promote the formation of adhesions—fibrous connections that eliminate the friction by locking the membranes together. Adhesions may also severely restrict the movement of the affected organ or organs and may compress blood vessels or nerves.

Pleuritis, or *pleurisy,* is an inflammation of the pleural cavities. At first the opposing membranes become drier and

scratch against one another, producing a sound known as a *pleural rub.* Adhesions seldom form between the serous membranes of the pleural cavities. More commonly, continued rubbing and inflammation leads to a gradual increase in fluid production to levels well above normal. Fluid then accumulates in the pleural cavities, producing a condition known as *pleural effusions.* Pleural effusions are also caused by heart conditions that elevate the pressure within the pulmonary blood vessels. Fluid then leaks into the alveoli and into the pleural spaces as well, compressing the lungs and making breathing difficult. This combination can be lethal.

Pericarditis is an inflammation of the pericardium. This condition typically leads to *pericardial effusion,* an abnormal accumulation of the fluid in the pericardial cavity. When sudden or severe, the fluid buildup can seriously reduce the efficiency of the heart and restrict blood flow through major vessels.

Peritonitis, an inflammation of the peritoneum, can follow infection of, or injury to, the peritoneal lining. Peritonitis is a potential complication of any surgical procedure in which the peritoneal cavity is opened or of a disease that perforates the walls of the intestines or stomach. Adhesions are common following peritoneal infections and may lead to constriction and blockage of the intestinal tract.

Liver disease, kidney disease, or heart failure can cause an accumulation of fluid in the peritoneal cavity. Called *ascites* (a-SĪ-tēz), this accumulation creates a characteristic abdominal swelling. The pressure and distortion of internal organs by the excess fluid can lead to symptoms such as heartburn, indigestion, shortness of breath, and low back pain.

■ CANCER STATISTICS, CLASSIFICATION, AND TREATMENT

HA p. 80

In the United States, the lifetime risk of developing some form of cancer is 50 percent for males and 33 percent for females. In 2004, an estimated 556,500 people will die of some form of cancer, making it second only to heart disease as a cause of mortality in the U.S. population. Because both population size and average age increased in the 20th century, just comparing the numbers of deaths from year to year can be misleading. The relevant statistics are better presented in terms of the cancer rate per 100,000 population, age-adjusted to match the most recent census data (2000), and compared to statistics on other causes of death. For example, from 1975 to 2000, the United States had an increase in cancer incidence of 12.5 percent, while the cancer death rate over this time did not increase, and the absolute death rate declined at least 18 percent. For more-detailed data analyses, visit the Website of the National Center for Health Statistics at http://www.cdc.gov/nchs.

▫ DETECTION AND INCIDENCE OF CANCER

Physicians who specialize in the identification and treatment of cancers are called **oncologists** (on-KOL-ō-jists; *onkos,* mass). Pathologists and oncologists classify cancers according to their cellular appearance and their sites of origin. More than a hundred kinds have been described, but broad categories are used that indicate the location of the primary tumor. A tumor is defined as a "new growth" resulting from

TABLE 8   Benign and Malignant Tumors in the Major Tissue Types

Tissue	Description
Epithelia	
Carcinomas	Any cancer of epithelial origin
Adenocarcinomas	Cancers of glandular epithelia
Angiosarcomas	Cancers of endothelial (vascular) cells
Mesotheliomas	Cancers of mesothelial cells
Connective tissues	
Fibromas	Benign tumors of fibroblast origin
Lipomas	Benign tumors of adipose tissue
Liposarcomas	Cancers of adipose tissue
Leukemias	Cancers of blood-forming tissues
Lymphomas	Cancers of lymphoid tissues
Chondromas	Benign tumors in cartilage
Chondrosarcomas	Cancers of cartilage
Osteomas	Benign tumors in bone
Osteosarcomas	Cancers of bone
Muscle tissues	
Myxomas	Benign muscle tumors
Myosarcomas	Cancers of skeletal muscle tissue
Cardiac sarcomas	Cancers of cardiac muscle tissue
Leiomyomas	Benign tumors of smooth muscle tissue
Leiomyosarcomas	Cancers of smooth muscle tissue
Neural tissues	
Gliomas	Cancers of neuroglial origin
Neuroblastomas	Cancers of neuronal origin

necessary for the correct diagnosis of cancer. Information is usually obtained by the histological examination of a tissue sample, or *biopsy,* typically supplemented by medical imaging and blood studies. A biopsy is one of the most significant diagnostic procedures, because it permits a direct look at the tumor cells. Not only do malignant cells have an abnormally high rate of mitosis, but they are also structurally distinct from healthy body cells.

If the tissue appears cancerous, other important questions must be answered, including the following: What is the measurable size of the primary tumor? Has the tumor invaded surrounding tissues? Has the cancer already metastasized to develop secondary tumors? Are any regional lymph nodes affected? The answers to these questions are combined with observations from the physical exam, the biopsy results, and information from any imaging procedures to arrive at an accurate diagnosis and prognosis.

In an attempt to develop a standard system, national and international cancer organizations have devised the *TNM system* for staging (identifying the stage of progression of) cancers. The letters stand for *tumor (T)* size and invasion, *lymph node (N)* involvement, and degree of *metastasis (M):*

- Tumor size is graded on a scale of 0 to 4. T0 indicates the absence of a primary tumor, and the largest dimensions and greatest amount of invasion are categorized as T4.
- Lymph nodes filter the tissue fluids from nearby capillary beds. The fluid, called *lymph,* then returns to the general lymphatic circulation. Once cancer cells have entered the lymphatic system, they can spread very quickly throughout the body. Lymph node involvement is graded on a scale of 0 to 3. A designation of N0 indicates that no lymph nodes have been invaded by cancer cells. A classification of N1 to N3 indicates the involvement of increasing numbers of lymph nodes:

N1   indicates the involvement of a single lymph node less than 3 cm in diameter.

N2   includes one medium-sized (3–6 cm) node or multiple nodes smaller than 6 cm.

N3   indicates the presence of a single lymph node larger than 6 cm in diameter, regardless of whether other nodes are involved.

- Metastasis is graded on a scale of 0 to 1. M0 indicates that there is no evidence of metastasis, whereas M1 indicates that the cancer cells have produced secondary tumors in other portions of the body.

This grading system provides a general overview of the progression of the disease. For example, a tumor classified as T1N1M0 has a better prognosis than one classified as T4N2M1; the latter tumor would be much more difficult to treat. The grading system alone does not provide all the information needed to plan treatment, however, because different types of cancer progress in different ways. Therapies must vary accordingly. Thus, *leukemia,* a cancer of the blood-forming tissues, is treated differently from colon cancer. We will consider specific treatments in discussions dealing with cancers that affect individual body systems; the next section provides a general overview of the strategies used to treat cancer.

uncontrolled cell division. A tumor can be *malignant* or *benign* and may *metastasize* (spread) rapidly or very slowly. Only malignant tumors are called cancers. Table 8 summarizes information about benign and malignant tumors (cancers) associated with the major tissues of the body.

A statistical profile of cancer incidences and survival rates in the United States is shown in Table 9. The numbers from other countries are different. For example, *bladder cancer* is common in Egypt, *stomach cancer* in Japan, and *liver cancer* in Africa. Variations in the combination of genetic factors and dietary, infectious, and other environmental factors are thought to be responsible for these differences.

■ CLINICAL STAGING AND TUMOR GRADING

The detection of a cancer often begins during a routine physical examination, when the physician notices an abnormal lump or growth. Many laboratory and diagnostic tests are

| | | | Five-Year Survival Rates | |
| | | | Diagnosis Date | |
Site	Estimated New Cases (2004)	Estimated Deaths (2004)	1974–76	1992–98
Digestive tract				
Esophagus	14,250	13,300	5%	13%
Stomach	22,710	11,780	15%	22%
Colon and rectum	148,940	56,730	50%	62%
Respiratory tract				
Lung and bronchus	173,770	160,440	12%	15%
Urinary tract				
Kidney and renal pelvis	35,710	12,480	52%	62%
Urinary bladder	60,240	12,710	73%	82%
Reproductive system				
Breast	217,440	40,580	75%	86%
Ovary	25,580	16,090	37%	53%
Testis	8,980	360	79%	95%
Prostate gland	230,110	29,900	67%	97%
Nervous system	18,400	12,690	22%	32%
Skin (melanoma only)	55,100	7,910	80%	89%

TABLE 9   Cancer Incidences and Survival Rates in the United States

Data courtesy of the American Cancer Society.

CANCER TREATMENT

It is unfortunate that the media tend to describe cancer as though it were one disease rather than many. This simplistic perspective fosters the belief that some dietary change, air ionizer, or wonder drug will be found that can prevent or cure the affliction. No single, universally effective cure for cancer is likely, because there are too many separate causes, underlying mechanisms, and individual differences.

The goal of cancer treatment is to achieve remission. A tumor in **remission** either ceases to grow or decreases in size. The treatment of malignant tumors must accomplish one of these two objectives to produce remission:

1. **The surgical removal or destruction of individual tumors.** Tumors containing malignant cells can be surgically removed or destroyed by radiation, heat, or freezing. These techniques are highly effective if the treatment is undertaken before metastasis has occurred. For this reason, early detection is important in improving survival rates for all forms of cancer.

2. **The killing of metastasized cells throughout the body.** This is much more difficult and potentially dangerous, because healthy tissues are likely to be damaged at the same time. At present, the most widely approved treatments are chemotherapy and radiation.

Chemotherapy may involve the administration of drugs that will either kill the cancerous tissues or prevent mitotic divisions. These drugs typically affect stem cells in normal tissues, and the side effects are usually unpleasant. For example, because some chemotherapy slows the regeneration and maintenance of epithelia of the skin and digestive tract, patients often lose their hair and experience nausea and vomiting. Several drugs are often administered simultaneously or in sequence, because, over time, cancer cells can develop a resistance to a single drug. Chemotherapy is used in the treatment of many kinds of metastasized cancer.

Massive doses of total body irradiation are sometimes used to treat advanced cases of *lymphoma,* a cancer of the immune system. In this rather drastic procedure, enough radiation is administered to kill all the blood-forming cells in the body. After treatment, new blood cells must be provided by a bone marrow transplant. In later sections dealing with the lymphatic system, we will discuss marrow transplants, lymphomas, and other cancers of the blood.

An understanding of molecular mechanisms and cell biology is leading to new approaches that may revolutionize cancer treatment. One approach focuses on the fact that cancer cells are ignored by the immune system. In **immunotherapy**, substances are administered that help the immune system recognize and attack cancer cells. More elaborate experimental procedures involve the creation of customized antibodies by the gene-splicing techniques. The resulting antibodies are specifically designed to attack the tumor cells in each particular patient. Although this technique shows promise, it remains difficult, costly, and very labor-intensive.

A second approach is targeted "designer" cancer drugs. One type of cancer, chronic myelogenous leukemia, involves the activity of an abnormal enzyme. A drug has been developed that inactivates this enzyme but has no effect on normal enzymes, and thus does not affect normal cells. In early trials of the drug imatinib (*Gleevac*), complete remission occurred in up to 95 percent of CML patients treated. Studies are now in progress to determine whether Gleevac could be effective against other types of cancer.

■ CANCER AND SURVIVAL

Advances in chemotherapy, radiation procedures, and molecular biology have produced significant improvements in the survival rates of several types of cancer patients. However, the improved survival rates indicated in Table 9 reflect advances not only in therapy, but also in early detection. Much of the credit goes to increased public awareness and concern about cancer. In general, the odds of survival increase markedly if the cancer is detected early, especially before it undergoes metastasis. The American Cancer Society has identified seven "warning signs" that mean that it's time to consult a physician. These signs are presented in Table 10.

■ TISSUE STRUCTURE AND DISEASE
HA p. 80

Physicians who specialize in the study of disease processes are called **pathologists** (pa-THOL-o-jists). Diagnosis, rather than treatment, is usually the main focus of their activities. In their analyses, pathologists integrate anatomical and histological observations to determine the nature and severity of a disease. Because disease processes affect the histological organization of tissues and organs, **biopsies**, or tissue samples, often play a key role in their diagnoses.

TABLE 10 Seven Warning Signs of Cancer
Change in bowel or bladder habits
A sore that does not heal
Unusual bleeding or discharge
Thickening or lump in breast or elsewhere
Indigestion or difficulty in swallowing
Obvious change in a wart or mole
Nagging cough or hoarseness

Figure 7● diagrams the histological changes induced in the respiratory epithelium by one relatively common irritating stimulus, cigarette smoke. The normal respiratory epithelium is shown in Figure 7a●. The first abnormality to be observed in a smoker is **dysplasia** (dis-PLĀ-zē-uh), a change in the shape, size, and organization of tissue cells. Dysplasia is generally a response to chronic irritation or inflammation, and the changes are reversible. The normal trachea (windpipe) and its branches are lined by a pseudostratified ciliated columnar epithelium. The

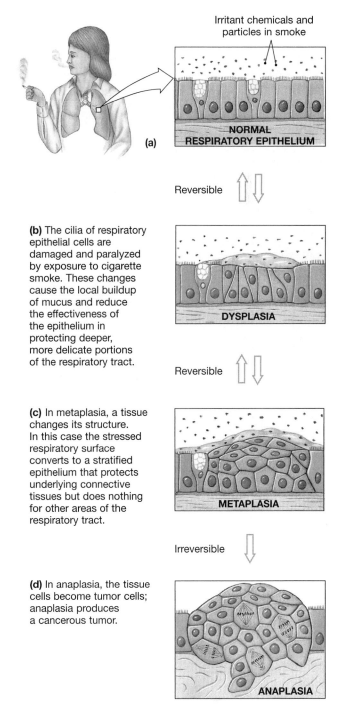

(a)

Irritant chemicals and particles in smoke

NORMAL RESPIRATORY EPITHELIUM

Reversible

(b) The cilia of respiratory epithelial cells are damaged and paralyzed by exposure to cigarette smoke. These changes cause the local buildup of mucus and reduce the effectiveness of the epithelium in protecting deeper, more delicate portions of the respiratory tract.

DYSPLASIA

Reversible

(c) In metaplasia, a tissue changes its structure. In this case the stressed respiratory surface converts to a stratified epithelium that protects underlying connective tissues but does nothing for other areas of the respiratory tract.

METAPLASIA

Irreversible

(d) In anaplasia, the tissue cells become tumor cells; anaplasia produces a cancerous tumor.

ANAPLASIA

● **FIGURE 7**
Changes in a Tissue under Stress

cilia move a mucous layer that traps foreign particles and moistens incoming air. The drying and chemical effects of smoking first paralyze the cilia, halting the movement of mucus (Figure 7b●). As mucus builds up, the individual coughs to dislodge it (the well-known "smoker's cough"). Dysplasia increases the risk of cancer formation in that tissue; in a tissue not subject to abnormal stresses, dysplasia may be the first indication of a developing cancer.

Epithelia and connective tissues may undergo more radical changes in structure, caused by the division and differentiation of stem cells. **Metaplasia** (me-tuh-PLĀ-zē-uh) is a structural change that dramatically alters the character of the tissue. In our example, over time heavy smoking causes the epithelial cells to lose their cilia altogether.

As metaplasia progresses, the epithelial cells produced by stem cell divisions no longer differentiate into ciliated columnar cells. Instead, they form a stratified squamous epithelium that provides greater resistance to drying and chemical irritation (Figure 7c●). This epithelium protects the underlying tissues more effectively, but it eliminates the moisturization and cleaning properties of the epithelium. Cigarette smoke will now have an even greater effect on more delicate portions of the respiratory tract. Fortunately, metaplasia is reversible, and unless a malignant tumor has formed, the epithelium will gradually return to normal if the individual quits smoking.

In **anaplasia** (a-nuh-PLĀ-zē-uh), tissue organization breaks down. Tissue cells change size and shape, typically becoming unusually large or small (Figure 7d●) and losing most if not all resemblance to mature tissue cells. Anaplasia is characteristic of many cancers; it occurs in smokers who develop one form of lung cancer. In anaplasia, the cells divide more frequently but not all divisions proceed in the normal way. Many of the tumor cells have abnormal chromosomes. Unlike dysplasia and metaplasia, anaplasia is irreversible.

CHAPTER 4

THE INTEGUMENTARY SYSTEM

The structures of the integumentary system include the skin and its accessory organs, such as hair, nails, and several types of exocrine glands. The integumentary system has a variety of functions, including protection of the underlying tissues, maintenance of body temperature, excretion of salts and water in sweat, cutaneous sensation, and the production of vitamin D_3.

The skin is the most visible organ of the body. As a result, abnormalities are easily recognized. A bruise, for example, typically creates a swollen and discolored area where blood has leaked into tissues surrounding damaged blood vessels. Changes in color, tone, texture and the overall condition of the skin commonly accompany illness or disease. These changes can assist in diagnosis. For instance, extensive bruising without obvious cause may indicate a blood-clotting disorder, and yellowish skin and mucous membranes may signify *jaundice*, a sign that often points to some type of liver disorder. The general condition of the skin can also be significant: In addition to color changes, changes in skin flexibility, elasticity, dryness, or sensitivity commonly follow malfunctions of other organ systems.

EXAMINATION OF THE SKIN

HA p. 86

In examining a patient, dermatologists (skin specialists) use a combination of investigative questions ("What has been in contact with your skin lately?" or "How does it feel?") and physical examination to arrive at a diagnosis. The condition of the skin is carefully observed. Notes are made about the presence of **lesions**, which are changes in skin structure caused by trauma or disease processes. Lesions are also called **skin signs**, because they are measurable, visible abnormalities of the skin surface. Figure 8● diagrams the most common skin signs and related disorders.

The distribution of lesions can be an important clue to the source of the problem. For instance, in *shingles (herpes zoster),* blisters on the skin occur in the area(s) innervated by peripheral sensory nerves. A ring of slightly raised, scaly (papular) lesions is typical of fungal infections that may affect the trunk, scalp, or nails. Examples of skin infections and allergic reactions are given in Table 11, with descriptions of the related skin signs. We consider skin lesions caused by trauma later, in the section titled "Trauma to the Skin" (p. 27). Table 11 summarizes signs on the surface of the skin, but signs involving the accessory organs of the skin can also be important. For instance:

- Nails have a characteristic shape that can change due to an underlying disorder. An example is *clubbing* of the fingernails, commonly a sign of *emphysema* or *congestive heart failure.* In these conditions, the fingertips broaden and the nails become distinctively curved.

- The condition of the hair can be an indicator of a person's overall health. For example, depigmentation and coarseness of hair occur in the protein deficiency disease *kwashiorkor.*

4

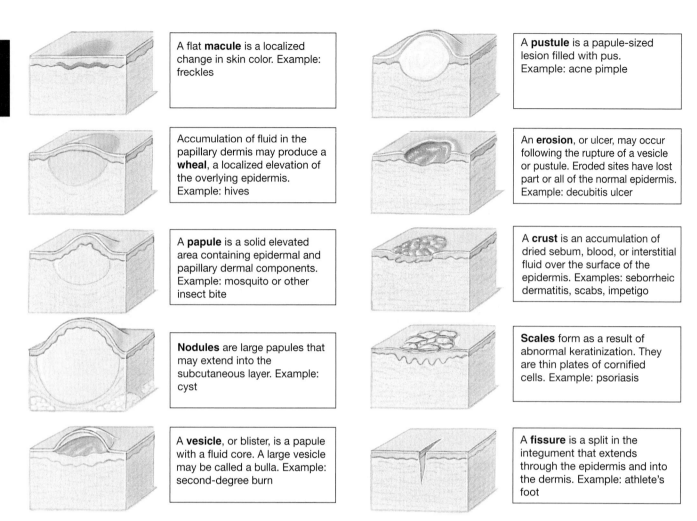

A flat **macule** is a localized change in skin color. Example: freckles

Accumulation of fluid in the papillary dermis may produce a **wheal**, a localized elevation of the overlying epidermis. Example: hives

A **papule** is a solid elevated area containing epidermal and papillary dermal components. Example: mosquito or other insect bite

Nodules are large papules that may extend into the subcutaneous layer. Example: cyst

A **vesicle**, or blister, is a papule with a fluid core. A large vesicle may be called a bulla. Example: second-degree burn

A **pustule** is a papule-sized lesion filled with pus. Example: acne pimple

An **erosion**, or ulcer, may occur following the rupture of a vesicle or pustule. Eroded sites have lost part or all of the normal epidermis. Example: decubitis ulcer

A **crust** is an accumulation of dried sebum, blood, or interstitial fluid over the surface of the epidermis. Examples: seborrheic dermatitis, scabs, impetigo

Scales form as a result of abnormal keratinization. They are thin plates of cornified cells. Example: psoriasis

A **fissure** is a split in the integument that extends through the epidermis and into the dermis. Example: athlete's foot

● **FIGURE 8**
Skin Signs

TABLE 11   Skin Signs of Various Disorders

Cause	Examples	Resulting Skin Lesions
Viral infections	Chickenpox	Lesions begin as macules and papules, but develop into vesicles
	Measles (rubeola)	A maculopapular rash that begins at the face and neck and spreads to the trunk and limbs
	Erythema infectiosum (fifth disease)	A maculopapular rash that begins on the cheeks (slapped-cheek appearance) and spreads to the limbs
	Herpes simplex	Raised vesicles that heal with a crust
Bacterial infections	Impetigo	Vesiculopustular lesions with exudate and yellow crusting
Fungal infections	Ringworm	An annulus (ring) of scaly papular lesions with central clearing
Parasitic infections	Scabies	Linear burrows with a small, very pruritic papule at one end
	Lice (pediculosis)	Dermatitis: excoriation (scratches) due to pruritus (itching)
Allergies to medication	Penicillin	Wheals (urticaria or hives)
Food allergies	Eggs, certain fruits	Wheals
Environmental allergies	Poison ivy	Vesicles

■ DIAGNOSING SKIN CONDITIONS

Table 12 lists several major types of vascular lesions. A single skin lesion, such as a *vesicle*, can have multiple causes. This is one of the challenges facing dermatologists; the signs may be apparent, but the underlying causes may not. Making matters more difficult, many skin disorders produce the same uncomfortable sensations. For example, **pruritus** (proo-RĪ-tus), an irritating itching sensation, is an extremely common symptom associated with a variety of skin conditions. Questions about the patient's medical history, medications, possible sources of infection, and environmental exposure, as well as the presence of other signs, can be the key to making an accurate diagnosis.

Pain is another common symptom of many skin disorders. Although pain is unwelcome, cutaneous sensation is an important function of the integumentary system. Its importance is dramatically demonstrated in *Hansen's disease,* or *leprosy.* Hansen's disease is caused by a bacterium that has an affinity for cool regions of the body. The bacterium destroys cutaneous nerve endings that are sensitive to touch, pain, heat, and cold. Damage to the distal tissues then occurs and accumulates, because the individual is no longer aware of painful stimuli. We will consider Hansen's disease further in a later section. Long-standing, poorly controlled diabetes mellitus can cause similar damage, particularly to the feet, and is a leading cause of amputations.

Figure 9● is an overview of skin disorders. Diagnostic tests that may prove useful in distinguishing among them include the following:

- **The scraping and microscopic examination of affected tissue,** a process often performed to check for fungal infections.
- **The culturing of fluid** removed from a lesion to identify pathogens.
- **Biopsy** of affected tissue to view its cell structure.
- **Skin tests,** through which various types of skin or systemic disorders can be detected. In a skin test, a localized area of the skin is exposed to an inactivated pathogen, a portion of a pathogen, or a substance capable of producing an immunologic reaction in sensitive individuals. Exposure occurs by injection or surface application. For example, in a skin test for *tuberculosis,* a small quantity of tuberculosis antigens is injected *intradermally* (*intra,* within). If the individual has anti-tuberculin antibodies from past or current tuberculosis infection, *erythema* (*erythros,* red; a change in skin color) and swelling will occur at the site of the injection 24–72 hours later. *Patch testing* is used to check sensitivity to *allergens,* environmental agents that can cause allergic reactions. In a patch test, the allergen is applied to the surface of the skin. If erythema, swelling, or itching develops, the individual is sensitive to that allergen.

Table 13 summarizes information about common infections of the integumentary system.

■ DISORDERS OF KERATIN PRODUCTION   *HA p. 89*

Not all skin signs are the result of infection, trauma, or allergy; some are the normal response to environmental stresses. One common response is the excessive production of keratin, a process called **hyperkeratosis** (hī-per-ker-a-TŌ-sis). The most obvious effects—calluses and corns—are easily observed. Calluses are thickened patches that appear on already thick-skinned areas, such as the palms of the hands or the soles or heels of the foot, in response to chronic abrasion and distortion. Corns are more localized areas of excessive keratin production that form in areas of thin skin on or between the toes.

In **psoriasis** (so-RĪ-a-sis), stem cells in the stratum germinativum are unusually active, causing hyperkeratosis in specific areas, often the scalp, elbows, palms, soles, groin, or nails. Normally, an individual stem cell divides once every 20 days, but in psoriasis it may divide every day and a half. Keratinization is abnormal and typically incomplete by the time the outer layers are shed. The affected areas have red bases covered with vast numbers of small, silvery scales that continuously flake off. Psoriasis develops in 20–30 percent of individuals with an inherited tendency for the condition. Roughly 5 percent of the U.S. population has psoriasis to some degree, frequently aggravated by stress and anxiety. Most cases are painless and controllable, but not curable.

Xerosis (zē-RŌ-sis), or dry skin, is a common complaint of the elderly and people who live in arid climates. In xerosis, cell membranes in the outer layers of skin gradually deteriorate and the stratum corneum becomes more a collection of scales than a single sheet. The scaly surface is much more permeable than an intact layer of keratin, and the rate of insensible perspiration increases. In persons with severe xerosis, the rate of insensible perspiration may increase by up to 75 times.

TABLE 12   Examples of Vascular Lesions

Lesion	Features	Some Possible Causes
Ecchymosis (contusion)	Reddish purple, blue, or yellow bruising related to trauma	Trauma; blood-clotting disorder; some vitamin deficiencies; thrombocytopenia; increased tendency to bruise can be due to aging or sun damaged skin
Hematoma	Pooling and possible clotting of blood, forming a mass; associated with pain and swelling	Trauma, broken blood vessel
Petechiae	Small red-to-purple pinpoint dots appearing in clusters	Leukemia; septicemia (toxins in blood); thrombocytopenia
Erythema	Red, flushed color of skin due to dilation of blood vessels in the skin	*Extensive:* drug reactions; *Localized:* burns, contact dermatitis

4

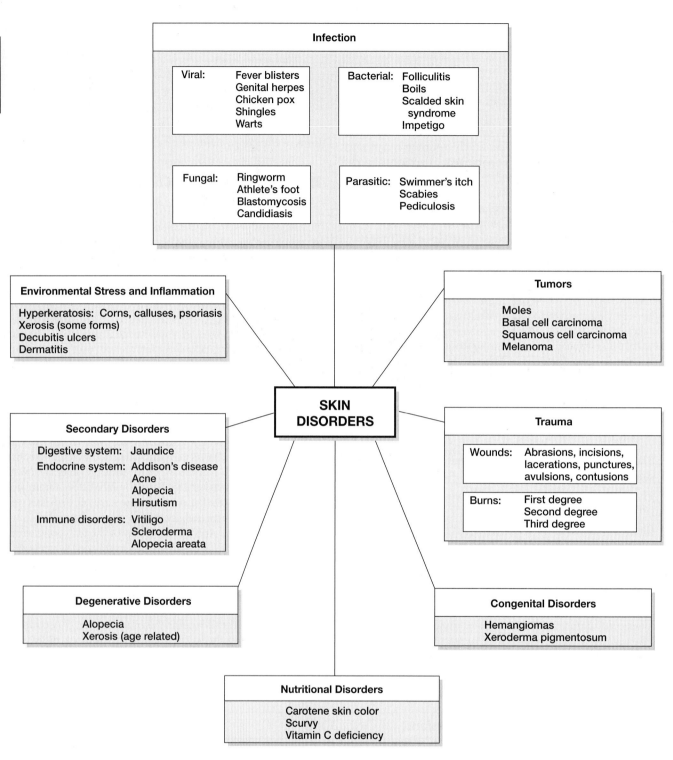

Infection

Viral: Fever blisters
Genital herpes
Chicken pox
Shingles
Warts

Bacterial: Folliculitis
Boils
Scalded skin
syndrome
Impetigo

Fungal: Ringworm
Athlete's foot
Blastomycosis
Candidiasis

Parasitic: Swimmer's itch
Scabies
Pediculosis

Environmental Stress and Inflammation

Hyperkeratosis: Corns, calluses, psoriasis
Xerosis (some forms)
Decubitis ulcers
Dermatitis

Tumors

Moles
Basal cell carcinoma
Squamous cell carcinoma
Melanoma

Secondary Disorders

Digestive system: Jaundice
Endocrine system: Addison's disease
Acne
Alopecia
Hirsutism
Immune disorders: Vitiligo
Scleroderma
Alopecia areata

SKIN DISORDERS

Trauma

Wounds: Abrasions, incisions,
lacerations, punctures,
avulsions, contusions

Burns: First degree
Second degree
Third degree

Degenerative Disorders

Alopecia
Xerosis (age related)

Congenital Disorders

Hemangiomas
Xeroderma pigmentosum

Nutritional Disorders

Carotene skin color
Scurvy
Vitamin C deficiency

● **FIGURE 9**
Disorders of the Integumentary System

Pressure on the skin, another form of stress, can produce *decubitis ulcers,* or *bedsores. Decubitis* means "to lie down"; an *ulcer* is a localized loss of epithelium. Decubitis ulcers form where dermal blood vessels are compressed against deeper structures such as bones or joints, so that local circulation is reduced enough to damage dependent tissues.

■ SKIN CANCERS   *HA p. 91*

Almost everyone has several benign lesions of the skin; freckles and moles are examples. Skin cancers are the most common form of cancer, and the most common skin cancers are caused by prolonged exposure to sunlight.

TABLE 13   Common Infectious Diseases of the Integumentary System

Disease	Organism (Name)	Description
Bacteria		
Folliculitis	*Staphylococcus aureus*	Infections of hair follicles may form pimples, furuncles (boils), and abscesses.
Scalded skin syndrome	*Staphylococcus aureus*	In infants; large areas of epidermis blister, peel, and leave oozing red areas.
Impetigo	*Staphylococci, Streptococci,* or both	Pustules form on skin, dry, and become yellow crusts; skin pigment may not reappear after healing.
Viruses		
Oral herpes	Herpes simplex 1	Vesicles (blisters), also called cold sores or fever blisters, form, usually on lips (vesicles heal, but may episodically recur)
Genital herpes	Herpes simplex 2	Lesions similar to those in oral herpes form on external genitalia; vesicles disappear and reappear.
Chicken pox (*varicella*)	Herpes varicella-zoster	In children; small, red macules form flaccid vesicles, which dry and become crusts.
Shingles (*zoster*)	Herpes varicella-zoster	In adults; distinctive vesicular lesions form a pattern along sensory nerves; severe, prolonged pain may follow attacks.
Warts	Human papillomaviruses	Rough papules form in the epidermis; genital warts may be sexually transmitted and may promote cervical cancer.
Fungi		
Ringworm (*tinea*)	*Epidermophyton, Microsporum,* and *Trichophyton*	Dry, scaly lesions form on the skin in different parts of the body, including the scalp (*tinea capitis*), body (*tinea corporis*), groin (*tinea cruris*), foot (*tinea pedis*, which may be moist), and nails (*tinea unguium*).
Blastomycosis	*Blastomyces dermatitidis*	Pustules and abscesses form in the skin; may affect other organs.
Candidiasis	*Candida albicans*	Normal inhabitant of the human body surface; may infect many organs; red, frequently moist lesions form in skin infections; nails may also become infected.
Parasites		
Swimmer's itch	*Schistosoma* worms (flukes)	Freshwater larval stages of schistosome worms (flukes) burrow into skin and cause itching.
Scabies	*Sarcoptes scabiei* (itch mite)	Itch mite burrows and lays eggs in skin, often in areas between fingers and at the wrists; entrance marked by tiny, scaly swellings that become red and intensely itchy.
Pediculosis	*Pediculus humanus* (human body louse)	Lice infestations on body and scalp; bites produce redness, dermatitis, and itching.
	Phthirus pubis (pubic louse)	"Crabs"; lice infestation of the pubic area; bites also cause itching.

A **basal cell carcinoma** (Figure 10a●) is a malignant cancer that originates in the germinativum (basal) layer. This is the most common skin cancer, and roughly two-thirds of these cancers appear in areas subjected to chronic UV exposure. These carcinomas have recently been linked to an inherited gene.

Squamous cell carcinomas are less common, but almost totally restricted to areas of sun-exposed skin. Metastasis seldom occurs in squamous cell carcinomas and almost never in basal cell carcinomas, and most people survive these cancers. The usual treatment involves surgical removal of the tumor,

and at least 95% of patients survive 5 years or more after treatment. (This statistic, the 5-year survival rate, is a common method of reporting long-term prognoses.) Compared with these common and seldom life-threatening cancers, **malignant melanomas** (mel-a-NŌ-maz) are extremely dangerous (Figure 10b●). In this condition, cancerous melanocytes grow rapidly and metastasize throughout the lymphatic system. The outlook for long-term survival is dramatically different, depending on when the condition is diagnosed. If the condition is localized, the 5-year survival rate is 90%; if widespread, the survival rate drops to 14%.

4

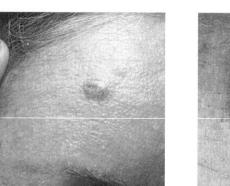
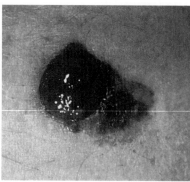

(a) Basal cell carcinoma　　　**(b) Melanoma**

Fair-skinned individuals who live in the tropics are most susceptible to all forms of skin cancer, because their melanocytes are unable to shield them from the ultraviolet radiation. Sun damage can be prevented by avoiding exposure to the sun during the middle hours of the day and by using clothing, a hat, and a sunblock (not a tanning oil or sunscreen)— a practice that also delays the cosmetic problems of sagging and wrinkling. Everyone who expects to be out in the sun for any length of time should choose a broad-spectrum sunblock with a sun protection factor (SPF) of at least 15; blonds, redheads, and people with very fair skin are better off with a sun protection factor of 20 to 30. (One should also remember the risks before spending time in a tanning salon or tanning bed.) The use of sunscreens has now become even more important as the ozone gas in the upper atmosphere is destroyed by our industrial emissions. Ozone absorbs UV before it reaches the earth's surface, and in doing so, it assists the melanocytes in preventing skin cancer. Australia, which is most affected by the depletion of the ozone layer near the South Pole (the "ozone hole"), is already reporting an increased incidence of skin cancers.

■ DERMATITIS　*HA p. 92*

Inflammation is a complex process that helps defend against pathogens and injury. Because skin contains an abundance of sensory receptors, inflammation can be very painful.

Dermatitis (der-muh-TĪ-tis) is an inflammation of the skin that involves primarily the papillary layer. The inflammation typically begins in a part of the skin exposed to infection or irritated by chemicals, radiation, or mechanical stimuli. Dermatitis may cause no discomfort, or it may produce an annoying itch, as in poison ivy. Sometimes the condition can be quite painful, and the inflammation can spread rapidly across the entire integument.

Dermatitis has many forms, some of them common:

- **Contact dermatitis** generally occurs in response to strong chemical irritants. It produces an itchy rash that may spread to other areas; poison ivy is an example.

- **Eczema** (EK-se-muh), also called atopic dermatitis, can be triggered by temperature changes, fungi, chemical irritants, greases, detergents, or stress. Hereditary factors, environmental factors, or both can promote its development.

- **Diaper rash** is a localized dermatitis caused by a combination of moisture, irritating chemicals from fecal or urinary wastes, and flourishing microorganisms.

- **Urticaria** (ur-ti-KAR-ē-uh), also called wheals or *hives,* is an extensive allergic response to a food, a drug, an insect bite, infection, stress, or some other stimulus.

TRAUMA TO THE SKIN　*HA p. 92*

Trauma is a physical injury caused by pressure, impact, distortion, or some other mechanical force. Trauma to the skin is common, and a number of terms are used to describe it. Such injuries generally affect all components of the integument, and each type of wound presents a different series of problems to clinicians attempting to limit damage and promote healing.

An **open wound** is an injury producing a break in the epithelium. The major categories of open wounds are illustrated in Figure 11●. **Abrasions** are the result of scraping (Figure 11a●). Bleeding may be slight, but a considerable area may be open to invasion by microorganisms. **Incisions** are linear cuts produced by sharp objects (Figure 11b●). Bleeding can be severe if deep vessels are damaged. The bleeding may help flush the wound, and closing the incision with bandages or stitches can limit the area open to infection while healing is underway. A **laceration** is a jagged, irregular tear in the surface of the skin produced by solid impact or by an irregular object (Figure 11c●). Tissue damage is extensive, and repositioning the opposing sides of the injury may be difficult. Despite the bleeding that generally occurs, lacerations are prone to infection. **Punctures** result when slender, pointed objects pierce the epithelium (Figure 11d●). Little bleeding results, and any microbes delivered under the epithelium in the process are likely to find conditions to their liking. In an **avulsion**, chunks of tissue are torn away by brute force in, for example, an auto accident or an explosion (Figure 11e●). Bleeding may be considerable, and even more serious internal damage may be present.

Closed wounds can affect any internal tissue, but because the epithelium is intact, the likelihood of infection is reduced. A **contusion** is a bruise causing bleeding in the dermis. "Black and blue" marks are familiar examples of contusions; most are not dangerous, but contusions of the head, such as "black eyes," may be the result of potentially life-threatening intracranial bleeding. Closed wounds caused by trauma severe enough to affect internal organs and organ systems are almost always serious threats to life.

Skin can regenerate effectively even after considerable damage. Burns, a type of trauma that can affect the epidermis, dermis, and deeper tissues, are discussed in a later section.

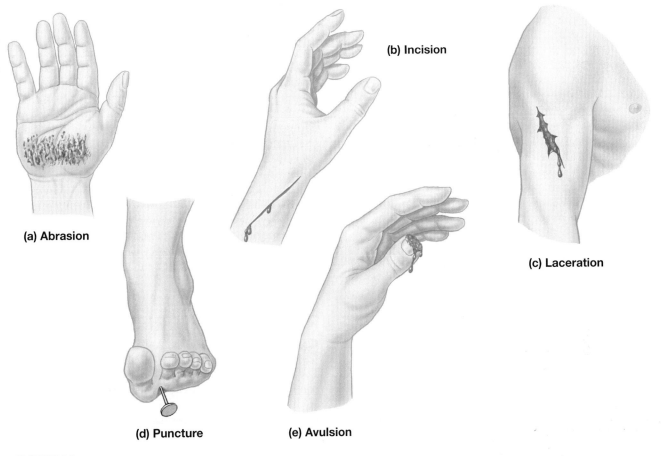

(b) Incision

(a) Abrasion

(c) Laceration

(d) Puncture

(e) Avulsion

● **FIGURE 11**
Major Types of Open Wounds

■ TUMORS IN THE DERMIS   *HA p. 92*

Tumors seldom develop in the dermis, and those that do appear are usually benign. Two forms of benign tumors called **hemangiomas** may appear among dermal blood vessels during embryonic development. Viewed from the surface, these form prominent *birthmarks*. A **capillary hemangioma** involves capillaries of the papillary layer. It usually enlarges after birth, but subsequently fades and disappears. **Cavernous hemangiomas,** or "port-wine stains," affect larger vessels in the dermis. Such birthmarks usually last a lifetime unless treated.

■ BALDNESS AND HIRSUTISM
HA p. 96

Hairs are dead, keratinized structures, so no amount of oiling, shampooing, or dousing with kelp extracts, vitamins, or nutrients will influence the follicle buried in the dermis. Skin conditions that affect follicles can contribute to hair loss. Temporary baldness can also result from exposure to radiation or to many of the toxic drugs used in cancer therapy. Two factors interact to cause most cases of baldness. A bald individual has a genetic susceptibility triggered by sufficiently large quantities of male sex hormones. Many women carry the genetic background for baldness, but unless major hormonal

abnormalities develop, as in certain endocrine tumors, hair loss does not occur. (In some women, however, scalp hair does thin after menopause.)

Male pattern baldness affects the top of the head and forehead first, only later reducing the density of hair along the sides (Figure 12●). Thus, hair follicles can be removed from the sides and implanted on the top or front of the head, tem-

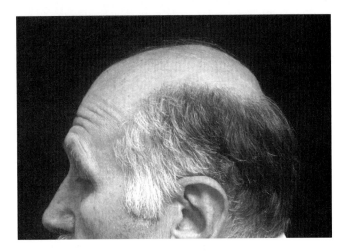

● **FIGURE 12**
Male Pattern Baldness

4

porarily delaying a receding hairline. This procedure is expensive, and not every hair transplant is successful. Finasteride is a prescription drug that blocks the production of one form of testosterone. In low doses, it slows the progression of male pattern baldness. *Minoxidil,* a drug originally marketed for the control of high blood pressure, appears to stimulate inactive hair follicles when rubbed onto the scalp. It is now available without a prescription as *Rogaine* and is most effective in preventing the progression of early hair loss.

Alopecia areata (al-ō-PĒ-shē-uh ar-ē-AH-ta) is a localized hair loss that can affect either gender. The cause is not known, and the severity and persistence of hair loss varies from case to case. The condition is associated with several disorders of the immune system. It has also been suggested that stress can promote alopecia areata in individuals who are genetically prone to baldness.

Hirsutism (HER-soot-izm; *hirsutus,* bristly) is hair growth on women that occurs in patterns generally characteristic of men. Because considerable overlap exists between the two genders in terms of normal hair distribution, and because racial and genetic differences are significant, the precise definition is more often a matter of personal taste than of objective analysis. Age and sex hormones may play a role, because hairiness increases late in pregnancy and menopause produces a change in body hair patterns.

Severe hirsutism is associated with abnormal amounts of androgen (male sex hormone) production in either the ovaries or the adrenal glands, and treatment of the underlying endocrine abnormality is indicated. Cosmetically unwanted follicles can be permanently "turned off" by plucking a growing hair and removing the papilla. *Electrolysis,* which destroys the follicle with a jolt of electricity, and lasers that damage or destroy pigmented follicles are more costly, as they require professional services, but the results are more reliable. Patients may also be treated with drugs that reduce or prevent androgen stimulation of the follicles.

■ FOLLICULITIS AND ACNE   *HA p. 97*

Although sebum has bactericidal (bacteria-killing) properties, under some conditions bacteria invade sebaceous glands. The presence of bacteria in sebaceous glands or follicles can produce **folliculitis** (fo-lik-ū-LĪ-tis), a local inflammation. If the duct of a sebaceous gland becomes blocked, a distinctive abscess called a **furuncle** (FUR-ung-kl), or boil, develops. The usual treatment for a furuncle is to cut it open, or "lance" it, so that drainage and healing can occur. Most individuals at puberty, especially those with a genetic tendency toward **acne**, have larger-than-average sebaceous glands. When the ducts of these glands become blocked, their secretions accumulate and the bacteria *P. acnes* colonizes the area. This produces a localized inflammation. Sex hormone production, which accelerates at puberty, stimulates the sebaceous glands. Their secretory output may be further encouraged by anxiety, stress, physical exertion, and certain foods or drugs.

The visible signs of acne are called **comedos** (ko-MĒ-dōz). Closed comedos ("whiteheads") contain accumulated, stagnant secretions. Open comedos ("blackheads") are open to the surface and contain more solid material. Although nei-

ther condition indicates the presence of dirt in the pores, washing may help keep superficial oiliness down.

Acne generally fades after sex hormone concentrations stabilize. Topical antibiotics, vitamin A derivatives such as *Retin-A,* or peeling agents may help reduce inflammation and minimize scarring. In cases of severe acne, the most effective treatment involves antibiotic drugs. However, oral antibiotic therapy has risks, including the development of antibiotic-resistant bacteria, so this therapy is not used unless other treatment methods have failed. Truly dramatic improvements in severe cases have been obtained with the prescription drug *Accutane.* This compound is structurally similar to vitamin A, and it reduces oil gland activity on a long-term basis. A number of side effects, such as dry skin, lips, and eyes, as well as depression, have been reported; these disappear when the treatment ends. The use of Accutane during the first month of pregnancy carries a 25-times-normal risk of inducing birth defects, so women must avoid pregnancy while using this drug.

REPAIRING INJURIES TO THE SKIN   *HA p. 101*

The skin can regenerate effectively, even after considerable damage has occurred, because stem cells persist in both the epithelial and connective tissue components. Germinative cell divisions replace lost epidermal cells, and mesenchymal cell divisions replace lost dermal cells. The process can be slow. When large surface areas are involved, problems of infection and fluid loss complicate the situation. The relative speed and effectiveness of skin repair vary with the type of wound involved. A slender, straight cut, or *incision,* may heal relatively quickly compared with a deep scrape, or *abrasion,* which involves a much greater surface area to be repaired.

Figure 13● illustrates the four stages in the regeneration of the skin after an injury. When damage extends through the epidermis and into the dermis, bleeding generally occurs (STEP 1). The blood clot, or **scab,** that forms at the surface temporarily restores the integrity of the epidermis and restricts the entry of additional microorganisms into the area (STEP 2). The bulk of the clot consists of an insoluble network of *fibrin,* a fibrous protein that forms from blood proteins during the clotting response. The clot's color reflects the presence of trapped red blood cells. Cells of the stratum germinativum undergo rapid divisions and begin to migrate along the edges of the wound in an attempt to replace the missing epidermal cells. Meanwhile, macrophages patrol the damaged area of the dermis, phagocytizing any debris and pathogens.

If the wound occupies an extensive area or involves a region covered by thin skin, dermal repairs must be underway before epithelial cells can cover the surface. Divisions by fibroblasts and mesenchymal cells produce mobile cells that invade the deeper areas of injury. Endothelial cells of damaged blood vessels also begin to divide, and capillaries follow the fibroblasts, enhancing circulation. The combination of blood clot, fibroblasts, and an extensive capillary network is called **granulation tissue.**

Over time, deeper portions of the clot dissolve, and the number of capillaries declines. Fibroblast activity leads to the

4

STEP 1

Bleeding occurs at the site of injury immediately after the injury, and mast cells in the region trigger an inflammatory response.

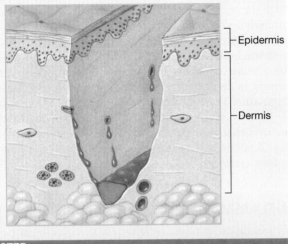

STEP 2

After several hours, a scab has formed and cells of the stratum germinativum are migrating along the edges of the wound. Phagocytic cells are removing debris, and more of these cells are arriving via the enhanced circulation in the area. Clotting around the edges of the affected area partially isolates the region.

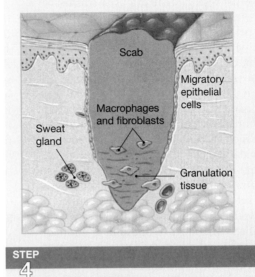

STEP 3

One week after the injury, the scab has been undermined by epidermal cells migrating over the meshwork produced by fibroblast activity. Phagocytic activity around the site has almost ended, and the fibrin clot is disintegrating.

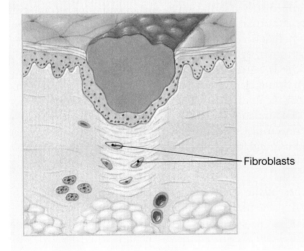

STEP 4

After several weeks, the scab has been shed, and the epidermis is complete. A shallow depression marks the injury site, but fibroblasts in the dermis continue to create scar tissue that will gradually elevate the overlying epidermis.

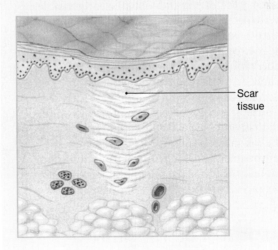

● **FIGURE 13**
Wound Healing

4

appearance of collagen fibers and typical ground substance (STEP 3). The repairs do not restore the integument to its original condition, however, because the dermis will contain an abnormally large number of collagen fibers and relatively few blood vessels. Severely damaged hair follicles, sebaceous or sweat glands, muscle cells, and nerves are seldom repaired, and they too are replaced by fibrous tissue. The formation of this rather inflexible, fibrous, noncellular **scar tissue** can be considered a practical limit to the healing process (STEP 4).

We do not know what regulates the extent of scar tissue formation, and the process is highly variable. For example, surgical procedures performed on a fetus do not leave scars, perhaps because damaged fetal tissues do not produce the same types of growth factors that adult tissues do. In some adults, most often those with dark skin, scar tissue formation may continue beyond the requirements of tissue repair. The result is a thickened mass of scar tissue that begins at the site of injury and grows into the surrounding dermis. This thick, raised area of scar tissue, called a **keloid** (KĒ-loyd), is covered by a shiny, smooth epidermal surface. Keloids most commonly develop on the upper back, shoulders, anterior chest, or earlobes. They are harmless; in fact, some aboriginal cultures intentionally produce keloids as a form of body decoration.

In fact, people in societies around the world adorn the skin with culturally significant markings of one kind or another. Tattoos, piercings, keloids and other scar patterns, and even high-fashion makeup are all used to "enhance" the appearance of the integument. Scarification is performed in several African cultures, resulting in a series of complex, raised scars on the skin. Polynesian cultures have long preferred tattoos as a sign of status and beauty. A dark pigment is inserted deep within the dermis of the skin by tapping on a needle, shark tooth, or bit of bone. Because the pigment is inert, if infection does not occur (a potentially serious complication), the markings remain for the life of the individual, clearly visible through the overlying epidermis. American popular culture has recently rediscovered tattoos as a fashionable form of body adornment. The colored inks that are commonly used are less durable, and older tattoos can fade or lose their definition.

Tattoos can now be partially or completely removed. The removal process takes time (10 or more sessions may be required to remove a large tattoo), and scars often remain. To remove the tattoo, an intense, narrow beam of light from a laser breaks down the ink molecules in the dermis. Each blast of the laser that destroys the ink also burns the surrounding dermal tissue. Although the burns are minor, they accumulate and result in the formation of localized scar tissue.

■ INFLAMMATION OF THE SKIN

HA p. 101

The epidermis provides significant protection from mechanical and chemical hazards because of its thick cell layers and keratin covering. Although the surface of the skin normally contains a variety of microorganisms, most of them are harmless as long as they remain outside the stratum corneum. Penetration of the superficial layers may produce familiar conditions such as "athlete's foot" (a fungal infection), and warts

and cold sores (both viral infections). To reach the underlying connective tissues, a bacterium must survive the bacteriocidal ingredients of sebum, avoid being flushed from the surface by the sweat gland secretions, penetrate the stratum corneum, squeeze between the junctional complexes of deeper layers, escape the Langerhans cells, and cross the basement membrane. Unfortunately, once there, the papillary layer of the dermis provides all the elements for the growth of microorganisms: warmth, darkness, moisture, and nutrients.

If the protective barriers are crossed, or if an injury breaks through the epidermis, foreign materials or pathogens reach the underlying tissues. Mast cells within the dermis then respond by triggering a powerful inflammatory response. Inflammation begins immediately after an injury and produces swelling, redness, heat, and pain. Inflammation in the skin is important in defending the body against injury and disease. Inflammation and regeneration are controlled at the tissue level. The two phases overlap; inflammation establishes a framework that guides the cells responsible for reconstruction, and repairs are under way well before cleanup operations have ended.

■ INFLAMMATION

Many stimuli can produce inflammation, including impact, abrasion, distortion, chemical irritation, infection by pathogenic organisms (such as bacteria or viruses), or extreme temperatures (hot or cold). The common factor is that each of these stimuli either kills cells, damages fibers, or injures the tissue in some other way. These changes alter the chemical composition of the interstitial fluid: Damaged cells release prostaglandins, proteins, and potassium ions, and the injury itself may have introduced foreign proteins or pathogens.

Immediately after the injury, tissue conditions become even more abnormal. **Necrosis** (ne-KRŌ-sis) is the tissue degeneration that occurs after cells have been injured or destroyed. This process begins several hours after the initial event, and the damage is caused by lysosomal enzymes. Lysosomes break down through autolysis, releasing digestive enzymes that first destroy the injured cells and then attack surrounding tissues. The accumulation of debris, fluid, dead and dying cells, and other necrotic tissue components that may result is known as **pus**. Pus often forms at an infection site in the dermis. An accumulation of pus in an enclosed tissue space is called an **abscess.**

These tissue changes trigger the inflammatory response by releasing chemicals that stimulate mast cells. When an injury occurs that damages fibers and cells, the stimulated mast cells release chemicals (**histamine** and **heparin**). Activation of mast cells has two primary effects (Figure 14●):

1. Histamine relaxes the smooth muscle tissue in the vessel walls, and the vessels enlarge, or **dilate.** This dilation increases the blood flow through the tissue, giving the region a reddish color and making it warm to the touch. The increased blood flow accelerates the delivery of nutrients and oxygen and the removal of waste products and toxic chemicals. It also brings white blood cells to the region. These

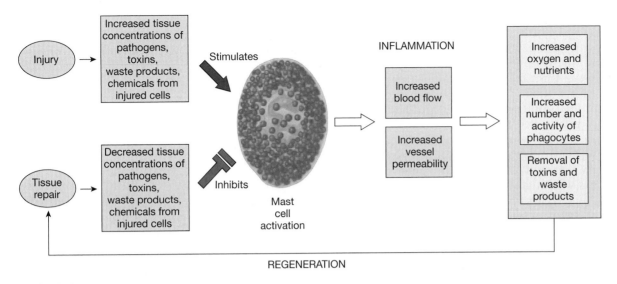

● **FIGURE 14**
Inflammation of the Skin

cells migrate into the injury site and assist in the defense and cleanup operations.

2. Histamine makes the endothelial cells of local capillaries more permeable.

Plasma, including blood proteins, diffuses into the injured tissue, and the area becomes swollen. Once in the tissue, some of the blood proteins combine to form large, insoluble fibers, known as **fibrin.** The heparin released by mast cells prevents the formation of fibrin at the injury site, but a dense fibrous meshwork, called a **clot,** surrounds the damaged area.

The clot walls off the inflamed region, isolating it and slowing the spread of cellular debris or bacteria into adjacent tissues.

The abnormal chemicals distributed by the increased blood flow attracts cells of the immune system to the site. Stimulated macrophages and newly arrived white blood cells defend the tissue and perform cleanup operations. Fixed macrophages, free macrophages, and microphages engulf debris and bacteria. Lymphocytes convert to plasma cells, producing antibodies that attack pathogens and foreign proteins. Usually the combination of physical attack, through phagocytosis, and chemical attack, through antibodies, succeeds in cleaning up the region and eliminating the inflammatory stimulus.

The combination of abnormal tissue conditions and chemicals released by mast cells also stimulate neurons which generate pain sensations. These sensations trigger immediate reflexive action (such as pulling the affected part of the body away from the source of the injury) that reduces the chances for further damage.

■ **REGENERATION**

When damage extends through the epidermis and into the dermis, bleeding usually occurs. The fibrin clot, or **scab,** that forms at the surface temporarily restores the integrity of the epidermis and restricts the entry of additional microorganisms and further blood loss. Cells of the stratum germinativum undergo rapid divisions and begin to migrate along the sides of the wound in an attempt to replace the missing epidermal cells. If the wound covers an extensive area or involves a region covered by thin skin, dermal repairs must be under way before epithelial cells can cover the surface. Fibroblast and mesenchymal cell divisions produce mobile cells that invade the deeper area of injury, migrating along fibrin strands. Endothelial cells of damaged blood vessels also begin to divide, and capillaries follow the fibroblasts, eventually uniting and providing a circulatory supply. The combination of fibrin clot, fibroblasts, and an extensive capillary network is called **granulation tissue.** Over time, the fibrin dissolves, and the number of capillaries declines. Fibroblast activity leads to the appearance of collagen fibers and typical ground substance.

While dermal repairs are in progress, **contraction** pulls the edges of the wound closer together. The mechanism of contraction is uncertain, but it is an essential part of the healing process when damage has been extensive. For example, after amputation of a finger or even the distal portion of a limb, over 90% of the exposed surface is covered by contraction of the wound edges, rather than by the divisions and migration of epithelial cells. Contraction distorts the adjacent surface, as if the skin were being stretched to cover the injury site. If contraction and epithelial cell migration cannot cover the wound, skin grafts, discussed in a later section, may be required.

■ **COMPLICATIONS OF INFLAMMATION**

When pus accumulates in an enclosed tissue space, the result is an abscess. In the skin, an abscess can form as pus builds up inside the fibrin clot that surrounds the injury site. If the cellular defenses succeed in destroying the invaders, the pus will be either absorbed or surrounded by a fibrous capsule, creating a cyst.

Erysipelas (er-i-SIP-e-las; *erythros,* red + *pella,* skin) is a widespread inflammation of the dermis caused by bacterial infection. If the inflammation spreads into the subcutaneous layer and deeper tissues, the condition is called **cellulitis** (sel-ū-LĪ-tis).

Erysipelas and cellulitis develop when bacterial invaders break through the fibrin wall. The bacteria involved produce large quantities of hyaluronidase, an enzyme that liquefies the ground substance, and fibrinolysin, an enzyme that breaks down fibrin and prevents clot and abscess formation. These are serious conditions that require prompt antibiotic therapy.

An **ulcer** is a localized shedding of an epithelium. **Decubitus ulcers,** also known as "bedsores," may afflict bedridden or mobile patients with circulatory restrictions, especially when splints, casts, or bedding continually press against superficial blood vessels. Such sores most often affect the skin near joints or projecting bones, where the vessels are constricted by pressure against hard underlying structures. The chronic lack of circulation kills epidermal cells, removing a barrier to bacterial infection, and eventually the dermal tissues deteriorate as well. (A comparable necrosis will occur in any tissues deprived of adequate circulation.) Bedsores can be prevented or treated by frequent changes in body position that vary the pressures applied to vulnerable skin sites.

BURNS AND GRAFTS *HA p. 101*

Burns result from exposure of the skin to heat, radiation, electrical shock, or strong chemical agents. The severity of the burn reflects the depth of penetration and the total area affected. Table 14 summarizes a classification of burns based on the depth of involvement.

First- and second-degree burns are also called **partial-thickness** burns because damage is restricted to the superficial layers of the skin. Accessory structures such as hair follicles and glands are usually unaffected. **Full-thickness burns,** or **third-degree burns,** destroy the epidermis and dermis, extending into subcutaneous tissues. These burns are actually less painful than second-degree burns, because sensory nerves are destroyed along with accessory structures, blood vessels, and other dermal components. Extensive third-degree burns cannot repair themselves, and the site remains exposed to potential infection.

Roughly 10,000 people die from burns each year in the United States. The larger the area burned, the more significant the effects on integumentary function. Burns that cover more than 20% of the skin surface represent serious threats to life. The extent of coverage of a burn is estimated on the basis of the percent surface area involved (Figure 15●) Burns are dangerous because they affect the following functions:

- **Fluid and electrolyte balance.** Even areas with partial-thickness burns lose their effectiveness as barriers to fluid and electrolyte losses. In full-thickness burns, the rate of fluid loss through the skin may reach five times the normal level.

- **Thermoregulation.** Increased fluid loss means increased evaporative cooling. More energy must be expended to keep body temperature within acceptable limits.

- **Protection from attack.** The burned epidermal surface, oozing tissue fluid, encourages bacterial growth. If the skin is broken at a blister or the site of a third-degree burn, infection is likely. Widespread bacterial infection, or **sepsis** (*septikos,* rotting), is the leading cause of death in burn victims.

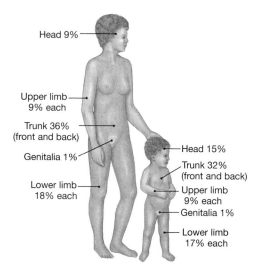

Head 9%
Upper limb 9% each
Trunk 36% (front and back)
Genitalia 1%
Lower limb 18% each
Head 15%
Trunk 32% (front and back)
Upper limb 9% each
Genitalia 1%
Lower limb 17% each

● FIGURE 15
A Method of Estimating the Surface Area Affected by Burns

TABLE 14	A Classification of Burns		
Classification	**Damage Report**		**Appearance and Sensation**
First-degree burn	*Killed:* superficial cells of epidermis.		Inflamed, tender. (example: mild sunburn)
	Injured: deeper layers of epidermis, papillary dermis.		
Second-degree burn	*Killed:* superficial and deeper cells of epidermis; dermis may be affected.		Blisters, pain. (example: touching a hot iron)
	Injured: damage may extend into reticular layer of the dermis, but many accessory structures unaffected.		
Third-degree burn	*Killed:* all epidermal and dermal cells.		Charred, no sensation at all.
	Injured: hypodermal and deeper tissues and organs.		

Because full-thickness burns cannot heal without aid, surgical procedures are necessary to encourage healing. In a skin graft, areas of intact skin are transplanted to cover the burn site. With the development of fluid replacement therapies, infection control methods, and grafting techniques, the recovery rate for severe burns has improved dramatically. At present, young patients with burns over 80% of the body have an approximately 50% chance of recovery.

Recent advances in cell culture techniques may improve survival rates further. It is now possible to remove a small section of skin and grow it under controlled laboratory conditions. Over time, the germinative cell divisions produce large sheets of epidermal cells that can then be used to cover the burn area. From initial samples the size of postage stamps, square yards of epidermis have been grown and transplanted onto body surfaces. Although questions remain concerning the strength and flexibility of the repairs, skin cultivation represents a substantial advance in the treatment of extensive burns (see the discussion on synthetic skin at the end of this section).

■ SYNTHETIC SKIN

Skin grafts are used if large areas of skin have been completely destroyed. Pieces of undamaged skin from other areas of the body are used. Usually a partial-thickness patch with the epidermis and part of the dermis is shaved off the donor site, cut in a mesh pattern to enlarge it, and tacked onto the burn area. An epithelium is reestablished at the donor site by epithelial cell migration from the edges and from follicles in the remaining dermis. After healing, accessory structures such as hair and sweat glands are missing in the grafted area, and flexibility is reduced. If the damaged area is large, there may not be enough normal skin available for grafting. *Epidermal culturing* can produce a new epithelial layer to cover a burn site. An epidermal sample from a burn patient can be cultured in a controlled environment that contains epidermal growth factors, fibroblast growth factors, and other stimulatory chemicals. This artificially produced epidermis can now be transplanted to cover the site of an injury. The larger the area that must be covered, the longer the culturing process continues. After three or four weeks of culturing, the cells obtained in the original sample can provide enough epidermis to cover the entire body surface of a typical adult. The absence of a dermis makes this grafted epidermis less flexible and more fragile than normal—a particular problem around joints and areas of friction.

The success of grafting is limited by the contamination of the wound site by bacteria while the epidermis is being cultured. Contamination is prevented by the use of a skin *allograft*. In this procedure, skin from a frozen cadaver is removed and placed over the wound as a temporary method of sealing the surface. Before the immune system of the patient attacks it, the graft will be partially or completely removed to provide a binding site for the epidermal transplant. After grafting, the complete reorganization and repair of the dermis and epidermis at the site of the injury takes approximately five years.

A second new procedure provides a model for dermal repairs that takes the place of normal tissue. A special synthetic skin, or "cellular wound dressing," is used. Several forms are available; most involve a biosynthetic dermis made of collagen (usually from cows) and/or a silicon polymer covered by a layer of dead or live human epidermal cells. This combination, available under trade names such as Integra (Transcyte) or Orcel, is placed over the burn site either alone or covered by a cultured epidermal layer. Over time, fibroblasts and epidermal cells of the recipient migrate among the grafted collagen fibers, and gradually replace the model framework with their own.

CHAPTER 5
OSSEOUS TISSUE AND SKELETAL STRUCTURE

AN INTRODUCTION TO THE SKELETAL SYSTEM AND ITS DISORDERS

The skeletal framework of the body is composed of at least 206 bones and the associated tendons, ligaments, and cartilages. The skeletal system has a variety of important functions, including the support of soft tissues, blood cell production, mineral and lipid storage, and, through its relationships with the muscular system, the support and movement of the body as a whole. Skeletal system disorders can thus affect many other systems. The skeletal system is in turn influenced by the activities of other systems. For example, weakness or paralysis of skeletal muscles will lead to a weakening of the associated bones and may cause changes in their relative positions.

■ THE DYNAMIC SKELETON   *HA p. 107*

Although the bones you study in the lab may seem to be rigid and permanent structures, the living skeleton is dynamic and undergoes continuous remodeling. The remodeling process involves (1) bone deposition by osteoblasts and (2) *osteolysis*, or dissolution of bone matrix, by osteoclasts. As indicated in Figure 16●, the net result of the remodeling varies with the following five factors:

1. **The age of the individual.** During development, bone deposition occurs faster than bone resorption; as the amount of bone increases, the skeleton grows. At maturity, bone deposition and resorption are in balance. As the aging process continues, the rate of bone deposition declines and the bones become less dense. This gradual weakening, called *osteopenia,* begins at age 30–40 and may ultimately progress to osteoporosis (p. 40).

2. **The applied physical stresses.** Heavily stressed bones become thicker and stronger, and lightly stressed bones become thinner and weaker. Skeletal weakness can therefore result from muscular disorders, such as *myasthenia gravis* (p. 54) and the *muscular dystrophies* (p. 53), and from conditions that affect motor neurons of the central nervous system, such as spinal cord injuries (Chapter 14), *demyelination disorders* (p. 65), and *multiple sclerosis* (p. 65).

3. **Circulating hormone levels.** Changing levels of growth hormone (GH), androgens and estrogens, thyroid hormones, parathyroid hormone, and calcitonin increase or decrease the rate of mineral deposition in bone. As a result, many disorders of the endocrine system affect the skeletal system.

4. **Rates of calcium and phosphate absorption and excretion.** For bone mass to remain constant, the rate of calcium and phosphate excretion, primarily at the kidneys, must be balanced by the rate of calcium and phosphate absorption at the digestive tract. Kidney failure, dietary calcium deficiencies, or problems that reduce calcium and phosphate absorption at the digestive tract will thus directly affect the skeletal system.

5. **Genetic or environmental factors.** Genetic or environmental factors can affect the structure of bone or the remodeling process. A number of abnormalities of skeletal

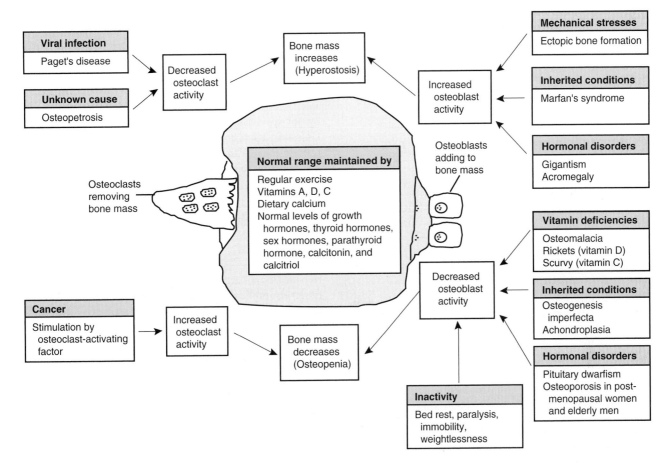

● **FIGURE 16**
Factors Affecting Bone Mass

development, such as *Marfan's syndrome* and *achondroplasia*
(p. 39), are inherited. When bone fails to form embryoni-
cally in certain areas, underlying tissues can be exposed and
associated functions can be altered. This type of abnormality occurs in a *cleft palate*, and in *spina bifida* (p. 42). Environmental stresses can alter the shape and contours of
developing bones. For example, some cultures used boards
to form an infant's skull to a shape considered fashionable,
and nerve damage from spina bifida may cause the bones of
an infant's feet to assume the equinovarus ("club foot")
position. Environmental forces can also result in the formation of bone in unusual locations. These *heterotopic bones*
(p. 36) may develop in a variety of connective tissues exposed to chronic friction, pressure, or mechanical stress. For
example, cowboys in the 19th century sometimes developed
heterotopic bones in the dermis of the thigh, from friction
with the saddle.

EXAMINATION OF THE SKELETAL
SYSTEM   *HA p. 111*

The bones of the skeleton cannot be seen without relatively sophisticated equipment. However, a number of physical signs can
assist in the diagnosis of a bone or joint disorder. Important factors noted in the physical examination include the following:

- **A limitation of movement or stiffness.** Many joint disorders, such as the various forms of arthritis, restrict movement or produce stiffness at one or more joints.

- **The distribution of joint involvement and inflammation.**
In a *monoarthritic* condition, only one joint is affected. In
a *polyarthritic* condition, several joints are affected
simultaneously.

- **Sounds associated with joint movement.** Bony crepitus
(KREP-i-tus) is a crackling or grating sound generated during the movement of an abnormal joint. The sound can result from the movement and collision of bone fragments
following a joint fracture or from friction at an arthritic
joint. If the periosteum is affected, movements generating
crepitus are very painful.

- **The presence of abnormal bone deposits.** The callus is a
thickened area of bone that develops at a fracture site as
part of the repair process. Abnormal bone deposits can also
develop around the joints in the fingers. These deposits are
called *nodules*. When palpated, nodules are solid and painless. Nodules, which can restrict movement, commonly
form at the interphalangeal joints of the fingers in osteoarthritis.

- **Abnormal posture.** Bone disorders that affect the vertebral
column can result in abnormal posture. The abnormality is
most apparent when the condition alters the normal spinal
curvature. Examples include *kyphosis, lordosis,* and *scoliosis*

(p. 42). A condition involving an intervertebral joint, such as a herniated disc, will also produce abnormal posture and movement, as can muscle and nerve problems.

Table 15 summarizes descriptions of the most important diagnostic procedures and laboratory tests that can be used to obtain information about the status of the skeletal system.

■ CLASSIFICATION OF SKELETAL DISORDERS

Figure 17● diagrams the major classes of skeletal disorders that affect the structure and function of bones (Figure 17a●) and joints (Figure 17b●). Some of these disorders, such as *osteosarcoma* and *osteomyelitis,* are the result of conditions that affect primarily the skeletal system; others, such as *acromegaly* and *rickets (HA p. 111),* result from problems originating in other systems.

Traumatic injuries, such as fractures or dislocations, and infections also damage cartilages, tendons, and ligaments. A somewhat different array of conditions affects the soft tissues of the bone marrow. Red bone marrow contains the stem cells for red blood cells, white blood cells, and platelets. Blood diseases characterized by blood cell overproduction (*polycythemia, leukemia,* p. 104) or underproduction (several *anemias,* pp. 99–101) result in bone marrow abnormalities.

■ SYMPTOMS OF BONE AND JOINT DISORDERS

A common symptom of a skeletal system disorder is pain. Because bone pain and joint pain are common symptoms associated with many bone and joint disorders, the presence of pain does not provide much help in identifying a specific

disorder. A person may be able to tolerate chronic, aching bone or joint pain and therefore not seek medical assistance until more definitive symptoms appear. By then, the condition is relatively advanced. For example, a symptom that may require immediate attention is a *pathologic fracture.* Pathologic fractures are the result of weakening of the skeleton by a disease process, such as *osteosarcoma* (a bone cancer). Such fractures can be caused by physical stresses that are easily tolerated by normal bones. Unless the usual course of the disease is understood, the minor trauma may be thought to have caused the fracture, when in fact it just revealed the underlying condition.

■ HETEROTOPIC BONE FORMATION
HA p. 112

Heterotopic (*hetero,* different + *topos,* place), or ectopic (*ektos,* outside), bones are bones that develop in unusual places. Such bones dramatically demonstrate the adaptability of connective tissues. Mesenchymal stem cells can develop into bone, cartilage, or even fat and muscle. Physical or chemical events can stimulate the development of osteoblasts in normal connective tissues. For example, sesamoid bones develop within tendons near points of friction and pressure. Bone can also form within a large blood clot at the site of an injury or within portions of the dermis subjected to chronic abuse. Other triggers include foreign chemicals and problems that affect calcium excretion and storage.

Almost any connective tissues can be affected. Ossification within a tendon or around joints can painfully interfere with movement. Bones can also form within the kidneys, between skeletal muscles, in the pericardium, in the walls of arteries, and around the eyes.

Myositis ossificans (mī-o-SĪ-tis os-SIF-i-kanz) involves the deposition of bone around skeletal muscles. A muscle injury can

---

TABLE 15   Examples of Tests Used in the Diagnosis of Bone and Joint Disorders

Diagnostic Procedure	Method and Result	Representative Uses
X-ray of bone and joint	Standard x-ray; film sheet with radiodense tissues in white on a black background	Detects fractures, tumors, dislocations, reduction in bone density, and bone infections (osteomyelitis)
Bone scans	Injected radiolabeled phosphate accumulates in bones, and radiation emitted is converted into an image	Especially useful in diagnosis of metastatic bone cancer; detects fractures, early infections, and some degenerative bone diseases
Arthrocentesis	Insertion of a needle into joint for aspiration of synovial fluid	See section on analysis of synovial fluid (below).
Arthroscopy	Insertion of fiber-optic tubing into a joint cavity; displays interior of joint	Detects abnormalities of the menisci, ligaments, and articular surfaces; useful in differential diagnosis of joint disorders
MRI	Standard MRI (p. 3); produces computer-generated images	Detects bone and soft tissue abnormalities; noninvasive
DEXA	Dual energy x-ray absorptiometry; measures changes in bone density as small as 1 percent. Uses very small amount of radiation	Quantitates and monitors loss of bone density in osteoporosis and osteopenia

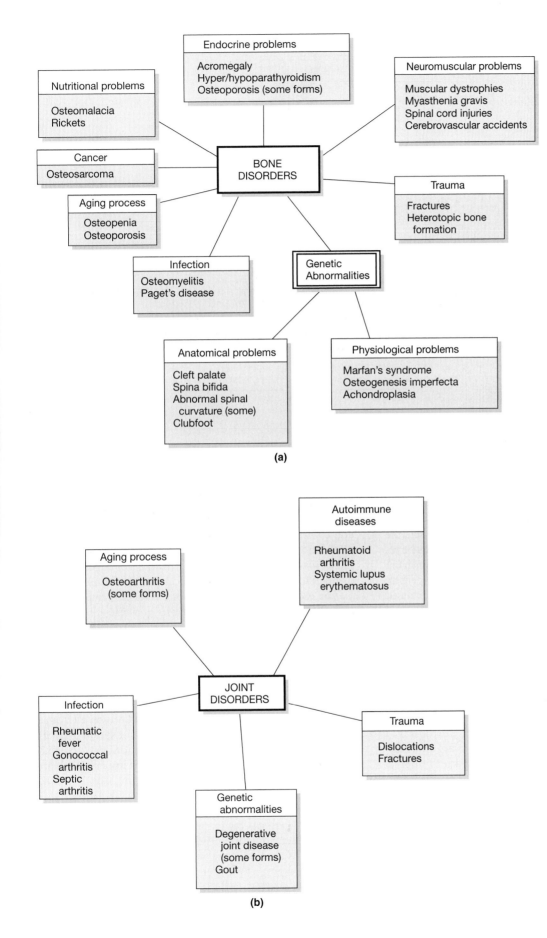

5

(a)

(b)

5

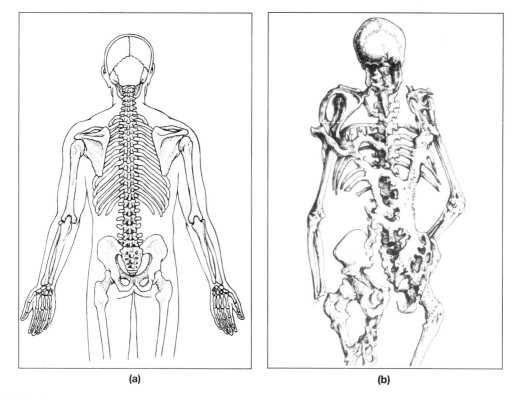

(a)   (b)

● **FIGURE 18**
Heterotopic Bone Formation. (a) The skeleton of a healthy adult male, posterior view.
(b) The skeleton of an adult male with advanced myositis ossificans.
[1974, L. B. Halstead Wykerman Publications, Ltd. (London)]

trigger a minor case. Severe cases have no known cause, but they certainly provide the most dramatic demonstrations of heterotopic bone formation. If the process does not reverse itself, the muscles of the back, neck, and upper limbs will gradually be replaced by bone. The extent of the conversion can be seen in Figure 18●. Figure 18a● shows the skeleton of a healthy adult male; Figure 18b● shows the skeleton of a 39-year-old man with advanced myositis ossificans. Several of the vertebrae have fused into a solid mass, and major muscles of the back, shoulders, and hips have undergone extensive ossification. Treatment can be problematic, as any surgical excision may trigger more ossification.

■ CONGENITAL DISORDERS OF THE SKELETON   *HA p. 115*

In **acromegaly** (*akron*, extremity + *megale*, great) an excessive amount of growth hormone is released after puberty, when most of the epiphyseal cartilages have already closed. Cartilages and small bones respond to the hormone, however, resulting in abnormal growth at the hands, feet, lower jaw, skull, and clavicle (Figure 19a●). Excessive growth resulting in *gigantism* occurs if there is hypersecretion of growth hormone before puberty. Inadequate production of growth hormone before puberty, by contrast, produces *pituitary dwarfism*. People with this condition are very short, but unlike achondroplastic dwarves (discussed below), their proportions are normal.

In **osteomalacia** (os-tē-o-ma-LĀ-shē-uh; *malakia*, softness) the size of the skeletal elements does not change, but their mineral content decreases, softening the bones. The osteoblasts work hard, but the matrix doesn't accumulate enough calcium salts. This condition, *rickets*, occurs in adults or children whose diet contains inadequate levels of calcium or vitamin D_3.

The excessive formation of bone is termed **hyperostosis** (hī-per-os-TŌ-sis). In *osteopetrosis* (os-tē-ō-pe-TRŌ-sis; *petros*, stone), the total mass of the skeleton gradually increases as a result of a decrease in osteoclast activity. Remodeling stops, and the shapes of the bones gradually change. Osteopetrosis in children produces a variety of skeletal deformities. The primary cause of this relatively rare condition is unknown.

There are more than 200 recognized inheritable disorders of connective tissues. Individual cases frequently result from a spontaneous mutation in either the oocyte or sperm of unaffected parents. Individuals with such conditions may transmit the disorder to their children. Osteogenesis imperfecta, Marfan's syndrome, and achondroplasia are examples of inherited conditions characterized by abnormal bone formation.

Osteogenesis imperfecta, appearing in roughly 1 individual in 20,000, affects the organization of collagen fibers. Osteoblast function is impaired, growth is abnormal, and in severe forms the bones are very fragile, leading to repeated fractures and progressive skeletal deformation. Fibroblast activity is also affected, and the ligaments and tendons can be very "loose," permitting excessive movement at the joints. There are four recognized types and large variations occur in

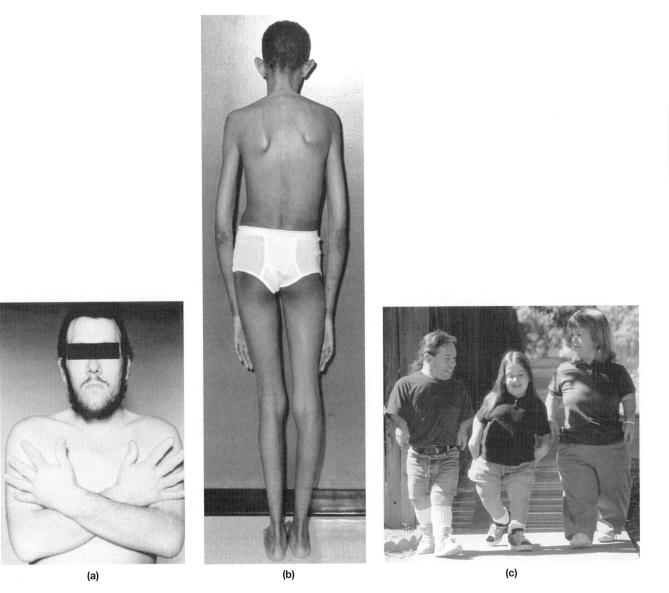

● FIGURE 19
Disorders of Bone Formation. (a) Acromegaly. **(b)** Marfan's syndrome. **(c)** Achondroplasia.

symptom severity. The lifetime fracture count can range from just a few to hundreds, with the worst form often fatal soon after birth.

    Marfan's syndrome is also linked to defective connective tissue structure. Extremely long and slender limbs, the most obvious physical indication of this disorder, result from excessive cartilage formation at the epiphyseal cartilages (Figure 19b●). An abnormality of a gene on chromosome 15 that affects the protein *fibrillin* is responsible. The skeletal effects are striking, but associated arterial wall weaknesses are more dangerous.

    Achondroplasia (ā-kon-drō-PLĀ-sē-uh) also results from abnormal epiphyseal activity. The child's epiphyseal cartilages grow unusually slowly, and the adult has short, stocky limbs (Figure 19c●). Although other skeletal abnormalities occur, the trunk is normal in size, and sexual and mental development remain unaffected. An adult with achondroplasia is known as an *achondroplastic dwarf*. The condition results from an

abnormal gene on chromosome 4 that affects a fibroblast growth factor. Most cases are the result of spontaneous mutations. If both parents have achondroplasia, the chances are that 25 percent of their children will be unaffected, 50 percent will be affected to some degree, and 25 percent will inherit two abnormal genes, leading to severe abnormalities and early death.

■ STIMULATION OF BONE GROWTH AND REPAIR   *HA p. 119*

Despite the body's considerable capacity for bone repair, not every fracture heals as expected. A *delayed union* is a repair that proceeds more slowly than anticipated. *Nonunion,* or no repair, can occur as a result of a complicating infection, continued movement, or other factors preventing complete callus formation.

Several techniques can induce bone repair. Surgical bone grafting is the most common treatment for nonunion. Each year in the United States, roughly 200,000 people receive *bone grafts*—transplants of bone—to stimulate bone repair. The bone fragments that serve as an "osteoconductive" scaffold are commonly taken from another part of the individual's body (an *autograft*), such as the iliac crest. However, because the inserted bone is ultimately destroyed and replaced, sterilized bone fragments from donors—cadavers (an *allograft*) or other species of animals (a *xenograft*)—can be used to establish a framework for the repair process. Thorough sterilization to prevent blood-borne diseases, including AIDS, is required. The calcium carbonate skeleton of corals has been sterilized and used, and synthetic "bio-ceramics" are being studied as well.

Steel plates, rods, and screws are sometimes used in severe fractures to help hold the ends in place while calluses form. This immobilizes the bone fragments and facilitates the repair process. Surgeons can also insert a shaped patch, made by mixing crushed bone and water, as an alternative to bone grafting.

Another approach involves the stimulation of osteoblast activity using strong electrical fields at the site of the injury. This procedure has been used to promote bone growth after fractures have failed to heal normally. Wires may be inserted into the skin, implanted in the adjacent bone, or wrapped around a cast. A success rate of about 80 percent can be achieved. There are experiments under way that attempt to use genetically engineered proteins to stimulate the conversion of osteoprogenitor cells into active osteoblasts. This approach is called "osteoinductive stimulation." Although results in animal experimentation have been encouraging, the technique has not yet been approved for human trials.

One advantage of autografts is the presence of bone-forming cells in the graft and around the injury site. An experimental method of inducing bone repair with these osteoprogenitor cells involves mixing the patients' bone marrow cells into a soft matrix of bone collagen and ceramic to form a composite graft. This combination is used like a putty at the fracture site. Mesenchymal cells in the marrow divide, producing chondrocytes that create a cartilaginous patch, which is later converted to bone by periosteal cells. In some cases, simply injecting marrow cells into the fracture site has been successful, even without the insertion of an autograft.

■ OSTEOPOROSIS AND AGE-RELATED SKELETAL ABNORMALITIES

HA p. 119

In **osteoporosis** (os-te-ō-por-Ō-sis; *porosus,* porous), there is a reduction in bone mass sufficient to compromise normal function. Our maximal bone density is reached in our early twenties and decreases as we age. Inadequate calcium intake in teenagers reduces peak bone density and increases the risk of osteoporosis. The distinction between the "normal" osteopenia of aging and the clinical condition of osteoporosis is a matter of degree.

Current estimates indicate that 29 percent of women between ages 45 and 79 can be considered osteoporotic. The increase in incidence after menopause has been linked to a decrease in the production of estrogens (female sex hormones). The incidence of osteoporosis in men of the same age is estimated at 18 percent.

The excessive fragility of osteoporotic bones commonly leads to breakage, and subsequent healing is impaired. Vertebrae may collapse, distorting the vertebral articulations and putting pressure on spinal nerves. Supplemental estrogens, dietary changes to elevate calcium levels in blood, exercise that stresses bones and stimulates osteoblast activity, and the administration of calcitonin by nasal spray appear to slow, but not prevent, the development of osteoporosis. The inhibition of osteoclast activity by drugs called *bisphosphonates,* such as *Fosamax,* can reduce the risk of spine and hip fractures in elderly women and improve bone density. For long-term use, exercise, dietary calcium, and bisphosphonates are currently preferred.

Osteoporosis can also develop as a secondary effect of some cancers. Cancers of the bone marrow, breast, or other tissues may release a chemical known as *osteoclast-activating factor.* This compound increases both the number and activity of osteoclasts and may produce severe osteoporosis.

Infectious diseases that affect the skeletal system become more common as individuals age. In part, this fact reflects the higher incidence of fractures, combined with slower healing and the reduction of immune defenses.

Osteomyelitis (os-tē-ō-mī-e-LĪ-tis; *myelos,* marrow) is a painful and destructive bone infection generally caused by bacteria. This condition, most common in people over age 50, can lead to dangerous systemic infections. Heredity and environmental factors, including the possibility of a viral infection, appear to be responsible for **Paget's disease**, also known as **osteitis deformans** (os-tē-Ī-tis dē-FOR-manz). This condition may affect up to 10 percent of the population over 70. Localized osteoclast activity accelerates, producing areas of acute osteoporosis, and osteoblasts produce abnormal matrix proteins. The result is a gradual deformation of the skeleton. Bisphosphonate treatment may slow the progression of the disease, as it does osteoporosis, by reducing osteoclast activity.

DISORDERS OF THE AXIAL SKELETON

■ SINUS PROBLEMS AND SEPTAL DEFECTS *HA p. 147*

When an irritant is introduced into the nasal passages, our bodies work to remove the irritant. Large particles or strong chemical agents can cause us to sneeze, expelling a sizable amount of air quickly and forcefully. With luck, this sweeps the offending particles or chemicals out with the air. Smaller particles or milder irritants trigger the production of large amounts of mucus by the epithelium of the paranasal sinuses. The mucus stream flushes the nasal surfaces clean, often removing irritants such as pepper, pollen grains, or dust.

A sinus infection, however, is another matter entirely. A viral or bacterial infection produces an inflammation of the mucous membrane of the nasal cavity. As swelling occurs, the communicating passageways narrow. Drainage of mucus slows, congestion increases, and the individual experiences headaches and a feeling of pressure within the facial bones. This condition of sinus inflammation and congestion is called **sinusitis**. The maxillary sinuses are commonly involved. Because gravity does little to assist mucus drainage from these sinuses, the effectiveness of the flushing action is reduced, and pressure on the sinus walls typically increases.

The relief of pain associated with sinusitis is the basis of a large over-the-counter (OTC) drug market in the United States. Every major pharmaceutical company has at least one product designed specifically to relieve sinus pressure. The active ingredients in these preparations are compounds that dry the epithelial linings, reduce pain, and restrict further swelling. Usually included are the antihistamine *chlorpheniramine maleate* and the nasal decongestant *pseudoephedrine HCl,* both taken orally, or *phenylephrine HCl* and *oxymetazoline HCl,* decongestants prepared in solutions for nose drops or sprays. Another formerly common oral decongestant and diet pill ingredient, phenylpropanolamine HCl, was recently taken off the market at the request of the FDA because it was linked to an increased risk of hemorrhagic stroke in young women. A relative newcomer to the market is a nasal spray containing an anti-inflammatory agent (*cromolyn*) that reduces swelling. Pain relievers such as acetaminophen or ibuprofen are included. With the exception of cromolyn, the ingredients and dosages differ very little from one "cold and sinus" product to the next. However, adult doses and concentrations are higher than in pediatric compounds, and some infant preparations are more concentrated than those marketed for older children. Careful attention to dosage guidelines with regard to age and weight (for children) and maximum daily dosage is required. Marketing and packaging play a major role in determining which OTC remedy dominates the market at any given time.

Temporary sinus problems may follow exposure of the mucous epithelium to chemical irritants or invading microorganisms, including viruses. Chronic sinusitis may accompany chronic allergies or occur as the result of a **deviated**

CHAPTER 6
THE AXIAL SKELETON

(nasal) **septum**. In this condition, the nasal septum has a bend in it, generally at the junction between the bony and cartilaginous regions. Septal deviation often blocks the drainage of one or more sinuses, producing chronic cycles of infection and inflammation. A deviated septum can result from developmental abnormalities or from injuries to the nose. Many boxers and hockey players have a deviated septum, as their sports subject them to numerous blows to the soft tissues of the nose. It can usually be corrected or improved by surgery.

■ KYPHOSIS, LORDOSIS, AND SCOLIOSIS   *HA p. 156*

The vertebral column has to move, balance, and support the trunk and head, with multiple bones and joints involved. Conditions or events that damage the bones, muscles, or nerves can result in distorted shapes and impaired function. In **kyphosis** (kī-FŌ-sis), the normal thoracic curvature becomes exaggerated posteriorly, producing a "round-back" appearance (Figure 20a●). This condition can be caused by (1) osteoporosis with compression fractures affecting the anterior portions of vertebral bodies, (2) chronic contractions in muscles that insert on the vertebrae, or (3) abnormal vertebral growth.

In **lordosis** (lor-DŌ-sis), or "swayback," both the abdomen and buttocks protrude abnormally (Figure 20b●). The cause is an anterior exaggeration of the lumbar curvature. This may result from abdominal wall obesity or weakness in the muscles of the abdominal wall.

Scoliosis (skō-lē-Ō-sis) is an abnormal lateral curvature of the spine (Figure 20c●) This lateral deviation can occur in one or more of the movable vertebrae. Scoliosis is the most common distortion of the spinal curvature. This condition may result from developmental problems, such as incomplete vertebral formation, or from muscular paralysis affecting one side of the back (as in some cases of polio). The "Hunchback of Notre Dame" suffered from severe scoliosis, which prior to the development of antibiotic therapies was often caused by a tuberculosis infection of the spine. In four out of five cases, the structural or functional cause of the abnormal spinal curvature is impossible to determine. This idiopathic scoliosis generally appears in girls during adolescence, when periods of growth are most rapid. Treatment may consist of a combination of exercises and braces that offer limited, if any, benefit. Severe cases can be treated through surgical straightening with implanted metal rods or cables.

■ SPINA BIFIDA   *HA p. 164*

During the third week of embryonic development, the vertebral arches form around the developing spinal cord. In the condition called *spina bifida* (SPĪ-nuh BI-fi-duh; *bifidus,* cut into two parts), the most common neural tube defect (NTD), a portion of the spinal cord develops abnormally such that the adjacent vertebral arches do not form. Because the vertebral arch is incomplete, the membranes (or *meninges*) that line the dorsal body cavity bulge outward. This is the most common developmental abnormality of the nervous system, occurring at a rate of up to 4 cases per 1000 births. (See the *Embryology Summary* in chapter 14 of the text for illustrations of this condition.) Both heredity and maternal diet, particularly the amount of folic acid present before and during early pregnancy, have been linked to NTDs. Women who may become pregnant are advised to take 400 micrograms of folic acid daily, and to assist in this, food in the United States containing wheat, rice, and corn has been fortified with folic acid since 1998. Probably as a result, the incidence of NTDs in the United States dropped 19 percent between 1998 and 2001.

The region affected and the severity of the condition vary widely. It is most common in the inferior thoracic, lumbar, or sacral region, typically involving 3-6 vertebrae. Variable degrees of paralysis occur distal to the affected vertebrae. Mild cases involving the sacral and lumbar regions may pass un-

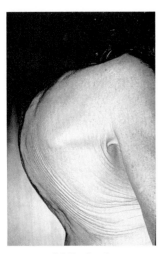
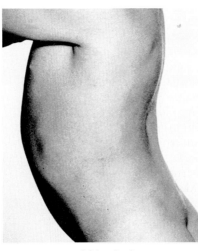
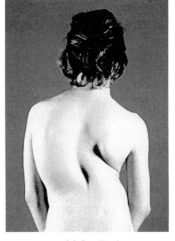

(a) Kyphosis   (b) Lordosis   (c) Scoliosis

● **FIGURE 20**
Abnormal Curvatures of the Spine

noticed, because neural function is not compromised significantly and "baby fat" may mask the fact that some of the spinous processes are missing. When spina bifida is detected, surgical repairs can close the gap in the vertebral wall. Severe cases, involving the entire spinal column and skull, reflect major problems with the formation of the spinal cord and brain. These neural problems usually kill the fetus before delivery; infants born with such developmental defects seldom survive more than a few hours or days.

■ THE THORACIC CAGE AND SURGICAL PROCEDURES   *HA p. 168*

Surgery on the heart, lungs, or other organs in the thorax often involves entering the thoracic cavity. The mobility of the ribs and the cartilaginous connections with the sternum allow the ribs to be temporarily moved out of the way. Special rib spreaders are used, which push them apart in much the same way that a jack lifts a car off the ground for a tire change. If more extensive access is required, the sternal cartilages can be cut and the entire sternum can be folded out of the way. Once replaced, the cartilages are reunited by scar tissue, and the ribs heal fairly rapidly.

After thoracic surgery, chest tubes may penetrate the thoracic wall to permit drainage of fluids. To install a **chest tube** or obtain a sample of pleural fluid, the wall of the thorax must be penetrated. This process, called **thoracentesis** (thō-ra-sen-TĒ-sis), involves the penetration of the thoracic wall along the superior border of one of the ribs. Penetration at this location avoids damaging vessels and nerves within the costal groove.

6

CHAPTER 7
THE APPENDICULAR SKELETON

DISORDERS OF THE APPENDICULAR SKELETON

■ PROBLEMS WITH THE ANKLE AND FOOT *HA pp. 200, 230*

The arches of the foot are usually present at birth. Sometimes, however, they fail to develop properly. In **congenital talipes equinovarus** *(clubfoot),* abnormal muscle development distorts growing bones and joints. One or both feet may be involved, and the condition can be mild, moderate, or severe. In most cases, the tibia, ankle, and foot are affected; the longitudinal arch is exaggerated, and the feet are turned medially and inverted. If both feet are involved, the soles face one another. This condition, which affects 2 in 1000 births, is roughly twice as common in boys as in girls. Prompt treatment with casts or other supports in infancy helps alleviate the problem, and fewer than half the cases require surgery. Kristi Yamaguchi, an Olympic gold medalist in figure skating, was born with clubfeet.

Someone with *flatfeet* loses or never develops the longitudinal arch. "Fallen arches" develop as tendons and ligaments stretch and become less elastic. Up to 40 percent of adults may have flatfeet, but no action is necessary unless pain develops. Individuals with abnormal arch development are most likely to suffer metatarsal injuries. Obese individuals and those who must constantly stand or walk on the job are likely candidates. Children have very mobile articulations and elastic ligaments, so they commonly have flexible, flat feet. Their feet look flat only while they are standing, and the arch appears when they stand on their toes or sit down. In most cases, the condition disappears as growth continues.

Claw feet are produced by muscular abnormalities. In individuals with a claw foot, the median longitudinal arch becomes exaggerated because the plantar flexors overpower the dorsiflexors. Causes include muscle degeneration and nerve paralysis. The condition tends to get progressively worse with age.

Even the normal ankle and foot are subjected to a variety of stresses during daily activities. In a *sprain,* a ligament is stretched to the point at which some of the collagen fibers are torn. The ligament remains functional, and the structure of the joint is not affected. The most common cause of a **sprained ankle** is a forceful inversion of the foot that stretches the lateral ligament. An ice pack is generally required to reduce swelling. With rest and support, the ankle should heal in about three weeks.

In more serious incidents, the entire ligament can be torn apart, or the connection between the ligament and the lateral malleolus can be so strong that the bone breaks instead of the ligament. In general, a broken bone heals more quickly and effectively than does a completely torn ligament. A dislocation may accompany such injuries.

In a **dancer's fracture**, the proximal portion of the fifth metatarsal is broken. Most such cases occur while the body weight is being supported by the longitudinal arch of the foot. A sudden shift in weight from the medial portion of the arch to the lateral, less elastic border breaks the fifth metatarsal close to its distal articulation.

JOINT DISORDERS

■ RHEUMATISM, ARTHRITIS, AND SYNOVIAL FUNCTION   *HA p. 208*

Rheumatism (ROO-ma-tizm) is a general term that indicates pain and stiffness affecting the skeletal system, the muscular system, or both. There are several major forms of rheumatism. **Arthritis** (ar-THRĪ-tis) includes all the rheumatic diseases that affect synovial joints. Arthritis always involves damage to the articular cartilages, but the specific cause varies. For example, arthritis can result from bacterial or viral infection, injury to the joint, metabolic problems, or autoimmune disorders.

Proper synovial function requires healthy articular cartilages. When an articular cartilage has been damaged, the matrix begins to break down; the exposed cartilage changes from a slick, smooth gliding surface to a rough feltwork of bristly collagen fibers. This feltwork drastically increases friction, damaging the cartilage further and producing pain. Eventually, the central area of the articular cartilage may completely disappear, exposing the underlying bone.

Fibroblasts are attracted to areas of friction, and they begin tying the opposing bones together with a network of collagen fibers. This network may later be converted to bone, locking the articulating elements into position. Such a bony fusion, called **ankylosis** (an-kē-LŌ-sis), eliminates the friction, but only at the drastic cost of making movement impossible.

The diseases of arthritis are usually considered either degenerative or inflammatory. **Degenerative diseases** begin at the articular cartilages, and modification of the underlying bone and inflammation of the joint occur secondarily. **Inflammatory diseases** start with the inflammation of synovial tissues, and damage later spreads to the articular surfaces. We will consider one example of each type.

Osteoarthritis (os-tē-ō-ar-THRĪ-tis), also known as **degenerative arthritis** or **degenerative joint disease** (DJD), generally affects older individuals. In the U.S. population, 25 percent of women and 15 percent of men over age 60 show signs of this disease. The condition seems to result from cumulative wear and tear on the joint surfaces and commonly affects weight-bearing joints including the knees, hips, lower back, and neck. The distal finger joints are also frequently involved.

Perhaps 30 percent of individuals with DJD have a gene that codes for an abnormal, weaker form of collagen that differs from the normal protein in only 1 of its 1000 amino acids. This collagen may lead to premature cartilage breakdown and arthritis.

Rheumatoid arthritis is a serious inflammatory condition that affects roughly 2.5 percent of the adult U.S. population. The cause is uncertain, although allergies, bacteria, viruses, and genetic factors have all been proposed. The patient's own immune system attacks the synovial membrane of joints, which become swollen and inflamed, a condition known as **synovitis** (si-no-VĪ-tis). The cartilaginous matrix begins to break down, and the process accelerates as dying cartilage cells release lysosomal enzymes. General anti-inflammatory medicines, such as aspirin and ibuprofen, or corticosteroids may help control inflammation and pain. Other "disease-modifying antirheumatic drugs" (DMARDs) may slow or stop joint damage. DMARDs include anti-cancer drugs such as methotrexate and newer biologic agents that target the immune system. Trials of combination therapy using several drugs may bring further improvement.

Advanced stages of inflammatory and degenerative forms of arthritis produce an inflammation that spreads into the surrounding area. Ankylosis, where a joint fuses, was relatively common in the past when complete rest was routinely prescribed for arthritis patients, but it is rarely seen today. Regular exercise, physical therapy, drugs that reduce inflammation (such as aspirin or ibuprofen), and immunosuppressants may slow the progression of the disease. Nutritional supplements such as *glucosamine* and *chondroitin sulfate* appear to have some anti-inflammatory effect (though smaller than that of aspirin), and a number of people report less pain while taking these supplements. Surgical procedures can realign or redesign the affected joint. In extreme cases involving the hip, knee, elbow, or shoulder, the defective joint can be replaced by an artificial one. Prosthetic (artificial) joints, such as those shown in Figure 21●, are weaker than natural ones, but elderly people seldom stress them to their limits.

■ DIAGNOSING AND TREATING INTERVERTEBRAL DISC PROBLEMS

HA p. 216

The most common sites for intervertebral disc problems are between vertebrae C_5 and C_6, L_4 and L_5, and L_5 and S_1. A clinician may be able to determine the location of the injured disc by noting the distribution of abnormal sensations and related spinal reflexes. For example, someone with a herniated disc at L_4–L_5 will experience pain in the hip, the groin, the posterior and lateral surfaces of the thigh, the lateral surface of the calf, and the superior surface of the foot. A herniation at L_5–S_1 produces pain in the buttocks, the posterior thigh, the posterior calf, and the sole of the foot. Most lumbar disc problems can be treated successfully with some combination of rest, muscle relaxants, painkillers, and physical therapy. Training to make proper body movements, especially when lifting and bending, can reduce the risk of relapse when combined with muscle strengthening exercises.

Surgery to relieve the symptoms may be required in some cases involving lumbar disc herniation. The primary method of treatment involves removing the offending disc, inserting material to restore the position of the vertebrae, and, if necessary, fusing the vertebral bodies together to prevent relative movement between them. Accessing the disc often requires that the laminae of the nearest vertebral arch be removed. For this reason, the procedure is known as a *laminectomy* (la-mi-NEK-to-mē). In cases in which the herniated portion of the disc does not extend far into the vertebral foramen, portions of the disc can be removed by a small tool that is guided to the site by radiological imaging. This minimally invasive procedure is faster than a laminectomy.

■ BURSITIS *HA p. 219*

When bursae become inflamed, causing pain in the affected area whenever the tendon or ligament moves, such a condition is termed **bursitis.** Inflammation can result from the friction associated with repetitive motion, pressure over the joint, irritation by chemical stimuli, infection, or trauma. Bursitis associated with repetitive motion often occurs at the shoulder; for example, golfers, pitchers, ceiling painters, and tennis players may develop bursitis, usually at the subscapular bursa. The most common pressure-related bursitis is a **bunion.** Bunions form over the base of the big toe as a result of the friction and distortion of the joint, frequently caused by tight shoes, frequently with pointed toes. There is chronic inflammation of the region, and as the wall of the bursa thickens, fluid builds up in

(a) Shoulder

(b) Hip

(c) Knee

● **FIGURE 21**
Artificial Joints

the surrounding tissues. The result is a firm, tender nodule. There are special names for bursitis at other locations; the names of these conditions indicate the occupations most often associated with them. In "housemaid's knee," which accompanies prolonged kneeling, the affected bursa lies between the patella (kneecap) and the skin. The condition of "student's elbow" is a form of bursitis that can result from propping your head above a desk while reading your anatomy textbook. Most of the symptoms of bursitis subside if the stimulus is removed, and a variety of anti-inflammatory drugs may help. In extreme cases the affected bursae can be surgically removed in a **bursectomy.**

■ HIP FRACTURES, AGING, AND ARTHRITIS   *HA p. 225*

Most hip fractures (~90 percent) are associated with osteoporosis, and they usually develop after a fall in individuals over age 60. The associated immobility leads to a 5–20 percent increase in deaths and a 15–25 percent decrease in the likelihood of independent living in the year following the fracture. Minimally displaced fractures of the femoral neck may be treated with the surgical placement of pins or lag screws, giving support that may promote an early return to normal activities. When the injury is severe the vascular supply to the joint may be damaged. As a result, two problems can develop:

1. **Avascular necrosis.** The mineral deposits in the bone of the pelvis and femur are turned over very rapidly, and osteocytes have high energy demands. In *avascular necrosis,* a reduction in blood flow to the femoral head first injures and then kills the osteocytes. When bone maintenance stops in the affected region, the matrix begins to break down.

2. **Degeneration of articular cartilages.** The chondrocytes in the articular cartilages absorb nutrients from the synovial fluid, which circulates around the joint cavity as the bones change position. A fracture of the femoral neck is generally followed by joint immobility and poor circulation to the synovial membrane. The combination results in a gradual deterioration of the articular cartilages of the femur and acetabulum.

In younger patients at risk for these complications, surgical repair with realignment of the bone, held in place by plates and screws, may preserve the articular surfaces. For older patients, replacement of the femoral head with a prosthetic head is often preferable.

If both the femoral head and the acetabulum are damaged, or if severe arthritis is present, both parts of the joint can be replaced. In a "total hip" replacement, the damaged portion of the femur is removed, and an artificial femoral head and neck are attached by a spike that extends into the marrow cavity of the shaft. The femoral head may be cemented in place, or carefully fitted so cement is not needed. Special cement is used to attach a new articular surface to the acetabulum. Joint replacement eliminates most pain and restores range of motion. Prosthetic joints (Figure 21●) are generally implanted in older people, who seldom subject them to severe stress.

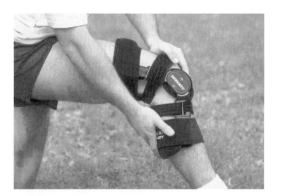

● **FIGURE 22**
A Knee Brace

Current models deteriorate in 10 to 15 years, so younger patients may need multiple replacements.

■ KNEE INJURIES   *HA p. 225*

Athletes place tremendous stresses on their knees. Ordinarily, the medial and lateral menisci move as the position of the femur changes. Placing a lot of weight on the knee while it is partially flexed can trap a meniscus between the tibia and femur, resulting in a break or tear in the cartilage. In the most common injury, the lateral surface of the leg is driven medially, tearing the medial meniscus. In addition to being quite painful, the torn cartilage may restrict movement at the joint. It can also lead to chronic problems and the development of a "trick knee"—a knee that feels unstable. Sometimes the meniscus can be heard and felt popping in and out of position when the knee is extended. To prevent such injuries, most competitive sports outlaw activities that generate side impacts to the knee, and athletes wishing to continue exercising with injured knees may use a brace that limits lateral movement (Figure 22●).

Other knee injuries involve tearing one or more stabilizing ligaments or damaging the patella. Torn ligaments can be difficult to correct surgically, and healing is slow. Rupture of the anterior cruciate ligament (ACL) is a common sports injury that affects women two to eight times as often as men. Twisting on an extended weight-bearing knee is frequently the cause. Nonsurgical treatment with exercise and braces is possible, but requires a change in activity patterns. Reconstructive surgery using part of the patellar tendon or an allograft from a cadaver tendon may allow a return to active sports.

The patella can be injured in a number of ways. If the leg is immobilized (as it might be in a football pileup) while you try to extend the knee, the muscles are powerful enough to pull the patella apart. Impacts to the anterior surface of the knee can also shatter the patella. Treatment of a fractured patella is difficult and time consuming. The fragments must be surgically removed and the tendons and ligaments repaired. The joint must then be immobilized. Total knee replacements are rarely performed on young people, but they are becoming increasingly common among elderly patients with severe arthritis.

8

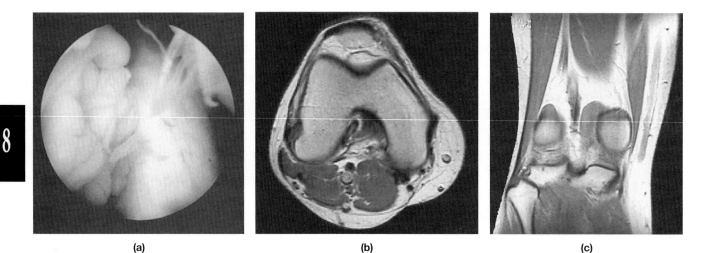

(a) (b) (c)

● **FIGURE 23**
Arthroscopy and MRI Scans of the Knee. (a) An arthroscopic view of a damaged knee, showing the torn edge of an injured meniscus. **(b)** A transverse and **(c)** a frontal MRI scan of the knee joint. For identification of the structures visualized in these images, see Plates 7a–f in the *Atlas*.

■ ARTHROSCOPIC SURGERY AND THE DIAGNOSIS OF JOINT INJURIES

HA p. 225

An **arthroscope** uses fiber optics to permit the exploration of a joint without major surgery. Optical fibers are thin threads of glass or plastic that conduct light. The fibers can be bent around corners, so they can be introduced into a knee or other joint and moved around, enabling the physician to see and diagnose problems inside the joint. Arthroscopic surgical treatment of the joint is possible at the same time. This procedure, called **arthroscopic surgery**, has greatly simplified the treatment of knee and other joint injuries. Figure 23a● is an arthroscopic view of the interior of an injured knee, showing

a damaged meniscus. Small pieces of cartilage can be removed and the meniscus surgically trimmed. A total **meniscectomy**, the removal of the affected cartilage, is generally avoided, because it leaves the joint prone to develop degenerative joint disease. New tissue-culturing techniques may someday permit the replacement of the meniscus or even the articular cartilage.

Arthroscopy is an invasive procedure with some risks. Magnetic resonance imaging (MRI) is a safe, noninvasive, and cost-effective method of viewing and examining soft tissues around the joint. It improves the diagnostic accuracy of knee injuries, and reduces the need for diagnostic arthroscopies. It can also help guide the arthroscopic surgeon. Figure 23b● and Figure 23c● are MRI views of the knee joint. Notice the image clarity and the soft-tissue details visible in these scans.

DISORDERS OF MUSCLE TISSUE

The muscular system includes more than 700 skeletal muscles that are directly or indirectly attached to the skeleton by tendons or aponeuroses. The muscular system produces movement as the contractions of skeletal muscles pull on the attached bones. Muscular activity does not always result in movement, however; it can also help stabilize skeletal elements and prevent movement. Skeletal muscles are also important in guarding entrances and exits of internal passageways, such as those of the digestive, respiratory, urinary, or reproductive systems, and in generating heat to maintain a stable body temperature.

THE DIAGNOSIS OF MUSCULAR SYSTEM DISORDERS

■ SIGNS OF MUSCULAR DISORDERS

Skeletal muscles normally contract only under the command of the nervous system. For this reason, clinical observation of muscular activity may provide indirect information about the nervous system, as well as direct information about the muscular system. The assessment of a patient's facial expressions, posture, speech, and gait can be an important part of a physical examination. Classical signs of muscle disorders include the following:

- **Gower's sign** is a distinctive method of standing up from a sitting or lying position on the floor. This sign is characteristic of young children with *muscular dystrophy* (p. 53). They move from a sitting position to a standing position by pushing the trunk off the floor with the hands and then moving the hands to the knees. The hands are then used as braces to force the body into the standing position. The extra support is necessary because the pelvic muscles are too weak to swing the weight of the trunk over the legs.

- **Ptosis** is a drooping of the upper eyelid. It may be seen in *myasthenia gravis* (p. 54), *botulism* (p. 54), and *myotonic dystrophy* (p. 53), or it may follow damage to cranial nerve III, which innervates the *levator palpebrae superioris muscle* of the eyelid.

- A **muscle mass**—an abnormal dense region within a muscle—is sometimes seen or felt in a skeletal muscle. A muscle mass results from torn muscle or tendon tissue, a hematoma, a parasitic infection such as *trichinosis* (p. 52), or bone deposition, as in *myositis ossificans* (p. 38).

- Abnormal contractions may indicate problems with the muscle tissue or its innervation. *Muscle spasticity* exists when a muscle has excessive tone. A *muscle spasm* is a sudden, strong, and painful involuntary contraction.

- **Muscle flaccidity** exists when the relaxed skeletal muscle appears soft and loose and its contractions are very weak or absent.

CHAPTER 9

MUSCLE TISSUE

• **Muscle atrophy** is skeletal muscle deterioration, or wasting, due to disuse, immobility, or interference with the normal innervation.

• Abnormal patterns of muscle movement, such as *tics, choreiform movements,* or *tremors,* and muscular paralysis are generally caused by nervous system disorders. We will describe these movements further in sections dealing with abnormal nervous system function.

■ SYMPTOMS OF MUSCULAR DISORDERS

Two common symptoms of muscular disorders are pain and weakness in the affected skeletal muscles. Possible causes of muscle weakness are diagrammed in Figure 24●.

Possible causes of muscle pain include the following:

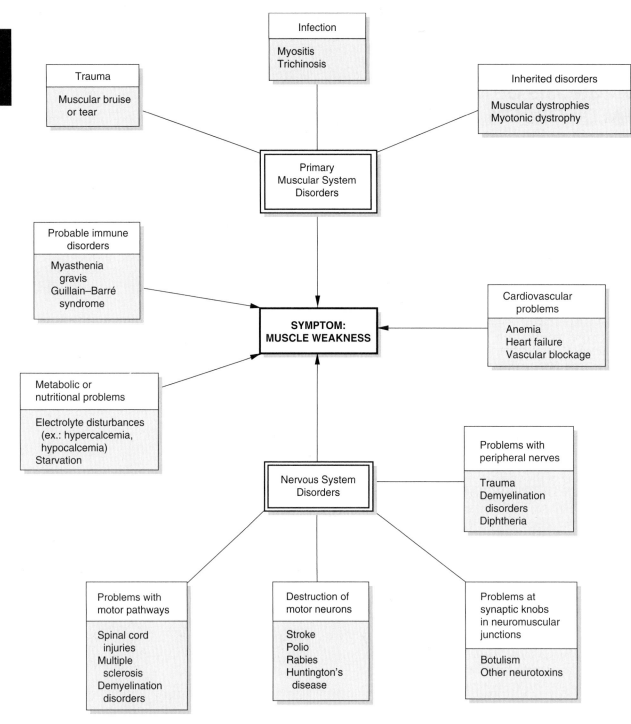

● **FIGURE 24**
Potential Causes of Muscle Weakness

- **Muscle trauma.** Examples of traumatic injuries to a skeletal muscle include a laceration, a deep bruise or crushing injury, a muscle tear, and a damaged tendon.

- **Muscle infection.** Skeletal muscles can be infected by viruses, as in some forms of myositis (muscle inflammation), or colonized by parasitic worms, such as those responsible for trichinosis (p. 52). These types of infections generally produce pain that is restricted to the affected muscles. Diffuse muscle pain can develop in the course of other infectious diseases, such as influenza.

- **Related problems with the skeletal system.** Muscle pain can result from skeletal problems such as arthritis (p. 00) or a sprained ligament near the point of muscle origin or insertion.

- **Problems with the nervous system.** Muscle pain can be related to the inflammation of sensory neurons or the stimulation of pain pathways in the central nervous system (CNS).

Muscle strength can be evaluated by applying an opposite force against a specific action. For example, an examiner might exert a gentle extending force on a patient's forearm while asking the patient to flex the elbow. Because the muscular and nervous systems are so closely interrelated, a single symptom, such as muscle weakness, can have a variety of causes (Figure 24●). Muscle weakness can also develop as a consequence of a condition that affects the entire body, such as anemia or starvation.

■ DIAGNOSIS OF MUSCULAR DISORDERS

Figure 25● provides an overview of muscular system disorders, and Table 16 provides information about representative diagnostic procedures that can be used to categorize such disorders.

DISRUPTION OF NORMAL MUSCLE ORGANIZATION   *HA p. 237*

A variety of disorders are characterized by a disruption in the structural organization of skeletal muscles. We will consider only a few representative examples: necrotizing fasciitis, characterized by a breakdown in the connective tissues of skeletal muscles; trichinosis, characterized by the colonization of muscles by parasites; and fibromyalgia and chronic fatigue syndrome, two muscle disorders of uncertain origin.

■ NECROTIZING FASCIITIS

Several bacteria produce enzymes such as *hyaluronidase* or *cysteine protease*. Hyaluronidase breaks down hyaluronic acid and disassembles the associated proteoglycans. Cysteine protease breaks down connective tissue proteins. The bacteria that produce these enzymes are dangerous, because they can spread rapidly by liquefying the matrix and dissolving the intercellular

9

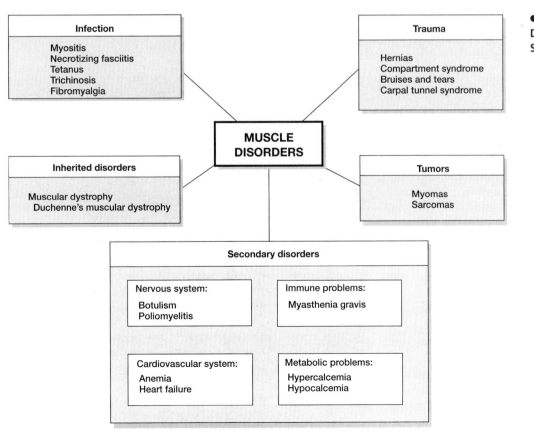

● **FIGURE 25**
Disorders of the Muscular System

TABLE 16   Examples of Tests Used in the Diagnosis of Muscle Disorders

Diagnostic Procedure	Method and Result	Representative Uses
Muscle biopsy	Removal of a small amount of affected muscle tissue	Identifies histologic muscle disease; also used to detect cyst formation or larvae to diagnose trichinosis.
Electromyography (EMG)	Insertion of a probe that transmits measurements of electrical activity in contracting muscles	Abnormal EMG readings occur in disorders such as myasthenia gravis, amyotrophic lateral sclerosis (ALS), and muscular dystrophy.
MRI	Standard MRI	Useful in the detection of muscle diseases and associated soft-tissue abnormalities.

9

cement that holds epithelial cells together. The *streptococci* are one group of bacteria that secrete both of these enzymes. *Streptococcus A* bacteria are involved in many human diseases, most notably "strep throat," an infection of the pharynx. In most cases, the immune response is sufficient to contain and ultimately defeat these bacteria before extensive tissue damage has occurred.

However, tabloid newspapers have a field day with stories of "killer bugs" and "flesh-eating bacteria." The details are horrific: Minor cuts become major open wounds, and interior connective tissues dissolve. This condition is called **necrotizing fasciitis.** Untreated, it is fatal. Even with rapid diagnosis, aggressive surgical removal of affected tissue, and antibiotics, the mortality rate is extremely high. The pathogen responsible is a strain of *Streptococcus A* that overpowers immune defenses and swiftly invades and destroys soft tissues. Moreover, the pathogens erode their way along the fascial wrapping that covers skeletal muscles and other organs. In most cases, myositis also occurs, followed by degeneration of the muscle tissue. Some form of highly aggressive infectious soft-tissue invasion occurs 75–150 times annually in the United States. The rapid development of extreme pain in an otherwise minor wound is one warning symptom.

■ TRICHINOSIS

Trichinosis (trik-i-NŌ-sis; *trichos,* hair + *nosos,* disease) results from infection by the parasitic nematode *Trichinella spiralis.* Symptoms include diarrhea, weakness, and muscle pain and are caused by the invasion of skeletal muscle tissue by larval worms, which create small pockets within the perimysium and endomysium (Figure 26●). Muscles of the tongue, eyes, diaphragm, chest, and legs are most often affected.

Larvae are common in the flesh of pigs, horses, dogs, and other mammals. The larvae are killed when the meat is cooked; people are most often exposed by eating undercooked infected pork. Once eaten, the larvae mature within the human intestinal tract, where they mate and produce eggs. The new generation of larvae then migrates through the body tissues to reach the muscles, where they complete their early development. The migration and subsequent settling produce a generalized achiness, muscle and joint pain, and swelling in infected tissues. An estimated 1.5 million Americans carry *Trichinella* in their muscles, and up to 300,000 new infections

occur each year. The mortality rate for people who have symptoms severe enough to require treatment is approximately 1 percent.

■ FIBROMYALGIA AND CHRONIC FATIGUE SYNDROME

Fibromyalgia (-*algia,* pain) is a disorder that has formally been recognized only since the mid-1980s. Although first described in the early 1800s, the condition is still somewhat controversial, because the symptoms cannot be definitively linked to any anatomical or physiological abnormalities. However, physicians now recognize a distinctive pattern of symptoms with the diagnostic criteria of widespread musculoskeletal pain for three months or more, and tenderness in 11 or more of 18 specific tender points. Sleep disorders, depression, and irritable bowel syndrome also occur alongside fibromyalgia.

Fibromyalgia may be the most common musculoskeletal disorder affecting women under age 40; from 3 to 6 million individuals in the United States may have this condition. The four most common tender points are (1) the medial surface of the knee, (2) the area distal to the lateral epicondyle of the humerus, (3) the area near the external occipital crest of the skull, and (4) the junction between the second rib and its costal cartilage. An additional clinical criterion is that the pains and stiffness cannot be explained by other mechanisms.

Most of the symptoms mentioned could be attributed to other problems. For example, chronic depression can, by itself, lead to fatigue and poor-quality sleep. As a result, the pattern of tender points is the diagnostic key to fibromyalgia. This symptom distinguishes fibromyalgia from **chronic fatigue syndrome** (CFS). The current symptoms accepted as a definition of CFS include (1) a sudden onset, generally following a viral infection, (2) disabling fatigue, (3) muscle weakness and pain, (4) sleep disturbance, (5) fever, and (6) enlargement of cervical lymph nodes. Roughly twice as many women as men are diagnosed with CFS.

Attempts to link either fibromyalgia or CFS to a viral infection, adrenal gland dysfunction, or some physical or psychological trauma have not been successful, and the causes remain unknown. For both conditions, treatment is at present limited to relieving symptoms when possible. For example, anti-inflammatory medications may help relieve pain, antidepressant

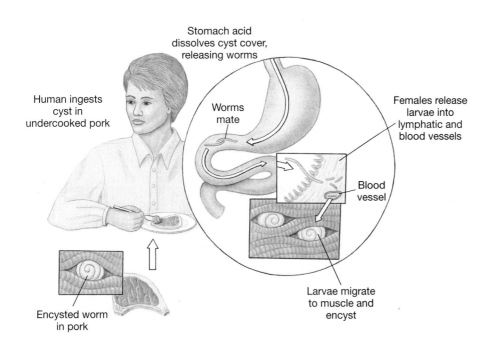

● **FIGURE 26**
The Life Cycle of *Trichinella spi-*
ralis

medications may improve sleep patterns and reduce depression, and exercise programs may help maintain a normal range of motion. Reassurance that fibromyalgia is not progressive, crippling, or life-threatening may help those who have it.

■ THE MUSCULAR DYSTROPHIES

HA p. 241

The **muscular dystrophies** (DIS-trō-fēz) are inherited diseases caused by a mutation in one of the many genes that affect muscle function. Progressive muscle weakness and deterioration occurs with variable severity depending on the type of mutation. One of the most common and best understood conditions is **Duchenne's muscular dystrophy (DMD)**, which may be inherited from the mother or, in 30 percent of cases, from spontaneous mutation. Symptoms of this form of muscular dystrophy start in childhood, commonly between ages 3 and 7. The condition is caused by mutation of the DMD gene, which codes for the protein dystrophin. This gene is located on the X chromosome and generally affects males (it can occur in females only if a very rare form of chromosome duplication occurs). DMD occurs at an incidence of roughly 30 per 100,000 male births. Having only 0–5 percent of the normal amount of dystrophin in muscle fibers, these children develop progressive muscular weakness and require wheelchairs by age 8 to 12. Most individuals die before age 20, due to respiratory paralysis or cardiac problems. Skeletal muscles are primarily affected, although for some reason the facial muscles continue to function normally. In later stages of the disease, the facial muscles and cardiac muscle tissue may also become involved.

Dystrophin is a large protein that attaches thin filaments of the sarcomeres to an anchoring protein on the sarcolemma, providing mechanical strength to the muscle fiber and connecting the myofibrils to the sarcolemma. In children with DMD, calcium channels remain open for an extended period, and calcium levels in the sarcoplasm rise to the point at which key proteins denature. Inflammation occurs and the muscle fiber then degenerates, resulting in elevated plasma levels of the muscle enzyme creatinine phosphokinase. Steroid treatment slows progression for up to three years at the price of significant side effects. Rats with a form of DMD treated by various forms of gene and stem cell therapy have sometimes improved, giving hope that an effective treatment may be possible.

Women carrying the defective gene are asymptomatic, but each of their sons will have a 50 percent chance of developing DMD; other techniques allow for carrier detection and fetal diagnosis by eight weeks' gestation.

■ MYOTONIC DYSTROPHY

Myotonic dystrophy is a form of muscular dystrophy that occurs in the United States with an incidence of 13.5 per 100,000 population. Symptoms may develop in infancy, but more commonly they develop after puberty. As with other forms of muscular dystrophy, adults developing myotonic dystrophy experience a gradual reduction in muscle strength and control. Problems with other systems, especially the cardiovascular and digestive systems, typically develop. There is no effective treatment.

The inheritance of myotonic dystrophy is unusual, because children of an individual with this condition commonly develop symptoms that are more severe than those of the parent. The increased severity of the condition appears to be related to the presence of multiple copies of a specific gene on chromosome 19. For some reason, the nucleotide sequence of that gene gets repeated several times, and the number can increase from generation to generation. This phenomenon has been called a "genetic stutter." The greater the number of copies, the more severe are the symptoms. It is not known why the stutter develops or how the genetic duplication affects the severity of the condition. Evidence indicates that the extra nucleotides interfere in some way with the transcription of an adjacent gene involved with the control of muscle tone.

■ PROBLEMS WITH THE CONTROL OF MUSCULAR ACTIVITY   *HA p. 245*

Another group of disorders interferes with normal neuromuscular communication by affecting either the nerve's ability to issue commands or the muscle's ability to respond. Anything that interferes with neural function or with excitation–contraction coupling will cause muscular paralysis. Two examples are worth noting:

1. **Botulism** results from the consumption of canned or smoked foods contaminated with a bacterial toxin. The toxin prevents the release of ACh at the synaptic terminals, leading to a potentially fatal muscular paralysis.

2. The progressive muscular paralysis of **myasthenia gravis** results from the loss of ACh receptors at the junctional folds. The primary cause is a misguided attack on the ACh receptors by the immune system. Genetic factors play a role in predisposing individuals to this condition.

3. The motor paralysis caused by **polio** is the result of viral damage to motor neurons.

■ BOTULISM

Botulinus (bot-ū-LĪ-nus) **toxin** prevents the release of acetylcholine (ACh) at synaptic terminals, thereby producing a severe and potentially fatal paralysis of the skeletal muscles. A case of botulinus poisoning is called **botulism**.[1] The toxin is produced by the bacterium *Clostridium botulinum*, which does not need oxygen to grow and reproduce. Because this bacterium can live quite well in a sealed can or jar, most cases of botulism are linked to improper canning or storing procedures, followed by failure to cook the food adequately before it is eaten. Home-canned tuna or beets, smoked fish, and cold soups are the foods most commonly linked to botulism. Boiling for a half-hour destroys both the toxin and the bacteria.

Symptoms generally begin 12–36 hours after a contaminated meal is eaten. The initial symptoms are typically disturbances in vision, such as double vision or a painful sensitivity to bright light. These symptoms are followed by other sensory and motor problems, including blurred speech and an inability to stand or walk. Roughly half of botulism patients experience intense nausea and vomiting. These symptoms persist for days to weeks, followed by a gradual recovery; some patients are still in recovery after a year.

The major risk of botulinus poisoning is respiratory paralysis and death by suffocation. Treatment is supportive: bed rest, observation, and, if necessary, the use of a mechanical respirator. In severe cases, an antitoxin and drugs that promote the release of ACh, such as *guanidine hydrochloride*, may be administered. The overall mortality rate in the United States is about 10 percent.

A commercial form of botulinum toxin, "Botox," is injected into selected muscles to treat problems such as spasms of the eyelid and neck muscles or unequal muscle strength in strabismus (crossed eyes). The effect is temporary and may need to be repeated. In recent years, Botox treatments have also been increasingly used by aging baby boomers to suppress facial wrinkles.

■ MYASTHENIA GRAVIS

Myasthenia gravis (mī-as-THĒ-nē-uh GRA-vis) is characterized by a general muscular weakness that tends to be most pronounced in the muscles of the arms, head, and chest. The first symptom is generally a weakness of the eye muscles and drooping eyelids. Facial muscles are commonly weak as well, and the individual develops a peculiar smile known as the "myasthenic snarl." As the disease progresses, weakness of the pharynx leads to problems with chewing and swallowing, and holding the head upright becomes difficult.

The muscles of the upper chest and upper limbs are next to be affected. All the voluntary muscles of the body may ultimately be involved. Severe myasthenia gravis produces respiratory paralysis, with a mortality rate of 5–10 percent. However, the disease does not always progress to such a life-threatening stage. For example, roughly 20 percent of people with the disease experience no symptoms other than eye problems.

The condition results from a decrease in the number of ACh receptors on the motor end plate. Before the remaining receptors can be stimulated enough to trigger a strong contraction, the ACh molecules are destroyed by cholinesterase. As a result, muscular weakness develops.

The primary cause of myasthenia gravis appears to be a malfunction of the immune system. Roughly 70 percent of individuals with myasthenia gravis have an abnormal thymus, an organ involved with the maintenance of normal immune function. In myasthenia gravis, the immune response attacks the ACh receptors of the motor end plate as if they were foreign proteins. For unknown reasons, 1.5 times as many women as men are affected. The typical age at onset is 20–30 for women, versus over 60 for men. Estimates of the incidence of this disease in the United States range from 2 to 10 cases per 100,000 population.

One approach to therapy involves the administration of drugs, such as *neostigmine,* that are termed **cholinesterase inhibitors**. As their name implies, these compounds are enzyme inhibitors; they tie up the active sites at which cholinesterase normally binds ACh. With cholinesterase activity reduced, the concentration of ACh at the synapse can rise enough to stimulate the surviving receptors and produce muscle contraction. Corticosteroid therapy is typically beneficial, as is surgical removal of the thymus.

---

[1] The disorder was described more than 200 years ago by German physicians who treated patients who were poisoned by dining on contaminated sausages. *Botulus* is the Latin word for "sausage."

■ POLIO

Because skeletal muscles depend on their motor neurons for stimulation, disorders that affect the nervous system can have an indirect effect on the muscular system. **Polio** is caused by the *poliovirus,* a virus that does not produce clinical symptoms in roughly 95 percent of infected individuals. The virus produces variable symptoms in the remaining 5 percent. Some individuals develop a nonspecific illness resembling the flu. Other individuals develop a brief *meningitis* (p. 67), an inflammation of the protective membranes that surround the CNS. In still another group of people, the virus attacks somatic motor neurons in the CNS.

In this third form of the disease, the individual develops a fever 7–14 days after infection. The fever subsides, but recurs roughly a week later, accompanied by muscle pain, cramping, and paralysis of one or more limbs. Respiratory paralysis may also occur, and the mortality rate of this form of polio is 2–5 percent in children and 15–30 percent in adults. If the individual survives, some degree of recovery generally occurs over a period of up to six months.

For unknown reasons, the survivors of paralytic polio may develop progressive muscular weakness 20–30 years after the initial infection. This *postpolio syndrome* is characterized by fatigue, muscle pain and weakness, and, in some cases, muscle atrophy. There is no treatment for the condition, although rest seems to help.

Polio has been almost completely eliminated from the U.S. population due to a successful immunization program. In 1954, 18,000 new cases occurred in the United States; there were 8 in 1976, and none since 1994. The World Health Organization reports that polio has been eradicated from the entire Western Hemisphere. Unfortunately, many parents today refuse to immunize their children against the poliovirus, because they assume that the disease has been "conquered." Failure to immunize is a mistake, because (1) there is still *no cure* for polio, (2) the virus remains in the environment in many areas of the world, and (3) up to 38 percent of children age 1–4 have not been immunized. A major epidemic could therefore develop very quickly if the virus were brought into the United States from another part of the world.

For years, the vaccine that has been used is the oral Sabin vaccine, preferred for its ease of administration and better immune stimulation than the injectable vaccine. However, unlike the injected vaccine, the oral vaccine carries a 1-in-1-million risk that the immunized person will develop polio. In 1996, the Centers for Disease Control and Prevention (CDC) recommended the use of either a combination of injected and oral vaccines or the injected vaccine alone.

■ DELAYED-ONSET MUSCLE SORENESS *HA p. 248*

You have probably experienced muscle soreness the day after a period of physical exertion (Figure 27●). Considerable controversy exists over the source and significance of this pain, which is known as *delayed-onset muscle soreness* (DOMS) and has several interesting characteristics:

- DOMS is distinct from the soreness you experience immediately after you stop exercising. The initial short-term soreness is probably related to the biochemical events associated with muscle fatigue.
- DOMS generally begins several hours after the exercise period ends and may last three or four days.
- The amount of DOMS is highest when the activity involves eccentric contractions. Activities dominated by concentric or isometric contractions produce less soreness.
- Levels of CPK and myoglobin are elevated in the blood, indicating damage to muscle cell membranes. The nature of the activity (eccentric, concentric, or isometric) has no effect on these levels, nor can the levels be used to predict the degree of soreness experienced.

Three mechanisms have been proposed to explain DOMS:

1. Small tears may exist in the muscle tissue, leaving muscle fibers with damaged membranes. The sarcolemma of each damaged muscle fiber permits the loss of enzymes, myoglobin, and other chemicals that may stimulate nearby pain receptors.

2. The pain may result from muscle spasms in the affected skeletal muscles. In some studies, stretching the muscle involved after exercise reduced the degree of soreness.

3. The pain may result from tears in the connective tissue framework and tendons of the skeletal muscle.

Some evidence supports each of these mechanisms, but it is unlikely that any one tells the entire story. For example, muscle fiber damage is certainly supported by biochemical findings, but if that were the only factor, the type of activity and the level of intracellular enzymes in the circulation would be correlated with the level of pain experienced, and this is not the case.

● **FIGURE 27**
Delayed-Onset Muscle Soreness. A rigorous workout can cause lingering pain whose origins are uncertain.

CHAPTER 10
THE AXIAL MUSCULATURE

DISORDERS OF THE AXIAL MUSCULATURE

■ HERNIAS   *HA p. 273*

When the abdominal muscles contract forcefully, pressure in the abdominopelvic cavity can increase dramatically. That pressure is applied to internal organs. If the individual exhales at the same time, the pressure is relieved because the diaphragm can move upward as the lungs collapse. But during vigorous isometric exercises or when lifting a weight while holding one's breath, pressure in the abdominopelvic cavity can rise to 106 kg/cm^2, roughly 100 times the normal pressure. A pressure that high can cause a variety of problems, including hernias. A **hernia** develops when a visceral organ or part of an organ protrudes abnormally through an opening in a surrounding muscular wall or partition. There are many types of hernias; here we will consider only *inguinal* (groin) *hernias* and *diaphragmatic hernias.*

Late in the development of male fetuses, the testes descend into the scrotum by passing through the abdominal wall at the **inguinal canals.** In adult males, the sperm ducts and associated blood vessels penetrate the abdominal musculature at the inguinal canals as the *spermatic cords,* on their way to the abdominal reproductive organs. In an inguinal hernia, the inguinal canal enlarges and the abdominal contents, such as a portion of the greater omentum, small intestine, or (more rarely) urinary bladder, enter the inguinal canal (Figure 28●). If the herniated structures become trapped or twisted, surgery may be required to prevent serious complications. Inguinal hernias are not always caused by unusually high abdominal pressures; injuries to the abdomen or inherited weakness or distensibility of the canal can have the same effect.

The esophagus and major blood vessels pass through openings in the diaphragm, the muscle that separates the

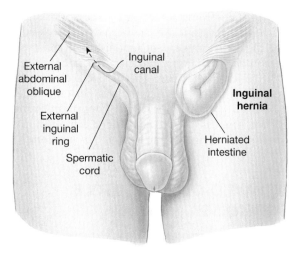

● FIGURE 28
An Inguinal Hernia

thoracic and abdominopelvic cavities. In a **diaphragmatic hernia**, abdominal organs slide into the thoracic cavity. If entry is through the *esophageal hiatus,* the passageway used by the esophagus, a *hiatal hernia* (hī-Ā-tal; *hiatus,* a gap or opening) exists. The severity of the condition depends on the location and size of the herniated organ or organs. Hiatal hernias are very common, and most go unnoticed, although they may increase the severity of gastric acid entry into the esophagus (gastroesophageal reflux disease, or GERD, commonly known as heartburn). Radiologists see them in about 30 percent of individuals whose upper gastrointestinal tracts are examined with barium-contrast techniques.

When clinical complications other than GERD develop, they generally do so because abdominal organs that have pushed into the thoracic cavity are exerting pressure on structures or organs there. Like inguinal hernias, a diaphragmatic hernia can result from congenital factors or from an injury that weakens or tears the diaphragm. If abdominal organs occupy the thoracic cavity during fetal development, the lungs may be poorly developed at birth.

10

CHAPTER 11
THE APPENDICULAR MUSCULATURE

DISORDERS OF THE APPENDICULAR MUSCULATURE

■ SPORTS INJURIES   *HA p. 287*

Sports injuries affect amateurs and professionals alike. A five-year study of college football players indicated that 73.5 percent experienced mild injuries, 21.5 percent moderate injuries, and 11.6 percent severe injuries during their playing careers. Contact sports are not the only activities that show a significant injury rate: A study of 1650 joggers running at least 27 miles a week reported 1819 injuries in a single year.

Muscles and bones respond to increased use by enlarging and strengthening. Poorly conditioned individuals are therefore more likely than people in good condition to subject their bones and muscles to intolerable stresses. Training is also important in minimizing the use of antagonistic muscle groups and in keeping joint movements within the intended ranges of motion. Planned warm-up exercises before athletic events stimulate circulation, improve muscular performance and control, and help prevent injuries to muscles, joints, and ligaments. Stretching exercises after an initial warm-up will stimulate blood flow to muscles and help keep ligaments and joint capsules supple. Such conditioning extends the range of motion and may prevent sprains and strains when sudden loads are applied.

Dietary planning can also be important in preventing injuries to muscles during endurance events, such as marathon running. Emphasis has commonly been placed on the importance of carbohydrates, leading to the practice of "carbohydrate loading" before a marathon. But while operating within aerobic limits, muscles also utilize amino acids extensively, so an adequate diet must include both carbohydrates and proteins.

Improved playing conditions, equipment, and regulations also play a role in reducing the incidence of sports injuries. Jogging shoes, ankle and knee braces, helmets, mouth guards, and body padding are examples of equipment that can be effective. The substantial penalties now earned for personal fouls in contact sports have reduced the numbers of neck and knee injuries.

Several traumatic injuries common to those engaged in active sports can also affect nonathletes, although the primary causes may be very different. A partial listing of activity-related conditions includes the following:

- **Bone bruise:** bleeding within the periosteum of a bone
- **Bursitis:** an inflammation of the bursae at joints
- **Muscle cramps:** prolonged, involuntary, and painful muscular contractions
- **Sprains:** tears or breaks in ligaments or tendons
- **Strains:** tears in muscles
- **Stress fractures:** cracks or breaks in bones subjected to repeated stresses or trauma
- **Tendinitis:** an inflammation of the connective tissue surrounding a tendon

We have discussed many of these conditions in related sections of the text.

Finally, many sports injuries would be prevented if people who engage in regular exercise would use common sense and recognize their personal limitations. It can be argued that some athletic events, such as the ultramarathon, place such excessive stresses on the cardiovascular, muscular, respiratory, and urinary systems that these events cannot be recommended, even for athletes in peak condition.

■ INTRAMUSCULAR INJECTIONS

HA p. 287

Drugs are commonly injected through hollow needles into tissues rather than directly into the bloodstream (accessing blood vessels may be technically more complicated). An **intramuscular (IM) injection** introduces a fairly large amount of a drug, which will then enter the circulation gradually. The drug is introduced into the mass of a large skeletal muscle. Uptake is generally faster and accompanied by less tissue irritation than when drugs are administered *intradermally* or *subcutaneously* (injected into the dermis or subcutaneous layer, respectively). Depending on the size of the muscle, up to 5 ml of fluid may be injected at one time, and multiple injections are possible. A decision on the injection technique and the injection site is based on the type of drug and its concentration.

For IM injections, the most common complications involve accidental injection into a blood vessel or piercing of a nerve. The sudden entry of massive quantities of a drug into the bloodstream can have fatal consequences, and damage to a nerve can cause motor paralysis or sensory loss., Thus, the site of injection must be selected with care. Bulky muscles that contain few large vessels or nerves are ideal sites. The gluteus medius muscle or the posterior, lateral, superior part of the gluteus maximus muscle is commonly selected. The deltoid muscle of the arm, about 2.5 cm (1 in.) distal to the acromion, is another effective site. Probably most satisfactory from a technical point of view is the vastus lateralis muscle of the thigh; an injection into this thick muscle will not encounter vessels or nerves, but may cause pain later when the muscle is used in walking. This is the preferred injection site in infants before they start walking, as their gluteal and deltoid muscles are relatively small (Figure 29●). The site is also used in elderly patients or others with atrophied gluteal and deltoid muscles.

● **FIGURE 29**
Intramuscular Injection in the Thigh

CHAPTER 12
SURFACE ANATOMY

For a discussion of the clinical relevance of surface anatomy, see the related sections on the diagnosis of disease in Chapter 1, "An Introduction to Clinical Anatomy."

AN INTRODUCTION TO THE NERVOUS SYSTEM AND ITS DISORDERS

The nervous system is a highly complex and interconnected network of neurons and supporting neuroglia. Neural tissue is extremely delicate, and the characteristics of the extracellular environment must be kept within narrow homeostatic limits for it to function well. When homeostatic regulatory mechanisms break down under the stress of genetic or environmental factors, infection, or trauma, symptoms of neurological disorders appear.

Hundreds of disorders can affect the nervous system. A *neurological examination* attempts to trace the source of a problem through an evaluation of the sensory, motor, behavioral, and cognitive functions of the nervous system. Figure 30● introduces several major categories of nervous system disorders. We will discuss many of these examples in the sections that follow. Table 17 summarizes representative infectious diseases of the nervous system.

■ THE SYMPTOMS OF NEUROLOGICAL DISORDERS *HA p. 332*

The nervous system has varied and complex functions, and the symptoms of neurological disorders are equally diverse. However, a few symptoms accompany a wide variety of disorders:

- **Headache.** Headache seems to be a universal experience, with 70 percent of people reporting at least one headache each year. Most of these are *tension-type headaches,* which are thought to be at least partially due to muscle tension, or *migraine headaches,* which have both neurological and cardiovascular origins (see p. 62). These conditions are rarely associated with life threatening problems.

- **Muscle weakness.** Muscle weakness can have an underlying neurological basis, as we noted in Figure 24●, p. 50. The examiner must determine the primary cause of the symptom for most effective treatment. Myopathies (muscle disease) must be differentiated from neurological diseases such as demyelinating disorders, neuromuscular junction dysfunction, and peripheral nerve damage.

- **Paresthesias.** Loss of feeling, numbness, or tingling sensations may develop after damage to (1) a sensory nerve (cranial or spinal nerve) or (2) sensory pathways in the central nervous system (CNS). The effects can be temporary or permanent. For example, a *pressure palsy* may last a few minutes, whereas the paresthesia that develops distal to an area of severe spinal cord damage will probably be permanent (p. 69).

■ THE NEUROLOGICAL EXAMINATION *HA p. 332*

During a physical examination, information about the nervous system is obtained indirectly by assessing sensory, motor, and intellectual functions. Examples of factors noted in the physical examination include the following:

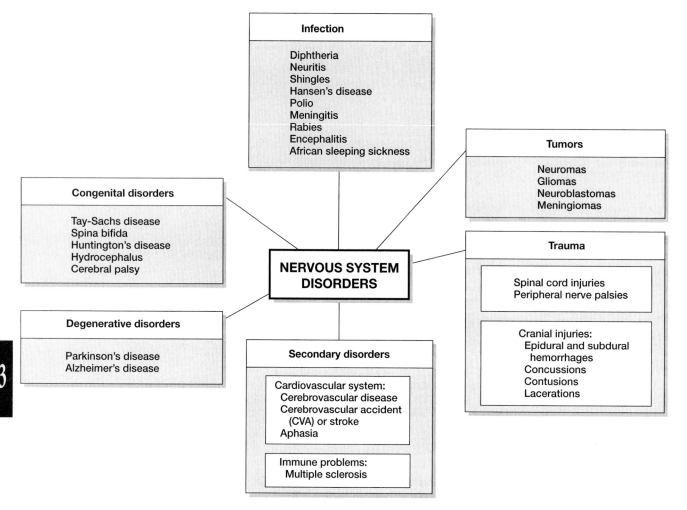

● **FIGURE 30**
Nervous System Disorders

- **State of consciousness.** There are many levels of con-sciousness, ranging from unconscious and incapable of being aroused, to fully alert and attentive, to hyperexcitable. We introduce the terms assigned to the various levels of con-sciousness in a later section.

- **Reflex activity.** The state of peripheral sensory and motor in-nervation can be checked by testing specific reflexes (p. 71). For example, the *knee-jerk reflex* will not be normal if dam-age has occurred in associated segments of the lumbar spinal cord, their spinal nerve roots, or the peripheral nerves in-volved in the reflex. Musculoskeletal abnormalities also af-fect reflex responses, but such abnormalities can usually be detected on further examination.

- **Abnormal speech patterns.** Normal speech involves intel-lectual processing, motor coordination at the speech centers of the brain, precise respiratory control, regulation of tension in the vocal cords, and adjustment of the musculature of the palate and face. Problems with the selection, production, or use of words commonly follow damage to the cerebral hemi-spheres, as in a stroke (p. 115).

- **Abnormal motor patterns.** An individual's posture, balance, and mode of walking, or gait, are useful indicators of the level of motor coordination and strength. Clinicians also ask about

abnormal involuntary movements that may indicate a *seizure*—a temporary disorder of cerebral function (p. 73).

A number of diagnostic procedures and laboratory tests can be used to obtain additional information about the status of the nervous system. Table 18 summarizes information about these procedures.

■ HEADACHES   *HA p. 332*

Almost everyone has experienced a **headache** at one time or another. Diagnosis and treatment pose a number of problems, primarily because, as we noted earlier, headaches can be produced by a wide variety of conditions (one source notes 316 established causes). The most common causes of headache appear to be either vascular or muscular problems.

Most headaches do not merit a visit to a neurologist. The majority are tension-type headaches with moderate pain that is pressing or tightening, poorly localized, and associated with muscle tension, such as tight neck muscles. Tension-type headaches do not have the associated features that define mi-graine headaches: throbbing, often unilateral severe pain, light

TABLE 17 Examples of Infectious Diseases of the Nervous System

Disease	Organism(s)	Description
Bacterial Diseases		
Hansen's disease (leprosy)	*Mycobacterium leprae*	Progresses slowly in coolest areas of body; invades nerves and produces sensory loss and motor paralysis; cartilage and bone may degenerate after unperceived trauma
Bacterial meningitis		Inflammation of the spinal or cranial meninges
	Haemophilus influenzae	Previously common meningitis; usually infects children (age 2 months–5 years); vaccine available and effective
	Neisseria meningitidis	Meningococcal meningitis; usually infects children and adults (age 5–40 years, young adults living closely together, military recruits and college students in dorms at higher risk); vaccines somewhat effective
	Streptococcus pneumoniae	Streptococcal meningitis; usually infects children and adults over age 40; high mortality rate (40 percent); vaccine effective
Brain abscesses	Various bacteria	Infection increases in size and compresses the brain
Viral Diseases		
Poliomyelitis	Polioviruses	Polio has different forms; only one attacks motor neurons, leading to paralysis of limbs and muscle atrophy; vaccine is available
Rabies	Rabies virus	Virus invades the central nervous system through peripheral nerves; untreated cases are fatal; treatment involves rabies antitoxin and vaccination; pre-exposure vaccine available for those at high risk of exposure
Encephalitis	Various encephalitis viruses	Inflammation of the brain; fever and headache; no vaccine is available; transmission occurs by mosquitoes; eastern equine encephalitis is most lethal (50–75 percent mortality rate)
Meningitis	Various viruses	Generally less severe than bacterial, usually resolves with symptomatic treatment
Parasitic Diseases		
African sleeping sickness	*Trypanosoma brucei*	Caused by a flagellated protozoan; infection occurs through bite of tsetse fly; infects blood, lymph nodes, and then nervous system; symptoms include headache, tiredness, weakness, and paralysis, before coma and death; no vaccine is available but drug treatment is effective

13

sensitivity, and nausea or vomiting. Other headaches develop secondarily due to the following problems:

- **CNS disorders,** such as infections (meningitis, encephalitis, rabies) or brain tumors
- **Trauma,** such as a blow to the head (p. 73)
- **Cardiovascular disorders,** such as a stroke (p. 94)
- **Metabolic disturbances,** such as low blood sugar

Evidence indicates that migraine headaches begin at a portion of the mesencephalon known as the *dorsal raphe*. Electrical stimulation of the dorsal raphe can produce changes in cerebral blood flow; several drugs with antimigraine action inhibit neurons there. The most effective drugs stimulate a class of serotonin receptors that are abundant in the dorsal raphe. The trigger for **tension-type headaches** probably involves a combination of factors, but sustained contractions of the neck and facial muscles are most commonly implicated. Tension headaches can last for days or can occur daily over longer periods. Some tension headaches may accompany severe depression or anxiety.

■ AXOPLASMIC TRANSPORT AND DISEASE *HA p. 340*

With a soft flutter of wings, dark shapes drop from the sky onto the backs of grazing cattle. Each shape is a small bat whose scientific name, *Desmodus rotundus,* is less familiar than the popular term, *vampire bat.* Vampire bats inhabit tropical and semitropical areas of North, Central, and South America. They range from the Texas coast to Chile and southern Brazil. These rather aggressive animals are true vampires, subsisting on a diet of fresh blood. Over the next hour, every bat in the flight—which may number in the hundreds—will consume about 65 ml of blood through small slashes in the skin of their prey.

As unpleasant as this blood collection may sound, it is not the blood loss that is the primary cause for concern. The major problem is that these bats can be carriers of the rabies virus. **Rabies** is an acute disease of the central nervous system. The rabies virus can infect any mammal, wild or domestic. With few exceptions, the result is death within three weeks. For

TABLE 18   Representative Diagnostic and Laboratory Tests for Nervous System Disorders

Diagnostic Procedure	Method and Result	Representative Uses
Lumbar puncture (LP) (spinal tap)	Needle aspiration of CSF from the subarachnoid space in the lumbar area of the spinal cord	See Analysis of CSF (below) for diagnostic.
Skull x-ray	Standard x-ray	Detects fractures, possible sinus involvement, and other bony abnormalities
Electroencephalography (EEG)	Electrodes placed on the scalp detect electrical activity of the brain, and EEG produces graphic record	Detects abnormalities in frequency and amplitude of brain waves due to cranial trauma or neurological disorders such as seizures
Computerized tomography (CT) scan of the brain	Standard CT; contrast media are commonly used	Detects tumors; cerebrovascular abnormalities; acute bleeding from strokes on aneurysms; scars; strokes; areas of edema
Cerebral angiography and digital subtraction angiography (DSA)	Dye is injected into an artery of the neck, and the movement of the dye is observed via serial x-rays; DSA transfers location of dye information to a computer for image enhancement	Detects abnormalities in the cerebral vessels, aneurysms or blockages
Positron emission tomography (PET) scan	Radiolabeled compounds injected into bloodstream accumulate at specific areas of the brain; the radiation emitted is monitored by a computer that generates a reconstructed image	Determines blood flow to specific regions of the brain; detects focal points of brain metabolic activity; diagnose mental illness; may help diagnosis of Parkinson's disease and Alzheimer's disease
Single-photon emission tomography (SPECT)	Similar to PET but technically easier	Similar to PET
Magnetic resonance imaging (MRI)	Standard MRI; contrast media are commonly used to enhance visualization	Detects brain tumors, hemorrhaging, edema, spinal-cord injury, and other soft tissue structural abnormalities

Laboratory Test	Normal Values	Significance of Abnormal Values
Pressure of CSF	200 cm H_2O	Pressure higher than 200 cm H_2O is considered abnormal, possibly indicating hemorrhaging, brain tumor, blood clots around the brain, or infection.
Color of CSF	Clear and colorless	Increased cloudiness suggests hemorrhage, faulty puncture technique, or infection.
Analysis of CSF		
Glucose in CSF	50–75 mg/dl	Decreased levels occur with CNS bacterial or fungal infections.
Protein in CSF	15–45 mg/dl	Elevated levels occur in some infectious processes, such as meningitis and encephalitis, also during inflammatory processes or brain tumor.
Cells present in CSF	No RBCs present; WBC count should be less than 5 per mm^3.	RBCs appear with subarachnoid hemorrhage and traumatic LP; increased number of neutrophils occurs in bacterial infections, such as bacterial meningitis; increase in lymphocyte numbers occurs in viral meningitis; cancer cells from a brain tumor may be shed into CSF.
Culture of CSF	Organism causing infectious process in brain or spinal cord can be cultured for identification and determination of sensitivity to antibiotics	Culture can determine causative agent in meningitis or brain abscess. Takes at least two days
Polymerase chain reaction (PCR)	negative	Identifies most viral and some bacterial meningitis. Results available same day

unknown reasons, bats can survive rabies infection for an indefinite period. That is why they are effective carriers of the disease. Because many bat species, including vampire bats, form dense colonies, a single infected individual can spread the disease throughout the entire colony.

Rabies is generally transmitted to people through the bite of a rabid animal. About 15,000 cases of human rabies occur each year worldwide, the majority of them the result of dog bites. Only about five of those cases, however, occur in the United States. In the United States, most cases of rabies are caused instead by the bites of raccoons, foxes, skunks, or bats because most dogs and cats are vaccinated against rabies.

Although these bites generally involve peripheral sites, such as the hand or foot, the symptoms are caused by CNS damage. The virus present at the site of the injury is absorbed by the synaptic knobs of peripheral nerves in the region. It then gets a free ride to the CNS, courtesy of retrograde flow. The incubation period is usually from one to seven weeks. During the first few days of illness, the individual may experience headache, fever, muscle pain, nausea, and vomiting. The afflicted person then enters the central nervous system phase marked by extreme excitability, hallucinations, muscle spasms, and disorientation. Swallowing becomes difficult, accounting for an early name for rabies: *hydrophobia* (literally, fear of water). The accumulation of saliva makes the individual appear to be "foaming at the mouth." Coma and death soon follow. Anyone exposed to the patient's saliva should receive preventive treatment.

Postexposure preventive treatment (PEP) was developed by Louis Pasteur in the late 1800s. The modern version of PEP must begin almost immediately after exposure and consists of an injection of immune globulin antibodies against the rabies virus, followed by a series of vaccinations against rabies. Without such treatment, rabies infection in humans is always fatal; but even this postexposure treatment may not be sufficient after a massive infection, which can lead to death in as little as four days. Individuals such as veterinarians and field biologists who are at high risk of exposure thus commonly take a pre-exposure series of vaccinations. These injections start the immune defenses and improve the effectiveness of the postexposure treatment.

Rabies is perhaps the most dramatic example of a clinical condition directly related to axoplasmic transport. However, many toxins, including heavy metals and toxins released by the tetanus bacteria, some pathogenic bacteria, and other viruses, use this mechanism to enter the CNS.

■ DEMYELINATION DISORDERS

HA p. 343

Demyelination disorders are linked by a common symptom: the destruction of myelinated axons in the CNS and peripheral nervous system (PNS). The mechanism responsible for this loss differs in each disorder. We will consider only the major categories:

- **Heavy-metal poisoning.** Chronic exposure to heavy-metal ions, such as arsenic, lead, and mercury, can lead to damage of neuroglia and to demyelination. As demyelination occurs, the affected axons deteriorate and the condition becomes irreversible. Historians note several examples of heavy-metal poisoning with widespread impact. For example, the contamination of drinking water with lead has been cited as one factor in the decline of the Roman Empire. In the 17th century, the great physicist Sir Isaac Newton is thought to have experienced several episodes of physical illness and mental instability brought on by his use of mercury in chemical experiments. Well into the 19th century, mercury used in the preparation of felt presented a serious occupational hazard for those employed in the manufacture of stylish hats. Over time, mercury absorbed through the skin and across the lungs accumulated in the CNS, producing neurological damage that affected both physical and mental functioning. (This effect is the source of the expression "mad as a hatter.") In the 1950s, Japanese fishermen and their families working in Minamata Bay, Japan, collected and consumed seafood contaminated with mercury discharged from a nearby chemical plant. Levels of mercury in their systems gradually rose to the point at which clinical symptoms appeared in hundreds of people. Making matters worse, pregnant women exposed to mercury had babies with severe, crippling birth defects. Less severe problems have affected children born to mothers in the midwestern United States who ate large amounts of fish during pregnancy. As a result, pregnant women are now advised to limit fish consumption. (For unknown reasons, the flesh of some species of fish contains relatively high levels of mercury.)

- **Diphtheria. Diphtheria** (dif-THĒ-rē-uh; *diphthera*, leather + *-ia*, disease) is a disease that results from a bacterial infection of the respiratory tract or the skin. In the case of respiratory infections, in addition to restricting airflow and damaging the respiratory surfaces, the bacteria produce a powerful toxin that injures the kidneys and adrenal glands, among other tissues. In the nervous system, diphtheria toxin damages Schwann cells and destroys myelin sheaths in the PNS. This demyelination leads to sensory and motor problems that can ultimately produce a fatal paralysis. The toxin also affects cardiac muscle cells, and heart enlargement and heart failure may occur. The fatality rate for untreated cases ranges from 35 to 90 percent, depending on the site of infection and the subspecies of bacterium. Because an effective vaccine (which is frequently combined with the tetanus vaccine) exists, cases are relatively rare in countries with adequate health care. Russia experienced an epidemic in the 1990s when vaccines were not widely available.

- **Multiple sclerosis. Multiple sclerosis** (skler-Ō-sis; *sklerosis*, hardness), or **MS,** is a disease characterized by recurrent incidents of demyelination that affects axons in the optic nerve, brain, and spinal cord. Common symptoms include partial loss of vision and problems with speech, balance, and general motor coordination, including bowel and urinary bladder control. The time between incidents and the degree of recovery vary from case to case. In about one-third of all cases, the disorder is progressive, and each incident leaves a greater degree of functional impairment. The average age at the first attack is 30–40 years; the incidence among women is 1.5 times that among men. In some patients, corticosteroid or interferon injections have slowed the progression of the disease.

13

■ GROWTH AND MYELINATION OF THE NERVOUS SYSTEM *HA p. 347*

The development of the nervous system continues for many years after birth. Nerve cells increase in number for the first year after delivery, and most of the important interconnections between neurons occur after birth, rather than before. Growth of the brain is completed by age 4, but the neurons have yet to be interconnected extensively. The myelination of axons may not be completed until early adolescence. The level of nervous system development limits mental and physical performance. For example, the degree of interconnection between neurons in the CNS affects intellectual abilities, and the myelination of axons improves coordination and control by decreasing the time between reception of a sensation and completion of a response. Because the early years are crucial for the development of normal interconnections, malnutrition in children under age 5 can lead to mental retardation and disability.

13

DISORDERS OF THE SPINAL CORD AND SPINAL NERVES

■ SPINAL MENINGITIS   *HA p. 352*

The warm, dark, nutrient-rich environment of the cranial or spinal meninges provides ideal conditions for a variety of bacteria and viruses. Microorganisms that cause brain infections (see Table 17, p. 63) may sometimes cause meningitis as well; the same bacteria may be associated with middle ear, sinus, throat, and lung infections. These pathogens may gain access to the meninges by traveling within blood vessels or by entering at sites of vertebral or cranial injury. Headache, chills, high fever, disorientation, and rapid heart and respiratory rates appear as higher centers are affected. Without treatment, delirium, coma, convulsions, and death may follow within hours to days.

The most common clinical assessment involves checking for a "stiff neck" by asking the patient to touch the chin to the chest. Meningitis affecting the cervical portion of the spinal cord results in a marked increase in the muscle tone of the extensor muscles of the neck. So many motor units become activated that flexion of the neck becomes painfully difficult, if not impossible.

The mortality rate for viral and bacterial forms of meningitis ranges from 1 to more than 50 percent, depending on the type of virus or bacteria, the age and health of the individual, and other factors. Bacterial meningitis can be combated with antibiotics, the maintenance of proper fluid and electrolyte balance, and other supportive care. Survivors may have permanent neurologic damage, including loss of sight or hearing, seizures, or mental retardation. The incidence of the most common forms of childhood bacterial meningitis, caused by *Haemophilus influenzae* (a bacterium despite its name) and streptococcal pneumonia, have been dramatically reduced by immunization.

■ SPINAL ANESTHESIA   *HA p. 355*

Injecting a local anesthetic around a nerve produces a temporary blockage of sensory and motor nerve function. This procedure can be done peripherally, as when skin lacerations are sewn up, or at sites around the spinal cord to obtain more widespread anesthetic effects. An *epidural block*—the injection of an anesthetic into the epidural space—has the advantage of (1) affecting only the spinal nerves in the immediate area of the injection, and (2) providing mainly sensory anesthesia. If a catheter is left in place, continued injection allows sustained anesthesia. Epidural anesthesia can be difficult to achieve in the upper cervical and midthoracic regions, where the epidural space is extremely narrow. It is more effective in the lower lumbar region, inferior to the conus medullaris, because the epidural space is somewhat broader.

CHAPTER 14
THE SPINAL CORD AND SPINAL NERVES

Caudal anesthesia involves the introduction of anesthetics into the epidural space of the sacrum. Injection at this site paralyzes and anesthetizes lower abdominal and perineal structures. Caudal anesthesia can be used to control pain during labor and delivery, but lumbar epidural anesthesia is often preferred.

Local anesthetics can also be introduced as a single dose into the subarachnoid space of the spinal cord. This procedure is commonly called "spinal anesthesia." However, the effects include both temporary muscle paralysis and sensory loss that tends to spread as the movement of cerebrospinal fluid distributes the anesthetic along the spinal cord. Problems with overdosing are seldom serious, because controlling the patient's position during administration can limit the distribution of the drug to a degree. Moreover, because the diaphragmatic breathing muscles are controlled by upper cervical spinal nerves, respiration continues even if all thoracic and abdominal segments have been paralyzed.

■ DAMAGE TO SPINAL TRACTS

14

HA p. 358

Damage to spinal tracts produces losses of sensation and/or motor control. The nature of the sensory and motor deficit depends on the site of damage and the specific tracts involved. We will consider three examples of conditions that produce sensory and motor problems by their impact on spinal tracts: *multiple sclerosis, polio,* and *spinal trauma.*

■ MULTIPLE SCLEROSIS

Multiple sclerosis (MS), introduced in the discussion of demyelination disorders (p. 65), is a disease that produces muscular paralysis and sensory losses through demyelination. Historically MS was diagnosed when neurologic abnormalities were "disseminated in space and time"—meaning that patients experienced episodic problems involving different areas of the nervous system. The initial symptoms, frequently involving some combination of paresthesias, clumsiness, and problems with vision and bladder control, appear as the result of myelin degeneration within the white matter of the lateral and posterior columns of the spinal cord, within the optic nerves, or along tracts within the brain. For example, spinal cord involvement can produce weakness, tingling sensations, and a loss of "position sense" for the limbs. For most patients, improvement occurs, but during subsequent attacks the effects become more widespread. The cumulative sensory and motor losses may eventually lead to a generalized muscular paralysis, with 50 percent of patients requiring help walking within 15 years after the initial occurrence of symptoms.

Evidence suggests that this condition is linked to problems in the immune system, caused by a combination of genetic and environmental factors, that results in the production of antibodies that attack myelin sheaths. Individuals with MS have lymphocytes that do not respond normally to foreign proteins. Because several viral proteins have amino acid sequences similar to those of normal myelin, it has been proposed that MS results from a case of mistaken identity. For unknown reasons, MS appears to be associated with cold and temperate climates. Relative lack of Vitamin D related to less sunlight may have some role. It has been suggested that individuals who develop MS may have an inherited susceptibility to a virus and that this susceptibility is exaggerated by environmental conditions. The yearly incidence in the United States averages around 50 cases for every 100,000 people in the population. Improvement has been noted in some MS patients treated with interferon, a peptide secreted by cells of the immune system; corticosteroid treatment of relapses has been linked to a slowdown in the progression of the disease.

■ POLIO

The viral disease *polio* causes paralysis by destroying somatic motor neurons. This disorder, introduced in Chapter 10 of the text, has been almost eliminated in the Western Hemisphere by an aggressive immunization program. Immunization continues because polio still occurs in other areas of the world. The disease could be brought into the United States at any time, leading to an epidemic among unimmunized persons.

■ SPINAL TRAUMA

Physical damage to the spinal cord by trauma from a severe auto crash or other accident can cause permanent paralysis because the damaged tracts seldom undergo even partial repair. Extensive damage to the spinal cord causes loss of motor and sensory function below the level of injury. Damage superior to the fifth cervical vertebra eliminates motor control (paralysis) and sensation of the upper and lower limbs, a condition termed *quadriplegia. Paraplegia,* the loss of motor control of the lower limbs, may follow damage to the thoracic spinal cord.

Less-severe injuries affecting the spinal cord or cauda equina produce symptoms of sensory loss or motor paralysis that reflect the specific nuclei, tracts, or spinal nerves involved. We will consider one example—the loss of peripheral sensation along the distribution of a spinal nerve—in a later section.

■ TECHNOLOGY AND MOTOR PARALYSIS   *HA p. 358*

Assistive technologies in the form of wheelchairs, modified cars and other tools, and accessible homes and buildings help paralyzed people live more varied lives. Computers and other electronic devices are making such technological approaches more effective and versatile.

When the spinal cord is injured above spinal segment C_3, respiratory paralysis results and patients require continuous mechanical ventilation. If the phrenic nerve, which arises from segments C_3 to C_5 and enervates the diaphragm, is not injured, respiratory cycles can be controlled by electrical stimulation of the phrenic nerve. The electrodes are implanted either along the phrenic nerve or intramuscularly on the diaphragm.

Research teams are experimenting with the use of computers to stimulate specific muscles and muscle groups electrically in a pre-determined sequence. The technique is called *functional electrical stimulation,* or FES. This approach commonly involves the implantation of a network of wires beneath the skin, with their tips in skeletal muscle tissue. The wires are connected to a small computer worn at the waist. The wires deliver minute electrical stimuli to the muscles, depolarizing their membranes and causing contractions. With this equipment and lightweight braces, quadriplegics have walked several hundred yards and paraplegics several thousand. The *Parastep* system, which uses a microcomputer controller and noninvasive surface electrodes, is also available now for exercise and limited ambulation. Using a network of wires woven into the fabric of close-fitting garments to create a set of electronic "hot pants," a paraplegic completed several miles of the 1985 Honolulu Marathon, and more recently another paraplegic walked down the aisle at her own wedding. However, muscular stimulation seems most beneficial in reducing the atrophy of unused muscles and improving overall fitness and health, rather than providing full mobility in paralyzed people.

Such technological solutions can provide only a degree of motor control without accompanying sensation. Everyone would prefer a biological procedure that would restore the functional integrity of the nervous system. The 1995 horseback-riding cervical spinal injury of actor Christopher Reeve has brought publicity and research funds to this area. Eight years after his injury, he had recovered limited sensation in much of his body, partial movement in one hand, and very slight movement of his limbs. An ongoing intensive physical therapy program, including electrical stimulation of his muscles, seems to have contributed to this small but significant improvement in his condition before his death in late 2004.

■ DAMAGE AND REPAIR OF PERIPHERAL NERVES *HA p. 360*

If a peripheral axon is damaged but not displaced, normal function may eventually return as the cut stump grows across the injury site, away from the soma and along its former path. The mechanics of this process were described at the close of Chapter 13 in the text. For normal function to be restored, several things must happen: The severed ends must be relatively close together (1–2 mm, or 0.04–0.08 in.); they must remain in proper alignment; and there must be no physical obstacles between them, such as the collagen fibers of scar

tissue. These conditions can be created in the laboratory, using experimental animals and individual axons or small fascicles. But in accidental injuries to peripheral nerves, the edges are likely to be jagged; intervening segments may be lost entirely; and elastic contraction in the surrounding connective tissues may pull the cut ends apart and misalign them.

Until recently, the surgical response would involve trimming the injured nerve ends, sewing them together neatly, and hoping for the best. This procedure was typically unsuccessful, in part because scalpels do not produce a smoothly cut surface and in part because the thousands of broken axons would never be perfectly aligned. Moreover, axons are not highly elastic, so if a large segment of the nerve was removed, crushed, or otherwise destroyed, there would be no way to bring the intact ends close enough to permit regeneration. In such instances, a **nerve graft** could be inserted, using a section from some other, less important peripheral nerve. The functional results were even less likely to be wholly satisfactory, because the growing axonal tips had to find their way across not one but two gaps, resulting in limited and random reinnervation. Nevertheless, any return of function was better than none!

Research now focuses on the physical and biochemical control of nerve regeneration, using the growth factors introduced earlier. Another promising strategy is the use of synthetic or biologic sleeves with nerve growth factors to guide and stimulate nerve growth. The sleeve may be a tube with an outer layer of silicone around a vein or an inner layer of cowhide collagen. When filled with muscle tissue or cultured Schwann cells (both of which seem to contain nerve growth factors) or saturated with purified growth factors, these grafts approach the performance of auto nerve grafts in animal tests.

■ PERIPHERAL NEUROPATHIES
HA p. 368

Peripheral neuropathies, or peripheral nerve palsies, are characterized by regional losses of sensory and motor function as a result of nerve trauma or compression. **Brachial palsies** result from injuries to the brachial plexus or its branches. **Crural palsies** involve the nerves of the lumbosacral plexus.

Palsies appear for several reasons. The *pressure palsies* are especially interesting; a familiar, but mild, example is the experience of having an arm or leg "fall asleep." The limb becomes numb, and afterward an uncomfortable "pins-and-needles" sensation, or **paresthesia**, accompanies the return to normal function. These incidents are seldom clinically significant, but they provide graphic examples of the effects of more serious palsies that can last for days to months. In **radial nerve palsy**, pressure on the back of the arm interrupts the function of the radial nerve, so the extensors of the wrist and fingers are paralyzed. This condition is also known as "Saturday night palsy," because falling asleep on a couch with your arm over the seat back (or beneath some-

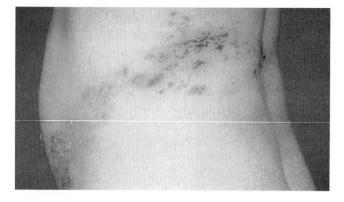

● **FIGURE 31**

Shingles. The side of a person with shingles. The skin eruptions follow the distribution of dermatomal innervation.

one's head) can produce the right combination of pressures. Students may also be familiar with **ulnar palsy**, which can result from prolonged contact between an elbow and a desk. The ring and little fingers lose sensation, and the fingers cannot be adducted. *Carpal tunnel syndrome* is a neuropathy resulting from compression of the median nerve at the wrist, where it passes deep to the flexor retinaculum with the flexor tendons. Repetitive flexion/extension at the wrist can irritate these tendon sheaths; the swelling that results is what compresses the median nerve.

Persons who carry large wallets in their hip pockets may develop symptoms of **sciatic compression** after they drive or sit in one position for extended periods. As nerve function declines, the individuals notice lumbar or gluteal pain, numbness along the back of the leg, and weakness in the leg muscles. Similar symptoms result from the compression of

nerve roots that form the sciatic nerve by a distorted lumbar intervertebral disc. This condition is termed **sciatica**, and one or both lower limbs may be affected, depending on the site of compression. Finally, sitting with your legs crossed can produce symptoms of a **peroneal palsy** *(fibular palsy)*. Sensory losses from the top of the foot and side of the leg are accompanied by a decreased ability to dorsiflex ("foot drop") or evert the foot.

■ SHINGLES AND HANSEN'S DISEASE
HA p. 368

In **shingles**, or *herpes zoster*, the *herpes varicella-zoster* virus attacks neurons within the dorsal roots of spinal nerves and sensory ganglia of cranial nerves. This disorder produces a painful rash whose distribution corresponds to that of the affected sensory nerves (Figure 31●). Shingles develops in people who were previously infected with the virus. The initial infection produces symptoms known as chickenpox. After this encounter, the virus remains dormant within neurons of the anterior gray horns of the spinal cord. It is not known what triggers the reactivation of this pathogen. Fortunately for those affected, attacks of shingles generally heal and leave behind only unpleasant memories.

Most people who contract shingles suffer just a single episode in their adult lives. However, the problem can recur in people with weakened immune systems, including those with AIDS or some forms of cancer. Treatment for shingles typically involves large doses of the antiviral drug *acyclovir (Zovirax)*.

The condition traditionally called **leprosy**, now more commonly known as **Hansen's disease**, is an infectious disease caused by a bacterium, *Mycobacterium leprae*. Leprosy is a

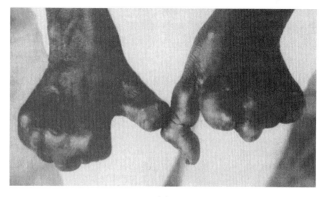

(a)

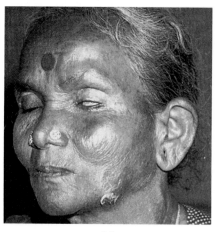

(b)

● **FIGURE 32**

Hansen's Disease. (a) The distal limbs are gradually deformed as untreated Hansen's disease progresses. **(b)** The disease also affects facial features, typically starting with degenerative changes around the eyes and at the nose and ears.

14

disease that progresses slowly, and symptoms may not appear for up to 30 years after infection. The bacterium invades peripheral nerves, growing faster in the cooler areas of the body, especially in the skin and distal limbs, producing sensory losses. Over time, motor paralysis develops, and the combination of sensory and motor loss can lead to recurring injuries and infections. The eyes, nose, hands, and feet may develop deformities as a result of neglected injuries (Figure 32a,b●). The disease has several forms; peripheral nerves are always affected, but some forms also involve extensive lesions of the skin and mucous membranes.

Only about 5 percent of those exposed to *Mycobacterium leprae* develop symptoms; people living in the tropics are at greatest risk. There are about 2000 cases of Hansen's disease in the United States and an estimated 12–20 million cases worldwide. The disease can be treated successfully with drugs such as rifampin and dapsone, and early treatment prevents deformities. Treated individuals are not infectious, and the practice of confining "lepers" in isolated compounds has thankfully been discontinued.

■ REFLEXES AND DIAGNOSTIC TESTING   *HA p. 373*

A neurological examination evaluates the sensory, motor, behavioral, and cognitive functions of the nervous system. The techniques involved were considered on p. 61; they range from asking questions to monitoring brain function through sophisticated scanning procedures.

Many somatic reflexes can be assessed through careful observation and the use of simple tools. The procedures are easy to perform, and the results can provide valuable information about the location of damage to the spinal cord or spinal nerves. By testing a series of spinal and cranial reflexes, the examiner can assess the function of sensory pathways and motor centers throughout the spinal cord and brain. Neurologists test many reflexes; only a few are so generally useful that physicians make them part of a standard physical examination. Representative examples are shown in Figure 33●, and Table 19 lists somatic reflexes that can be used in this way.

TABLE 19   Reflexes Used in Diagnostic Testing

Reflex	Stimulus	Afferent Nerve(s)	Spinal Segment	Efferent Nerve(s)	Normal Response
Superficial Reflexes					
Abdominal reflex	Light stroking of skin of abdomen	T7–T12, depending on region stroked	T7–T12, at level of arrival	Same as afferent	Contraction of abdominal muscles that pull navel toward the stimulus
Cremasteric reflex	Stroking of skin of upper thigh	Femoral nerve	L_1	Genitofemoral nerve	Contraction of cremaster, elevation of scrotum
Plantar reflex	Longitudinal stroking of sole of foot	Tibial nerve	S_1, S_2	Tibial nerve	Flexion at toe joints
Anal reflex	Stroking of region around the anus	Pudendal nerve	S_4, S_5	Pudendal nerve	Constriction of external anal sphincter
Stretch Reflexes					
Biceps reflex	Tap to tendon of biceps brachii muscle near its insertion	Musculocutaneous nerve	C_5, C_6	Musculocutaneous nerve	Flexion at elbow
Triceps reflex	Tap to tendon of triceps brachii muscle near its insertion	Radial nerve	C_6, C_7	Radial nerve	Extension at elbow
Brachioradialis reflex	Tap to forearm near styloid process of the radius	Radial nerve	C_5, C_6	Radial nerve	Flexion at elbow, supination, and flexion at finger joints
Patellar reflex	Tap to patellar tendon	Femoral nerve	$L_2–L_4$	Femoral nerve	Extension at knee
Ankle-jerk reflex	Tap to calcaneal tendon	Tibial nerve	S_1, S_2	Tibial nerve	Extension (plantar flexion) at ankle

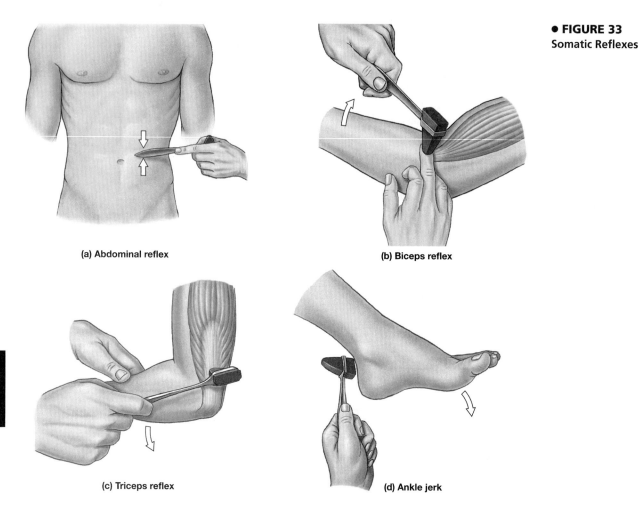

● **FIGURE 33**
Somatic Reflexes

(a) Abdominal reflex

(b) Biceps reflex

(c) Triceps reflex

(d) Ankle jerk

The *abdominal reflex* (Figure 33a●) is a superficial reflex that is normally present in adults. In this reflex, a light stroking of the skin produces a reflexive twitch in the abdominal muscles that moves the navel toward the stimulus. The reflex is facilitated by descending commands; it disappears after descending tracts have been damaged. The *patellar reflex* (see Figure14.18 in HA), *biceps reflex* (Figure 33b●), *triceps reflex* (Figure 33c●), and *ankle-jerk reflex* (Figure 33d●) are stretch reflexes controlled by specific segments of the spinal cord. Testing these reflexes provides information about the corresponding spinal segments. For example, a normal patellar reflex indicates that spinal nerves and spinal segments L$_2$–L$_4$ are undamaged.

■ ABNORMAL REFLEX ACTIVITY

In **hyporeflexia**, normal reflexes are weak but present. In **areflexia** (ā-rē-FLEK-sē-uh; *a-*, without), normal reflexes are completely absent. Hyporeflexia or areflexia may indicate temporary or permanent damage to skeletal muscles, dorsal or ventral nerve roots, spinal nerves, the spinal cord, or the brain.

Hyperreflexia occurs when higher centers maintain a high degree of facilitation along the spinal cord. Under these conditions, reflexes are easily triggered and the responses may be grossly exaggerated. This effect also results from compression of the spinal cord or diseases that target higher centers or descending tracts. One potential result of hyperreflexia is the appearance of alternating contractions in opposing muscles. When one muscle contracts, it stimulates the stretch receptors in the opposing muscle. The stretch reflex then triggers a contraction in that muscle, and this contraction stretches receptors in the original muscle. This self-perpetuating sequence, which can be repeated indefinitely, is called **clonus** (KLŌ-nus). In a hyperreflexive person, a tap on the patellar tendon will set up a cycle of kicks rather than just one or two.

A more extreme hyperreflexia develops if the motor neurons of the spinal cord lose contact with higher centers. In many cases, after a severe spinal injury, the individual first experiences a temporary period of areflexia known as spinal shock. When the reflexes return, they respond in an exaggerated fashion, even to mild stimuli. For example, the lightest touch on the skin may produce a massive withdrawal reflex with intense muscle spasms strong enough to break bones. In the **mass reflex**, the entire spinal cord becomes hyperactive for several minutes, issuing exaggerated skeletal muscle and visceral motor commands.

DISORDERS OF THE BRAIN AND CRANIAL NERVES

■ TRAUMATIC BRAIN INJURIES

HA p. 381

Traumatic brain injury (TBI) may result from harsh contact between the head and another object or from a severe jolt. Head injuries account for more than half the deaths attributed to trauma. Every year roughly 1.5 million cases of TBI occur in the United States. Approximately 50,000 people die, and another 80,000 have long-term disability. We presented the characteristics of spinal concussion, contusion, and laceration on p. 62; comparable descriptions apply to brain injuries.

Concussions may accompany even minor head injuries. A concussion may involve transient confusion with abnormal mental status, temporary loss of consciousness and some degree of amnesia (p. 77). Physicians examine concussed individuals quite closely and may x-ray or CT-scan the skull to check for fractures or cranial bleeding. Mild concussions produce a brief interruption of consciousness and little memory loss. Severe concussions produce extended periods of unconsciousness and abnormal neurological functions. Severe concussions are typically associated with **contusions** (bruises), hemorrhages, or **lacerations** (tears) of the brain tissue; the possibilities for recovery vary with the areas affected. Extensive damage to the reticular formation can produce a permanent state of unconsciousness, and damage to the lower brain stem generally proves fatal.

Wearing helmets during activities such as bike, horse, skateboard, or motorcycle riding; contact sports such as football and hockey; and when batting and base running in baseball provides protection for the brain. Seat belts give similar protection in the event of a motor vehicle accident. If a concussion does occur, restricted activities, including delay in return to the activity that led to the injury, is recommended.

■ SEIZURES AND EPILEPSIES

HA p. 388

A *seizure* is a temporary disorder of cerebral function, accompanied by abnormal, involuntary movements; unusual sensations; or inappropriate behavior. The individual may lose consciousness for the duration of the attack. There are many types of seizures. The terms **epilepsy** and *seizure disorder* refer to more than 40 conditions characterized by a recurring pattern of seizures over extended periods. In roughly 50 percent of patients, no obvious cause can be determined. Known causes include infection, brain trauma, brain damage from metabolic problems, stroke, genetic factors, and poisoning.

Seizures of all kinds are usually accompanied by a marked change in the pattern of electrical activity that can be detected in an electroencephalogram. The change begins in one portion of the cerebral cortex, but may thereafter spread to adjacent regions, potentially involving the entire cortical

CHAPTER 15

THE BRAIN AND CRANIAL NERVES

surface. The neurons at the site of origin of the change are abnormally sensitive. When they become active, they may facilitate and subsequently stimulate adjacent neurons. As a result, the abnormal electrical activity can spread across the entire cerebral cortex.

The extent of the cortical involvement determines the nature of the observed symptoms. A **focal seizure** affects a relatively restricted cortical area, producing sensory or motor symptoms or both. The individual generally remains conscious throughout the attack. If the seizure occurs within a portion of the primary motor cortex, the activation of pyramidal cells will produce uncontrollable movements. The muscles affected or the specific sensations experienced provide an indication of the precise region involved. In a **temporal lobe seizure**, the disturbance spreads to the sensory cortex and association areas, so the individual may experience unusual memories, sights, smells, or sounds as well. Involvement of the limbic system can also produce sudden emotional changes. The individual may lose consciousness at some point during the seizure.

Convulsive seizures are associated with uncontrolled muscle contractions. In a **generalized seizure**, the entire cortical surface is involved. Generalized seizures range from prolonged, major events to brief, almost unnoticed incidents. We will consider only two examples here: *grand mal* and *petit mal seizures*.

Some epileptic attacks involve powerful, uncoordinated muscular contractions that affect the face, eyes, and limbs as well as loss of consciousness. These are symptoms of a tonic-clonic seizure, often called a **grand mal seizure**. During a grand mal attack, the cortical activation begins at a single focus and then spreads across the entire surface. There may be no warning, but some individuals experience a vague apprehension or awareness that a seizure is about to begin. A sudden loss of consciousness follows, and the individual drops to the floor as major muscle groups go into tonic contraction. The body remains rigid for several seconds before a rhythmic series of contractions occurs in the limb muscles. Incontinence may occur. After the attack subsides, the individual may appear disoriented or may sleep for several hours. Muscles or bones subjected to extreme stresses may be damaged, and the person will probably be sore for days after the incident.

Petit mal seizures, are very brief (less than 10 seconds) and involve few motor abnormalities. Typically, the individual loses consciousness suddenly, with no warning. It is as if an internal switch were thrown and the conscious mind turned off. Because the individual is "not there" for brief periods during petit mal attacks, the incidents are also known as *absence seizures*. During the seizure, small-motor activities, such as fluttering of the eyelids or trembling of the hands, may occur.

Petit mal seizures generally begin between ages 6 and 14. They can occur hundreds of times a day, so the child lives each day in small segments separated by blank periods. The individual is aware of brief losses of consciousness that occur without warning, but, due to embarrassment, may not seek help. He or she becomes extremely anxious about the timing of future attacks. However, the motor signs are so minor that they tend to go unnoticed by family members, and the psychological stress caused by the condition is in many cases over-

looked. The initial diagnosis is typically made during counseling for learning problems. (You have probably taken an exam after you have missed 1 or 2 lectures out of 20. Imagine taking an exam after you have missed every third minute of every lecture.)

Both petit mal and grand mal forms of epilepsy can be treated with anticonvulsive drugs, such as barbiturates, phenytoin (*Dilantin*) or valproic acid (*Depakene*). More than twenty different anticonvulsants are available, with varying effectiveness, side effects, and other risks. Usually one drug suffices to control or eliminate seizures. Many people can discontinue medication after being seizure-free for two or three years. This is less likely if multiple drugs are needed, and very rarely brain surgery to affect the seizure focus may be required to control seizures.

■ CEREBELLAR DYSFUNCTION
HA p. 405

Cerebellar function can be altered permanently by trauma or a stroke or temporarily by drugs such as alcohol. The alterations can produce disturbances in motor control. In severe ataxia, balance problems are so great that the individual cannot sit or stand upright. Less-severe conditions cause an obvious unsteadiness and irregular patterns of movement. The individual typically watches his or her feet to see where they are going and controls ongoing movements by intense concentration and voluntary effort. Reaching for something becomes a major exertion, because the only information available must be gathered by sight or touch while the movement is taking place. Without the cerebellar ability to adjust movements while they are occurring, the individual becomes unable to anticipate the course of a movement over time. Most commonly, a reaching movement ends with the hand overshooting the target. This inability to anticipate and stop a movement precisely is called **dysmetria** (dis-MET-rē-uh; *dys-*, bad + *metron*, measure). In attempting to correct the situation, the person usually overshoots again, this time in the opposite direction, and so on. The hand oscillates back and forth until either the object can be grasped or the attempt is abandoned. This oscillatory movement is known as an **intention tremor**.

Clinicians check for ataxia by watching an individual walk in a straight line; the usual test for dysmetria involves touching the tip of the index finger to the tip of the nose or the examiner's fingertip. Because many drugs impair cerebellar performance, the same tests are used by police officers to check drivers suspected of driving while under the influence of alcohol or other drugs.

■ ANALYSIS OF GAIT AND BALANCE

To check a person's gait and balance, the examiner typically asks the person to walk a line, first as usual, then heel to toe. Heel-to-toe walking on a straight line will magnify any gait abnormalities or problems due to a loss of balance sensations. While the subject is walking, the examiner also watches how the feet are placed and how the arms swing back and forth.

The pattern of limb movement during walking is normally regulated by the cerebral nuclei. Problems with these nuclei, as in Parkinson's disease, will upset the pace and rhythm of the associated movements.

Another test (the *Romberg test*) involves standing with the feet together, first with the eyes open and then with the eyes closed. The Romberg test is used to check balance and equilibrium sensations because it reveals how much the individual is relying on visual information to fine-tune motor functions. If the person stands still with the eyes open, but sways or starts to fall with the eyes closed, there are problems with the balance pathways and perhaps the cerebellum as well. This result, called a *positive Romberg sign,* may indicate a stroke (or other cause of brain damage), multiple sclerosis, peripheral neuropathies, or any of several vestibular disorders.

■ CRANIAL NERVE TESTS   *HA p. 416*

A number of different tests are used to test the condition of specific cranial nerves:

- The olfactory nerve (I) is assessed by asking the subject to distinguish among various odors.
- Cranial nerves II, III, IV, and VI are assessed when vision and movement of the eyes are checked. Vision (cranial nerve II) is tested separately. For eye motion the person is asked to hold the head still and track the movement of the examiner's finger with the eyes. For the eyes to track the finger, the oculomotor muscles and associated cranial nerves must be functioning normally. For example, if the person cannot track with the right eye a finger that is moving from left to right, the right lateral rectus muscle or cranial nerve VI on the right side may be damaged.
- Cranial nerve V, which provides motor innervation to the muscles of mastication, can be checked by asking the person to clench the teeth. The jaw muscles are then palpated. If motor components of V on one side are damaged, the muscles on that side will be weak or flaccid. Sensory components of cranial nerve V can be tested by lightly touching areas of the forehead and side of the face.
- The facial nerve (VII) is checked by watching the muscles of facial expression or by asking the person to perform particular facial movements. Wrinkling the forehead, raising the eyebrows, pursing the lips, and smiling are controlled by the facial nerve. If a branch of VII has been damaged (see *Bell's palsy,* discussed in the text) the affected side will show muscle weakness or drooping. For example, the corner of the mouth may sag and fail to curve upward when the person smiles. Special sensory components of VII can be checked by placing solutions known to stimulate taste receptors on the anterior third of the tongue.
- The glossopharyngeal and vagus nerves (IX and X) can be evaluated by having the person say "ahh" or "ehh," and eliciting the gag reflex by touching the oropharynx with a tongue blade or swab. Examination of the soft-palate arches and uvula for normal movement is important.
- The accessory nerve (XI) can be checked by asking the person to shrug the shoulders. Atrophy of the sternocleidomastoid or trapezius muscles may also indicate problems with the accessory nerve.
- The hypoglossal nerve (XII) can be checked by having the person extend the tongue and move it from side to side.

15

CHAPTER 16
SENSORY AND MOTOR PATHWAYS AND HIGHER-ORDER FUNCTIONS

DISORDERS AFFECTING NEURAL INTEGRATION

■ AMYOTROPHIC LATERAL SCLEROSIS   *HA p. 432*

Demyelinating disorders (see p. 65) affect both sensory and motor neurons, producing losses in sensation and motor control. **Amyotrophic lateral sclerosis (ALS)** is a progressive disease that affects specifically motor neurons, leaving sensory neurons intact. As a result, individuals with ALS experience a loss of motor control, but have no loss of sensation or intellectual function. Motor neurons throughout the CNS are destroyed. Neurons involved with the innervation of skeletal muscles are the primary targets.

Symptoms of ALS generally do not appear until the individual is over age 40. ALS occurs at an incidence of three to five cases per 100,000 population worldwide. The disorder is somewhat more common among males than females. The pattern of symptoms varies with the specific motor neurons involved. When motor neurons in the cerebral hemispheres of the brain are the first to be affected, the individual experiences difficulty in performing voluntary movements and has exaggerated stretch reflexes. If motor neurons in other portions of the brain and the spinal cord are targeted, the individual experiences weakness, initially in one limb, but gradually spreading to other limbs and ultimately the trunk. When the motor neurons innervating skeletal muscles degenerate, a loss of muscle tone occurs. Over time, the skeletal muscles atrophy. The disease progresses rapidly, and the average survival after diagnosis is just three to five years. Because intellectual functions remain unimpaired, a person with ALS remains alert and aware throughout the course of the disease. This is one of the most disturbing aspects of the condition. Among well-known people who have developed ALS are baseball player Lou Gehrig and physicist Stephen Hawking.

The primary cause of ALS is uncertain; only 5–10 percent of ALS cases appear to have a genetic basis, with 5 percent of these genetic cases caused by a mutation in a gene that codes for an enzyme that protects the cell from harmful chemicals generated during metabolism. At the cellular level, it appears that the underlying problem is at the postsynaptic membranes of motor neurons. Treatment with *riluzole*, a drug that suppresses the release of glutamate (a neurotransmitter), has delayed the onset of respiratory paralysis and extended the life of ALS patients. The Food and Drug Administration (FDA) has approved this drug for clinical use.

■ TAY–SACHS DISEASE   *HA p. 432*

Tay–Sachs disease is a genetic abnormality involving the metabolism of *gangliosides*, important components of nerve cell membranes. People with this disease lack an enzyme needed to break down one particular ganglioside, which accumulates within the lysosomes of CNS neurons and causes them to deteriorate. Affected infants seem normal at birth, but within six months neurological problems begin to appear. The progress of symptoms typically includes muscular weakness, blindness, seizures, and death, usually before age 4. No effective treatment exists, but prospective parents can be tested to determine whether they are carrying the gene responsible for this condition. The disorder is most prevalent in one ethnic group, the Ashkenazi Jews of Eastern Europe.

■ HUNTINGTON'S DISEASE   *HA p. 437*

Huntington's disease (*Huntington's chorea*) is an inherited disease marked by a progressive deterioration of mental abilities and pronounced movement disorders. There are racial differences in the incidence of this condition; Caucasians have by far the highest incidence, three to four cases per million population.

In Huntington's disease the basal nuclei show degenerative changes, as do the frontal lobes of the cerebral cortex. The basic problem is the destruction of neurons secreting ACh and GABA (an inhibitory neurotransmitter) in the basal nuclei. The first signs of the disease usually appear in early adulthood. As you would expect in view of the areas affected, the symptoms involve difficulties in performing voluntary and involuntary patterns of movement and a gradual decline in intellectual abilities leading eventually to dementia. Screening tests can now detect the presence of the gene for Huntington's disease, which is located on chromosome 4. However, no effective treatment is available.

■ AMNESIA   *HA p. 439*

Amnesia, or memory loss, occurs suddenly or progressively, and recovery is complete, partial, or nonexistent, depending on the nature of the problem. In **retrograde amnesia** (*retro-*, behind), the individual loses memories of past events. Some degree of retrograde amnesia commonly follows a head injury; after a car wreck or a football tackle, many people are unable to remember the moments preceding the accident. In **anterograde amnesia** (*antero-*, ahead), an individual may be unable to store additional memories, but earlier memories are intact and accessible. The problem appears to involve an inability to generate long-term memories. At least two drugs—*diazepam (Valium)* and *Halcion*—have been known to cause brief periods of anterograde amnesia. Brain injuries can cause more prolonged memory problems. A person with permanent anterograde amnesia lives in surroundings that are always new. Magazines can be read, chuckled over, and reread a few minutes later with equal pleasure, as if they had never been seen before. Clinicians must introduce themselves at every meeting, even if they have been treating the patient for years. Some degree of **post-traumatic amnesia (PTA)** commonly occurs after a head injury. The duration of the amnesia varies with the severity of the injury. This condition combines the characteristics of retrograde and anterograde amnesias; the individual can neither remember the immediate past nor consolidate recent memories of the present.

■ ALZHEIMER'S DISEASE   *HA p. 440*

Alzheimer's disease is a chronic, progressive illness characterized by memory loss and impairment of higher-order cerebral functions including abstract thinking, judgment, and personality. It is the most common cause of **senile dementia**, or *senility*. Symptoms may appear at age 50–60 or later, although the disease occasionally affects younger individuals. Alzheimer's disease has widespread impact. An estimated 4 million people in the United States have Alzheimer's—including roughly 3 percent of those from age 65 to 70, with the number doubling for every five years of aging until nearly 50 percent of those over age 85 have some form of the condition. Over 230,000 victims require nursing home care, and Alzheimer's disease causes more than 53,000 deaths each year.

Most cases of Alzheimer's disease are associated with large concentrations of neurofibrillary tangles and plaques in the nucleus basalis, hippocampus, and parahippocampal gyrus. These brain regions are directly associated with memory processing. It remains to be determined whether these deposits cause Alzheimer's disease or are secondary signs of ongoing metabolic alterations with an environmental, hereditary, or infectious basis.

In Down syndrome (which results from an extra copy of chromosome 21 and is discussed further on p. 174) and in some inherited forms of Alzheimer's disease, mutations affecting genes on either chromosome 21 or a small region of chromosome 14 lead to increased risk of the early onset of the disease. Other genetic factors certainly play a major role. The late-onset form of Alzheimer's disease has been traced to a gene on chromosome 19 that codes for proteins involved in cholesterol transport.

Diagnosis involves excluding metabolic and anatomical conditions that can mimic dementia, a detailed history and physical, and an evaluation of mental functioning. Initial symptoms are subtle: moodiness, irritability, depression, and a general lack of energy. These symptoms are often ignored, overlooked, or dismissed. Elderly relatives are viewed as eccentric or irascible and are humored whenever possible.

As the condition progresses, however, it becomes more difficult to ignore or accommodate. An individual with Alzheimer's disease has difficulty making decisions, even minor ones. Mistakes—sometimes dangerous ones—are made, through either bad judgment or forgetfulness. For example, the person might light the gas burner, place a pot on the stove, and go into the living room. Two hours later, the pot, still on the stove, melts and starts a fire.

As memory losses continue, the problems become more severe. The individual may forget relatives, his or her home address, or how to use the telephone. The memory loss commonly starts with an inability to store long-term memories, followed by the loss of recently stored memories. Eventually, basic long-term memories, such as the sound of the individual's own name, are forgotten. The loss of memory affects both intellectual and motor abilities, and a person with severe Alzheimer's disease has difficulty performing even the simplest motor tasks. Although by that time victims are relatively unconcerned about their mental state or motor abilities, the condition can continue to have devastating emotional effects on the immediate family.

Individuals with Alzheimer's disease show a pronounced decrease in the number of cortical neurons, especially in the frontal and temporal lobes. This loss is correlated with inadequate ACh

16

production in the *nucleus basalis* of the cerebrum. Axons leaving that region project throughout the cerebral cortex; when ACh production declines, cortical function deteriorates.

There is no cure for Alzheimer's disease, but a few medications and supplements slow its progress in many patients and reduce the need for nursing home care. The antioxidants vitamin E and ginkgo biloba and the B vitamins of folate, B_6, and B_{12} help some patients and may delay or prevent the disease. Drugs that increase glutamate levels (a neurotransmitter in the brain) also give some additional benefit. Various toxicities and side effects determine what combination of drugs is used. In mice, a vaccine has reduced tangles and plaques in the brain and improved maze-running ability. A preliminary trial of a human vaccine was stopped because cases of immune encephalitis developed in some treated patients. Modification of the vaccine may eliminate this problem, allowing further study of this new approach.

16

DISORDERS OF THE AUTONOMIC NERVOUS SYSTEM

■ HYPERSENSITIVITY AND SYMPATHETIC FUNCTION *HA p. 450*

Two interesting clinical conditions result from the disruption of normal sympathetic functions. In **Horner's syndrome**, the sympathetic postganglionic innervation to one side of the face becomes interrupted. The interruption may be the result of an injury, a tumor, or some progressive condition such as multiple sclerosis. The affected side of the face becomes flushed as vascular tone decreases. Sweating stops in the region, and the pupil on that side becomes markedly constricted. Other symptoms include a drooping eyelid and an apparent retreat of the eye into the orbit.

Primary Raynaud's phenomenon, also called **Raynaud's disease**, most commonly affects young women. In this condition, for unknown reasons, the sympathetic system temporarily orders excessive peripheral vasoconstriction of small arteries, usually in response to cold temperatures. The hands, feet, ears, and nose become deprived of their normal blood circulation and the skin in these areas changes color, becoming initially pale and then developing blue tones. A red color ends the cycle as normal blood flow returns. The symptoms may spread to adjacent areas as the disorder progresses. Most cases do not cause tissue damage, although in rare cases prolonged decreased blood flow may distort the skin and nails, even progressing to skin ulcers or the more extensive tissue death of dry gangrene.

Behavioral changes such as avoiding cold environments or wearing mittens and other protective clothing can usually reduce the frequency of occurrence. Stopping smoking and avoiding drugs that can cause vasoconstriction may also be beneficial. Drugs that prevent vasoconstriction (vasodilators) can be used if preventive steps prove ineffective.

Trauma such as frostbite or the cumulative damage caused by the chronic use of vibrating machinery can cause symptoms of Raynaud's syndrome. Symptoms may also appear in individuals with arterial diseases and in connective tissue disorders such as scleroderma, rheumatoid arthritis, and systemic lupus erythematosis (SLE). These secondary forms of Raynaud's syndrome usually improve with treatment of the underlying disorder.

A regional **sympathectomy** (sim-path-EK-tō-mē), cutting the fibers that provide sympathetic innervation to the affected area, may occasionally be beneficial. Under normal conditions, sympathetic tone provides the effectors with a background level of stimulation. After the elimination of sympathetic innervation, peripheral effectors may become extremely sensitive to norepinephrine and epinephrine. This hypersensitivity can produce extreme alterations in vascular tone and other functions after stimulation of the adrenal medullae. If the sympathectomy involves cutting the postganglionic fibers, hypersensitivity to circulating norepinephrine and epinephrine may eliminate the beneficial effects. The prognosis improves if the preganglionic fibers are transected, because the ganglionic neurons will continue to release small

CHAPTER 17
THE AUTONOMIC NERVOUS SYSTEM

quantities of neurotransmitter across the neuromuscular or neuroglandular junctions. This release keeps the peripheral effectors from becoming hypersensitive.

■ BIOFEEDBACK   *HA p. 461*

Although conscious thought processes affect the ANS, you normally do not perceive the effect because visceral sensory information does not reach your cerebral cortex. Even when a conscious mental process triggers a physiological shift, such as a change in blood pressure, sweat gland activity, skin temperature, or muscle tone, the information never arrives at your primary sensory cortex. *Biofeedback* is an attempt to bridge this gap. In this technique, a person's autonomic processes are monitored, and a visual or auditory signal is used to alert the person when a particular change takes place. These signals let the individual know when ongoing conscious thought processes have triggered a desirable change in autonomic function. For example, when biofeedback is used to regulate blood pressure, a light or tone informs the person when blood pressure drops.

With practice, some people can learn to recreate the thoughts, sensations, or mood that will lower their blood pressure. This action is possible because the cerebral cortex (conscious thought), thalamus (sensory feedback), and limbic system (emotions) all affect the hypothalamic centers that control autonomic activity.

Biofeedback techniques have been used to promote conscious control of blood pressure, heart rate, circulatory pattern, skin temperature, brain waves, and so forth. Putting one's hands in warm water upon entering a cold room may desensitize the ANS and prevent primary Raynaud's phenomenon. By reducing stress, lowering blood pressure, and improving circulation, these techniques can reduce the severity of clinical symptoms. In the process, they also lower the risks for serious complications, such as heart attacks or strokes, in individuals with high blood pressure. Unfortunately, not everyone can learn to influence their own autonomic functions, and the combination of variable results and expensive equipment has kept biofeedback unsuitable for widespread application.

17

SENSORY DISORDERS

■ ACUTE AND CHRONIC PAIN   *HA p. 468*

Pain management poses a number of problems for clinicians. Painful sensations can result from tissue damage or sensory nerve irritation; it can originate where it is perceived, be referred from another location, or represent a false signal generated along the sensory pathway. The treatment differs in each case, and an accurate diagnosis is an essential first step.

Acute pain is the result of tissue injury; the cause is apparent, and local treatment of the injury is typically effective in relieving the pain. The most effective solution is to stop the damage, end the stimulation, and suppress the painful sensations at the site of injury. Pain sensations are suppressed when topical or locally injected anesthetics inactivate nociceptors in the immediate area. Analgesic drugs can also be administered. They work in many different ways; we will consider only a few examples here.

Tissue injury result in damage to cell membranes. A fatty acid called arachidonic acid escapes from injured membranes. In interstitial fluid, an enzyme called *cyclo-oxygenase* converts arachidonic acid molecules to prostaglandins, which then stimulate nociceptors in the area, Aspirin, ibuprofen, and related analgesics reduce inflammation and suppress pain by blocking the action of cyclo-oxygenase .

Chronic pain is more difficult to categorize and treat. It includes (1) injury-related pain that persists after tissue structure has been repaired; (2) pain from a chronic disease, such as cancer; and (3) pain without any apparent cause. Chronic pain in part reflects permanent facilitation of the [pain pathways and the creation of a reverberating "pain memory." Complex psychological and physiological components are also involved. For example, many chronic-pain patients develop a tolerance for pain medications, and insomnia and depression are common complaints. Chronic pain can be helped by antidepressants, which affects neurotransmitter levels, and antiseizure medicines. Counseling may help the person focus attention outward rather than inward; the outward focus can lessen the perceived level of pain and reduce the amount of pain medication required. Curiously, developing a second, acute source of pain, such as a herpes-zoster infection, can reduce the perception of preexisting chronic pain.

In some cases, chronic pain and severe acute pain can be suppressed by inhibition of the central pain pathway. Analgesics related to morphine reduce pain by mimicking the action of endorphins. The perception of pain may be altered, although the pain remains. For example, patients on morphine report being aware of painful sensations, but they are not distressed by them. Surgical steps can be taken to control severe pain; for instance, (1) the sensory innervation of an area can be destroyed by an electrical current, (2) the dorsal roots carrying the painful sensations can be cut (*a rhizotomy*), (3) the ascending tracts in the spinal cord can be severed (*a tractotomy*), or (4) thalamic or limbic centers can be stimulated or destroyed. These options, listed in order of increasing degree of effectiveness, surgical complexity, and associated risk, are used only when other methods of pain control have failed to provide relief.

CHAPTER 18
THE GENERAL AND SPECIAL SENSES

When used to control pain, the Chinese technique of acupuncture involves the insertion of fine needles at specific locations. These needles are either heated or twirled by the therapist. Several theories have been proposed to account for the positive effects, but non is widely accepted or proven. If has been suggested that the pain relief may follow the release of endorphins. It is not know how acupuncture stimulates endorphin release; the acupuncture points do not correspond to the distribution of any of the major peripheral nerves.

Many other aspects of pain generation and control remain a mystery. Up to 30 percent of patients who receive nonfunctional medication subsequently experience a significant reduction on pain. It has been suggested that this *placebo effect results* from endorphins release triggered by the expectation of pain relief. Although the medication has no direct effect, the indirect effect can be quite significant and complicates the evaluation of analgesic medications.

ASSESSMENT OF TACTILE SENSITIVITIES *HA p. 469*

Regional sensitivity to light touch can be checked by gentle contact with a fingertip or a slender wisp of cotton. The **two-point discrimination test** provides a more detailed sensory map of tactile receptors. Two fine points of a bent paper clip or another object are applied to the skin surface simultaneously. The subject then describes the contact. When the points fall within a single receptive field, the individual will report only one point of contact. A normal individual loses two point discrimination at 1mm (0.04 in.) on the surface of the tongue, st 2–3 mm (0.08–0.12 in.) on the lips, at 3–5 mm (0.12–.020 in.) on the backs of the hands and feet, and at 4–7 cm (1.6–2.75 in.) over the general body surface. Vibration receptors are tested by applying the base of a tuning fork to the skin. Damage to an individual spinal nerve produces insensitivity to vibration along the paths of the related sensory nerves. If the sensory loss results from spinal-cord damage, the injury site can typically be located by walking the tuning fork down the spinal column, resting its base on the vertebral spines.

Descriptive terms are used to indicate the degree of sensitivity in the area considered. *Anesthesia* implies a total loss of sensation; the individual cannot perceive touch, pressure, pain, or temperature sensations in that area. *Hypesthesia* is a reduction in sensitivity, and paresthesia is the presence of abnormal sensations, such as the pins-and-needles sensation when an arm or leg "falls asleep" due to pressure on a peripheral nerve.

TICKLE AND ITCH SENSATIONS

HA p. 469

Tickle and itch sensations are closely related to the sensations of touch and pain. The receptors involved are free nerve endings, and he information is carried by unmyelinated Type C fibers. Tickle sensations, which are usually (but not always) described as pleasurable, are produced by a light touch that moves across the skin. Psychological factors are involved in the interpretation of tickle sensations, and tickle sensitivity differs greatly among individuals. Itching is probably produced by the stimulation of the same receptors. Specific "itch spots" can be mapped in the skin, the inner surfaces of the eyelids, and other mucous membranes of the nose. Itch sensations are absent from other mucous membranes and from deep tissue and viscera. Itching is extremely unpleasant, even more unpleasant that pain. Individuals with extreme itching will scratch even when pain is the result. Itch receptors can be stimulated by the injection of histamine or proteolytic enzymes into the epidermis and superficial dermis. The precise receptor mechanism is unknown.

VERTIGO, MOTION SICKNESS, AND MÉNIÈRE'S DISEASE *HA p. 481*

The term **vertigo** describes an inappropriate sense of motion, usually a spinning sensation. This meaning distinguishes it from "dizziness," a sensation of light-headedness and disorientation. Vertigo can result from abnormal conditions in or stimulation of the inner ear or from problems elsewhere along the sensory pathway that carries equilibrium sensations. It can accompany CNS or other infections, and many people experience vertigo when they have a high fever.

Any event that sets endolymph in motion can stimulate the equilibrium receptors and produce vertigo. Placing an ice pack in contact with the area over the mastoid process of the temporal bone or flushing the external acoustic meatus with cold water may chill the endolymph in the outermost portions of the labyrinth and establish a temperature-related circulation of fluid. A mild and temporary vertigo is the result. The consumption of excessive quantities of alcohol or exposure to certain drugs can also produce vertigo by changing the composition of the endolymph or disturbing the hair cells of the inner ear.

Other causes of vertigo include viral infection of the vestibular nerve and damage to the vestibular nucleus or its tracts. Acute vertigo can also result from damage caused by abnormal endolymph production, as in Ménière's disease. Probably the most common cause of vertigo is *motion sickness*.

The exceedingly unpleasant signs and symptoms of **motion sickness** include headache, sweating, flushing of the face, nausea, vomiting, and various changes in mental perspective. (Sufferers may go from a state of giddy excitement to almost suicidal despair in a matter of moments.) It has been suggested that the condition results when central processing stations, such as the tectum of the mesencephalon, receive conflicting sensory information. Why and how these conflicting reports result in nausea, vomiting, and other symptoms are not known. Sitting belowdecks on a moving boat or reading in a car or airplane tends to provide the necessary conditions. Your eyes (which are tracking lines on a page) report that your position in space is not changing, but your semicircular ducts report that your body is lurching and turning. To counter this effect, seasick sailors watch the horizon rather than their immediate surroundings, so that their eyes will provide visual confirmation of the movements detected by their inner ears. It is not known why some individuals are almost immune to motion sickness, whereas others find travel by boat or plane almost impossible. Visual and equilibrium receptors are both involved; roughly half of the astronauts on the space shuttle suffer from "space sickness" upon glancing out a window and seeing Earth appearing at an unexpected angle. Fortunately for most of them, the nervous system adapts relatively quickly, and space sickness is seldom a problem after a day or two in orbit.

Drugs commonly administered to prevent motion sickness include dimenhydrinate (*Dramamine*), scopolamine, and promethazine. These compounds appear to depress activity at the vestibular nuclei. Sedatives, such as prochlorperazine (*Compazine*), may also be effective. Scopolamine can be administered across the skin surface by using an adhesive patch (*Transderm Scop*).

In **Ménière's disease**, distortion of the membranous labyrinth of the inner ear by high fluid pressures may rupture the membranous wall and mix endolymph and perilymph. The receptors in the vestibule and semicircular canals then become highly stimulated. The individual may be unable to start a voluntary movement because he or she is experiencing intense spinning or rolling sensations. In addition to the vertigo, the person may "hear" unusual sounds as the cochlear receptors are activated.

■ TESTING AND TREATING HEARING DEFICITS *HA p. 483*

As noted in the text *conductive deafness* results from conditions in the outer or middle ear that block the normal transfer of vibrations from the tympanic membrane to the oval window. An external acoustic meatus plugged with accumulated wax or trapped water can cause a temporary hearing loss. Scarring or perforation of the tympanic membrane and immobilization of one or more of the auditory ossicles are more serious causes of conductive hearing loss.

In **nerve deafness**, the problem lies within the cochlea or somewhere along the auditory pathway. The vibrations reach the oval window and enter the perilymph, but either the receptors cannot respond or their response cannot reach its central destinations. Causes of nerve deafness include the following:

• Very loud (high-intensity) sounds can produce nerve deafness by breaking stereocilia off the surfaces of the hair cells. (The reflex contraction of the tensor tympani and stapedius muscles in response to a dangerously loud noise occurs in less than 0.1 second, but this may not be fast enough to prevent damage.) As trauma occurs repeatedly over a lifetime, with every loud noise, high-frequency hearing loss gradually develops, and hearing losses are common among older individuals. Wearing ear protection around loud machinery, gunfire, and music is advised.

• Drugs such as the aminoglycoside antibiotics (neomycin and gentamicin) can diffuse into the endolymph and kill hair cells. Because hair cells and sensory nerves can also be damaged by bacterial infection, the potential side effects must be balanced against the severity of infection.

In the most common hearing test, an individual listens to sounds of varying frequency and intensity generated at irregular intervals. A record is kept of the responses, and the graphed record, or **audiogram**, is compared with that of an individual with normal hearing (Figure 34●). **Bone conduction tests** are used to discriminate between conductive

deafness and nerve deafness. If you put your fingers in your ears and talk quietly, you can still hear yourself because the bones of the skull conduct the sound waves to the cochlea, bypassing the middle ear. In a bone conduction test, the physician places a vibrating tuning fork against the skull. If the subject hears a louder sound when the tuning fork is in contact with the mastoid process next to the ear being tested, than when it is held near the entrance to the external acoustic meatus, the problem must lie with the external or middle ear on that side (that is, it is a conductive problem). If the tuning fork is held against the top of the skull in the center of the head, both ears normally hear the sound equally. If the person hears a louder sound in one ear, the problem must be at the receptors or along the auditory pathway in the other ear (that is, it is a nerve problem).

Several effective treatments exist for conductive deafness. A hearing aid overcomes the loss in sensitivity by increasing the volume of received sounds. Surgery may patch the tympanic membrane or free damaged or immobilized

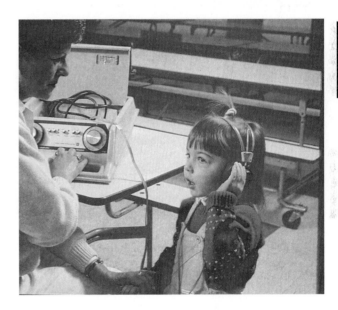

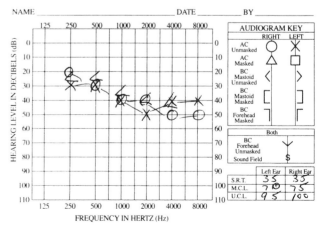

● **FIGURE 34**
An Audiogram

auditory ossicles. Artificial ossicles may also be implanted if the original ones are damaged beyond repair.

Few treatments are available for nerve deafness. Mild conditions may be overcome by the use of a hearing aid if some functional hair cells remain. In a **cochlear implant**, a small battery-powered device is inserted beneath the skin behind the mastoid process. Small wires run through the round window to reach the cochlear nerve; the implant "hears" a sound and stimulates the nerve directly. Increasing the number of wires (channels) and varying their implantation sites make it possible to create a number of different-frequency sensations. Those sensations do not approximate normal hearing: There is as yet no way to target the specific afferent fibers responsible for the perception of a particular sound. Instead, a random assortment of afferent fibers are stimulated, and the individual learns to recognize the meaning and probable origin of the perceived sound. Radio personality Rush Limbaugh received a multichannel cochlear implant in 2002, and congenitally deaf children given implants have learned normal speech. Older people who cannot hear with regular hearing aids are also benefiting from cochlear implants.

A new approach involves inducing the regeneration of hair cells of the organ of Corti. Researchers working with mammals other than humans have been able to induce hair cell regeneration both in cultured hair cells and in live animals. This is a very exciting area of research, and there is hope that it may ultimately lead to an effective treatment for human nerve deafness.

■ SCOTOMAS AND FLOATERS
HA p. 492

Abnormal blind spots, or **scotomas**, that are fixed in position may result from compression of the optic nerve, damage to retinal photoreceptors, or central damage along the visual pathway. Scotomas are abnormal, permanent features of the visual field. Most readers will probably be more familiar with floaters, small spots that drift across the field of vision. Floaters are common, temporary phenomena that result from blood cells or cellular debris within the vitreous body. They can be detected by staring at a blank wall or a white sheet of paper.

18

CHAPTER 19

AN INTRODUCTION TO THE ENDOCRINE SYSTEM AND ITS DISORDERS *HA p. 504*

THE ENDOCRINE SYSTEM

The endocrine system provides long-term regulation and adjustment of homeostatic mechanisms and a variety of body functions. For example, the endocrine system is responsible for the regulation of fluid and electrolyte balance, cell and tissue metabolism, growth and development, and reproductive functions. The endocrine system also helps the nervous system respond to stressful stimuli.

The endocrine system is composed of nine major endocrine glands and several other organs, such as the heart and kidneys, that have nonendocrine functions as well. The hormones secreted by these endocrine organs are distributed by the circulatory system to target tissues throughout the body. Each hormone affects a specific set of target tissues that may differ from those affected by other hormones. The selectivity is based on the presence or absence of hormone-specific receptors in the target cell's cell membrane, cytoplasm, or nucleus. As researchers learn more about how cells interact, they are discovering that tissues contain a variety of molecules secreted by cells to affect their neighbors and coordinate tissue activities. The terms *local hormones* and *cytokines* have been used to describe these molecules. Although it was initially thought that their effects were limited to their tissues of origin, it is now clear that many have more widespread impact. How their effects are coordinated with those of the "traditional" hormones of the endocrine system has yet to be determined.

■ A CLASSIFICATION OF ENDOCRINE DISORDERS *HA p. 505*

Homeostatic regulation of circulating hormone levels primarily involves negative feedback control mechanisms. The feedback loop features an interplay between the endocrine organ and its target tissues. An endocrine gland may release a particular hormone in response to one of three types of stimuli:

1. *Some hormones are released in response to variations in the concentrations of specific substances in body fluids.* Parathyroid hormone, for example, is released when calcium levels decline.

2. *Some hormones are released only when the gland cells receive hormonal instructions from other endocrine organs.* For example, the rate of production and release of triiodothyronine (T_3) and tetraiodothyronine (T_4, thyroxine) by the thyroid gland is controlled by thyroid-stimulating hormone (TSH) from the anterior pituitary gland. The secretion of TSH is in turn regulated by the release of thyrotropin-releasing hormone (TRH) from the hypothalamus.

3. *Some hormones are released in response to neural stimulation.* The release of epinephrine and norepinephrine from the adrenal medullae during sympathetic activation is an example.

Endocrine disorders can therefore develop due to abnormalities in the endocrine gland, the endocrine or neural regulatory mechanisms, or the target tissues. Figure 35● provides an overview of the major classes of endocrine disorders. In the

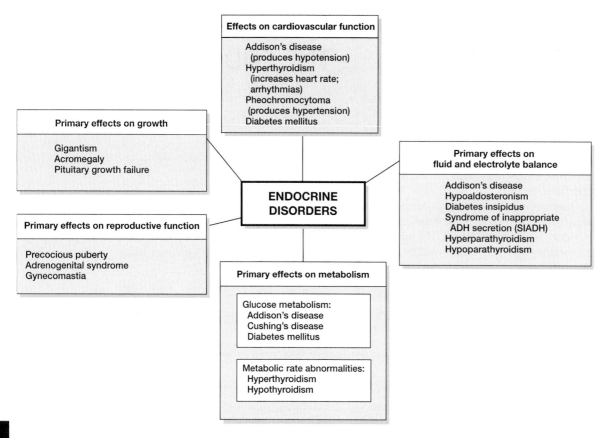

● FIGURE 35
Disorders of the Endocrine System

19

discussions that follow, we will first consider *primary disorders* that originate in an endocrine gland itself and that may result in hormone overproduction (*hypersecretion*) or underproduction (*hyposecretion*). We will use the thyroid gland as an example.

■ DISORDERS DUE TO ENDOCRINE GLAND ABNORMALITIES

Most endocrine disorders are the result of problems within the endocrine gland itself. Causes of hyposecretion include the following:

- **Metabolic factors.** Hyposecretion may result from a deficiency in some key substrate needed to synthesize the hormone in question. For example, hypothyroidism can be caused by inadequate dietary iodine levels or by exposure to drugs that inhibit iodine transport or utilization at the thyroid gland.
- **Physical damage.** Any condition that interrupts the normal circulatory supply or that physically damages the endocrine cells may cause them to become inactive immediately or after an initial surge of hormone release. If the damage is severe, the gland can become permanently inactive. For instance, temporary or permanent hypothyroidism can result from infection or inflammation of the gland (*thyroiditis*), from the interruption of normal blood flow, or from exposure to radiation as part of treatment for cancer of the thyroid gland or adjacent tissues. The thyroid gland can also be damaged in an *autoimmune disorder* that results in the production of antibodies that attack and destroy normal follicle cells.

- **Congenital disorders.** An individual may be unable to produce normal amounts of a particular hormone because (1) the gland itself is too small, (2) the required enzymes are abnormal, (3) the receptors that trigger secretion are relatively insensitive, or (4) the gland cells lack the receptors normally involved in stimulating secretory activity.

■ DISORDERS DUE TO ENDOCRINE OR NEURAL REGULATORY MECHANISM ABNORMALITIES

Endocrine disorders can result from problems with other endocrine organs involved in the negative feedback control mechanism. For example:

- **Secondary hypothyroidism** can be caused by inadequate TSH production at the pituitary gland or by inadequate TRH secretion at the hypothalamus.
- **Secondary hyperthyroidism** can be caused by excessive TRH or TSH production. Secondary hyperthyroidism may develop in individuals with tumors of the pituitary gland.

■ DISORDERS DUE TO TARGET TISSUE ABNORMALITIES

Endocrine abnormalities can also be caused by the presence of abnormal hormonal receptors in target tissues. In such a case, the gland involved and the regulatory mechanisms are normal, but the peripheral cells are unable to respond to the circulating hormone. The best example of this type of abnormality is *type 2 diabetes,* in which peripheral cells do not respond normally to insulin.

SYMPTOMS AND DIAGNOSIS OF ENDOCRINE DISORDERS *HA p. 505*

Knowledge of the individual endocrine organs and their functions makes it possible to predict the symptoms of specific endocrine disorders. For example, thyroid hormones increase basal metabolic rate, body heat production, perspiration, and heart rate. An elevated metabolic rate, increased body temperature, weight loss, nervousness, excessive perspiration, and an increased or irregular heartbeat are symptoms of hyperthyroidism. Conversely, a low metabolic rate, decreased body temperature, weight gain, lethargy, dry skin, and a reduced heart rate typically accompany hypothyroidism.

The next step in the diagnosis of an endocrine disorder, after obtaining a patient's medical history, is the physical examination. Several disorders produce characteristic physical signs that reflect abnormal hormone activities. Examples include the following:

- **Cushing's disease**, which results from an oversecretion of glucocorticoids by the adrenal cortex. As the condition progresses, the normal pattern of fat distribution in the body shifts. Adipose tissue accumulates in the abdominal area, the lower cervical area (causing a "buffalo hump"), and the face (producing a "moonface"), but the limbs become relatively thin.
- **Acromegaly**, which results from the oversecretion of growth hormone in adults. In this condition, the facial features become distorted due to excessive cartilage and bone growth, and the lower jaw protrudes, a sign known as *prognathism.* The hands and feet also become enlarged.
- **Adrenogenital syndrome**, which results from the oversecretion of androgens by the adrenal glands in females. Hair growth patterns change to resemble that of males, and *hirsutism* (p. 27) develops.
- **Hypothyroidism**, which can produce a distinctively enlarged thyroid gland, or *goiter.*
- **Hyperthyroidism**, which can produce protrusion of the eyes, or *exophthalmos.*

These signs are very useful, but many other signs and symptoms related to endocrine disorders are less definitive. For example, *polyuria,* or increased urine production, can result from hyposecretion of ADH (*diabetes insipidus*) or the hyperglysuria caused by *diabetes mellitus;* a symptom such as hypertension (high blood pressure) can be caused by a variety of cardiovascular or endocrine problems. In these instances, many diagnostic decisions are based on blood and other tests, which can confirm the presence of an endocrine disorder by detecting abnormal levels of circulating hormones or metabolic products resulting from hormone action, followed by tests that determine whether the primary cause of the problem lies with the endocrine gland, the regulatory mechanism(s), or the target tissues. Often, a pattern of several different test results leads to the diagnosis. Table 20 provides an overview of important anatomical tests used in the diagnosis of endocrine disorders.

■ DIABETES INSIPIDUS *HA p. 507*

There are several different forms of **diabetes**, all characterized by excessive urine production (**polyuria**). Although diabetes can be caused by physical damage to the kidneys, most forms are the result of endocrine abnormalities. The two most important forms are *diabetes insipidus,* considered here, and *diabetes mellitus,* considered later.

Diabetes insipidus develops when the posterior lobe of the pituitary gland no longer releases adequate amounts of antidiuretic hormone (ADH). Water conservation at the kidneys is impaired, and excessive amounts of water are lost in the urine. As a result, an individual with diabetes insipidus is constantly thirsty, but the fluids consumed are not retained by the body. Mild cases may not require treatment, as long as fluid and electrolyte intake keep pace with urinary losses. In severe diabetes insipidus the fluid losses can reach 10 liters per day, and a fatal dehydration will occur unless treatment is provided. Administering a synthetic form of ADH, *desmopressin acetate* (DDAVP), in a nasal spray concentrates the urine and reduces urine volume. The drug enters the bloodstream after diffusing through the nasal epithelium. It is also an effective treatment for bed-wetting if used at bedtime.

■ GROWTH HORMONE ABNORMALITIES *HA p. 509*

Growth hormone stimulates muscular and skeletal development. Until the epiphyseal cartilages close, it causes an increase in height, weight, and muscle mass, In extreme cases, *gigantism* can result. In **acromegaly** (*akron,* great), an excessive amount of GH is released after puberty, when most of the epiphyseal cartilages have already closed. Cartilages and small bones respond to the hormone, however, resulting in abnormal growth of the hands, feet, lower jaw, skull, and clavicle. Figure 19.12 in the text shows a typical acromegalic individual.

Children who are unable to produce adequate concentrations of GH have *pituitary growth failure,* a condition introduced in the text. The steady growth and maturation that typically precede and accompany puberty do not occur in these individuals, who have short stature, slow epiphyseal growth, and larger-than-normal adipose tissue reserves.

Normal growth patterns can be restored by the injected administration of GH. Before the advent of gene splicing and recombinant DNA techniques, GH had to be carefully extracted and purified from the pituitary glands of cadavers, at considerable expense and risk of infectious disease. Now genetically manipulated bacteria are used to produce human GH in commercial quantities. The current availability of purified human growth hormone has led to its use under medically questionable circumstances. For example, it is now being praised as an "anti-aging" miracle cure. Although GH injections do slow or even reverse the losses in

19

TABLE 20 Representative Diagnostic Procedures for Disorders of the Endocrine System

Diagnostic Procedure	Method and Result	Representative Uses
Pituitary Gland		
Wrist and hand x-ray film	Standard x-rays of epiphyseal cartilages for estimation of "bone age," based on the time of closure of epiphyseal cartilages	Compares a child's bone age and chronological age; a bone age greater than two years behind the chronological age suggests possible growth hormone deficiency with hypopituitarism or pituitary growth failure.
X-ray study of sella turcica	Standard x-ray of the sella turcica, which houses the pituitary gland	Determine (with increasing accuracy (and cost) the size of the pituitary gland; detect pituitary tumors
CT scan of pituitary gland	Standard cross-sectional CT; contrast media may be used	
MRI of pituitary gland	Standard MRI	
Thyroid Gland		
Thyroid scanning	A dose of radionucleotide accumulates in the thyroid, giving off detectable radiation to create an image of the thyroid	Determines size, shape, and abnormalities of the thyroid gland; detects presence of nodules and/or tumors; may detect hyperactive or hypoactive areas; may determine cause of a mass in the neck
Ultrasound examination of thyroid	Sound waves reflected off internal structures are used to generate a computer image	Detects thyroid cysts or tumors, enlarged lymph nodes, or abnormalities in the shape or size of the thyroid gland
Radioactive iodine uptake (RAIU) test	Radioactive iodine is ingested and trapped by the thyroid; detector determines the amount of radioiodine taken up over a period of time	Determines hyperactivity or hypoactivity of the thyroid gland; frequently done at the same time as thyroid scan
Parathyroid Glands		
Ultrasound examination of parathyroid glands	Standard ultrasound	Determines structural abnormalities of the parathyroid gland, such as enlargement
Adrenal Glands		
Ultrasound of adrenal gland	Standard ultrasound	Determines abnormalities in adrenal gland size or shape; may detect tumors
CT scan of adrenal gland	Standard cross-sectional CT	Determines abnormalities in adrenal gland size or shape; may detect tumor (pheochromocytoma)
Adrenal angiography	Injection of radiopaque dye for examination of the vascular supply to the adrenal gland	Detects tumors and hyperplasia

bone and muscle mass that accompany aging, continuous use is required to maintain benefits and little is known about adverse side effects that may accompany long-term use of the hormone in adults. GH is also being sought by some parents of short but otherwise healthy children. These parents view short stature as a handicap that merits treatment, and what height is considered "too short" varies with the local population norms. Whether we are considering GH treatment of adults or children, it is important to remember that GH and the somatomedins affect many different tissues and have widespread metabolic effects. For example, children exposed to GH may grow faster, but their body fat content declines drastically, and sexual maturation is delayed. The decline is associated with metabolic changes in many organs. The range and significance of these metabolic side effects are now the subjects of long-term studies. Since growth hormone is a protein that is digested and rendered ineffective if taken orally, diet supplements advertised as supplying growth hormone are probably (and fortunately) ineffective.

■ THYROID GLAND DISORDERS

HA p. 509

Normal production of thyroid hormones controls the background rate of cellular metabolism. These hormones exert their primary effects on metabolically active tissues and organs, including skeletal muscles, the liver, and the kidneys. The inadequate production of thyroid hormones is called **hypothyroidism**.

Hypothyroidism typically results from some problem that involves the thyroid gland, rather than a problem with the pituitary production of TSH. In primary hypothyroidism, TSH levels are elevated, because the pituitary gland attempts to stimulate thyroid activity, but levels of T_3 and T_4 are depressed In a fetus or infant, hypothyroidism produces *cretinism* (see Figure 19.11 in the text), a condition marked by inadequate skeletal and nervous system development and a metabolic rate as much as 40 percent below normal levels. Significant mental retardation may occur. The condition affects approximately one birth out of every 5000. Hypothyroidism developing later in childhood retards growth, delays puberty, and may affect mental development.

The signs and symptoms of adult hypothyroidism, collectively known as *myxedema* (miks-e-DE-muh), include subcutaneous swelling, dry skin, hair loss, low body temperature, muscular weakness, and slowed reflexes. Adults with hypothyroidism are lethargic and unable to tolerate cold temperatures. Hypothyroidism may also be associated with enlargement of the thyroid gland, producing a distinctive swelling called a goiter (see Figure 19.11 in the text). The enlargement generally indicates an increased thyroid follicle size, but thyroxine release may be increased or decreased, depending on the cause of the goiter. Most goiters develop when the thyroid gland is unable to synthesize and release adequate amounts of thyroid hormones. Under continuing TSH stimulation, thyroglobulin production accelerates and the thyroid follicles enlarge. One type of goiter occurs if the thyroid fails to obtain enough iodine to meet its synthetic requirements. Goiters from inadequate dietary iodide are very rare in the United States, in part due to the addition of iodine to table salt, but these conditions can be relatively common in poorer countries, especially landlocked ones (seafood is a good source of iodine). Administering iodine may not solve the problem entirely: The sudden availability of iodine can temporarily produce symptoms of hyperthyroidism as the stored thyroglobulin becomes available.

In the absence of iodine deficiency, the usual therapy for hypothyroidism and/or goiter involves the administration of synthetic thyroid hormone, thyroxine, which has a negative feedback effect on the hypothalamus and pituitary gland, thus inhibiting the production of TSH. Over time, the resting thyroid may return to its normal size. Treatment of chronic hypothyroidism, such as the hypothyroidism that follows radiation exposure or autoimmune thyroiditis, generally involves the administration of thyroxine to maintain normal blood concentrations.

Hyperthyroidism, or *thyrotoxicosis,* occurs when thyroid hormones are produced in excessive quantities, and may be associated with slight enlargement of the thyroid gland. The metabolic rate climbs, and the skin may be flushed and moist with perspiration. Blood pressure and heart rate increase, and the heartbeat may become irregular.. The effects on the central nervous system make the individual restless, excitable, and subject to insomnia and shifts in mood and emotional states. Despite the drive for increased activity, the person has limited energy and fatigues easily. *Graves' disease* is a form of hyperthyroidism that develops when antibodies are produced

that attack the thyroid gland. This autoimmune condition results in the release of excessive amounts of thyroid hormones, accompanied by goiter formation and other signs of hyperthyroidism. Protrusion of the eyes, or **exophthalmos** (eks-ahf-THAL-mos), may also appear for unknown reasons. Graves' disease has a genetic autoimmune basis and affects many more women than it does men. Treatment may involve the use of antithyroid drugs, surgical removal of portions of the glandular mass, or destruction of part of the gland by exposure to radioactive iodine.

Hyperthyroidism may also result from inflammation or, rarely, thyroid tumors. In extreme cases, the individual's metabolic processes accelerate out of control. During a *thyrotoxic crisis,* or "thyroid storm," the individual experiences a high fever, a rapid heart rate, and dangerous malfunctioning of a variety of physiological systems.

■ DISORDERS OF PARATHYROID FUNCTION *HA p. 511*

When the parathyroid glands secrete inadequate or excessive amounts of parathyroid hormone, calcium concentrations move outside normal homeostatic limits. Inadequate parathyroid hormone production, a condition called **hypoparathyroidism**, leads to low Ca^{2+} concentrations in body fluids. The most obvious symptoms involve neural and muscle tissues: The nervous system becomes more excitable, and the affected individual may experience *hypocalcemic tetany,* a dangerous condition characterized by prolonged muscle spasms that initially involve the limbs and face.

Hypoparathyroidism with hypocalcemia can develop after neck surgery, especially a thyroidectomy, if the parathyroid glands are damaged. In many other cases, the primary cause of the condition is uncertain. Parathyroid hormone (PTH) is extremely costly; because supplies are limited, PTH administration is not used to treat hypoparathyroidism, despite its probable effectiveness. As an alternative, a dietary combination of vitamin D and calcium can be used to elevate body fluid calcium concentrations, because vitamin D stimulates the absorption of calcium ions across the lining of the digestive tract.

In **hyperparathyroidism**, Ca^{2+} concentrations become abnormally high. Calcium salts in the skeleton are mobilized, and over time bones are weakened. On x-rays, the bones have a light, airy appearance, because the dense calcium salts no longer dominate the tissue. Central nervous system (CNS) function is depressed, thinking slows, memory is impaired, and the individual often experiences emotional swings and depression. Nausea and vomiting occur, and in severe cases the person becomes comatose. Muscle function deteriorates and skeletal muscles become weak. Other tissues are typically affected as calcium salts crystallize in joints, tendons, and the dermis; calcium deposits may produce masses called *kidney stones,* which block filtration and conduction passages in the kidneys and urinary tract. Hyperparathyroidism most commonly results from a tumor of the parathyroid gland. Treatment involves the surgical removal of the overactive tissue. Fortunately, humans have four parathyroid glands, and the secretion of even a portion of one gland can maintain normal calcium concentrations.

19

■ DISORDERS OF THE ADRENAL CORTEX   *HA p. 513*

Clinical problems related to the adrenal gland vary with the zone involved. The problems may result from changes in the functional capabilities of the adrenal cells (primary conditions) or disorders that affect the regulatory mechanisms (secondary conditions).

In **hypoaldosteronism**, the zona glomerulosa fails to produce enough aldosterone, generally either as an early sign of adrenal insufficiency or because the kidneys are not releasing adequate amounts of renin. Low aldosterone levels lead to excessive losses of water and sodium ions at the kidneys, and the water loss in turn leads to low blood volume and a fall in blood pressure. The resulting changes in electrolyte concentrations eventually cause dysfunctions in neural and muscular tissues.

Hypersecretion of aldosterone results in **aldosteronism**, or *hyperaldosteronism.* Under continued aldosterone stimulation, the kidneys retain sodium ions in exchange for potassium ions that are lost in urine. Hypertension and hypokalemia occur as extracellular potassium levels decline, increasing the concentration gradient for potassium ions across cell membranes. This increase leads to an acceleration in the rate of potassium diffusion out of the cells and into interstitial fluids. The reduction in intracellular and extracellular potassium levels eventually interferes with the function of excitable membranes, especially cardiac muscle cells, neurons, and kidney cells.

Addison's disease (see Figure 19.11 in the text) results from inadequate stimulation of the zona fasciculata by the pituitary hormone ACTH (adrenocorticotropic hormone) or from the inability of the adrenal cells to synthesize the necessary hormones, generally from adrenal cell loss caused by autoimmune problems or infection. Affected individuals produce insufficient levels of glucocorticoids. They become weak and lose weight, owing to a combination of appetite loss, hypotension, and hypovolemia. They cannot adequately mobilize energy reserves, and their blood glucose concentrations may fall sharply within hours after a meal. Stresses cannot be tolerated, and a minor infection or injury can lead to a sharp and even fatal decline in blood pressure. A particularly interesting symptom is the increased production of the pigment melanin in the skin. The ACTH molecule and the melanocyte-stimulating hormone (MSH) molecule are similar in structure, and at high concentrations ACTH stimulates the MSH receptors on melanocytes. President John F. Kennedy had this disorder.

Addison's disease is treated by replacement corticosteroid drugs (cortisone, prednisone, and others). However, chronic use or higher doses of corticosteroids to treat inflammatory conditions, such as rheumatoid arthritis or asthma, carries the risk of suppressing ACTH secretion and *causing* a secondary form of Addison's disease. For this reason corticosteroids are used for these conditions only when they are unresponsive to other treatments.

Cushing's disease (see Figure 19.11 in the text) results from the overproduction of glucocorticoids. The symptoms resemble those of a protracted and exaggerated response to stress. Glucose metabolism is suppressed, lipid reserves are mobilized, and peripheral proteins are broken down. Lipids and amino acids are mobilized in excess of the existing demand. The energy reserves are shuffled around, and the distribution of body fat changes. Adipose tissues in the cheeks and around the base of the neck become enlarged at the expense of other areas, producing a "moonfaced" appearance. The demand for amino acids falls most heavily on the skeletal muscles, which respond by breaking down their contractile proteins. This response reduces muscular power and endurance. The skin becomes thin and may develop *stria,* or stretch marks.

If the primary cause of Cushing's disease is ACTH oversecretion at the anterior lobe of the pituitary gland, the most common source is a *pituitary adenoma* (a benign tumor of glandular origin). Microsurgery can be performed through the sphenoid bone to remove the adenomatous tissue. Some oncology centers use pituitary radiation rather than surgery. Several pharmacological therapies act at the hypothalamus, rather than at the pituitary gland, to prevent the release of corticotropin-releasing hormone (CRH). The drugs used are serotonin antagonists, gamma-aminobutyric acid (GABA) transaminase inhibitors, or dopamine agonists. Alternatively, a bilateral adrenalectomy (the removal of the adrenal glands) can be performed, but further complications may arise as the adenoma in the pituitary gland enlarges. Cushing's disease also results from the production of ACTH outside the pituitary gland; for example, the condition may develop with one form of lung cancer (oat cell carcinoma). The removal of the causative tumor in some cases relieves the symptoms.

As mentioned in the discussion of Addison's disease, the chronic administration of large doses of steroids is sometimes required to treat severe asthma, arthritis, and certain cancers or to prevent transplanted organs from being rejected. Prolonged use of such large doses can produce symptoms similar to those of Cushing's disease, if not Addison's disease, which provides another reason to avoid such treatment if at all possible.

The zona reticularis ordinarily produces a negligible amount of androgens. If a tumor forms there, androgen secretion may increase dramatically, producing symptoms of **adrenogenital syndrome**. In women, this condition leads to the gradual development of male secondary sex characteristics, including body and facial hair patterns, adipose tissue distribution, and muscle development. Tumors affecting the zona reticularis of males can result in the production of large quantities of estrogens. Affected males develop enlarged breasts (a condition called **gynecomastia** (*gynaikos,* woman + *mastos,* breast), and, in some cases, other female secondary sex characteristics.

■ DISORDERS OF THE ADRENAL MEDULLA   *HA p. 514*

The overproduction of epinephrine by the adrenal medullae may reflect chronic sympathetic activation. A **pheochromocytoma** (fē-ō-krō-mō-sī-TŌ-muh) is a tumor that produces catecholamines in massive quantities. The tumor generally develops within an adrenal medulla, but may also involve other sympathetic ganglia. The most dangerous symptoms are episodes of rapid and irregular heartbeat and high blood pressure; other symptoms include uneasiness, sweating, blurred vision, and headaches. The condition is rare, and surgical removal of the tumor is the most effective treatment.

■ DIABETES MELLITUS   *HA p. 516*

Diabetes mellitus is a condition caused by inadequate production of, or sensitivity to, insulin. In the absence of insulin regulation, blood glucose levels skyrocket, yet peripheral tissues become glucose-starved because they are unable to transport glucose into the cytoplasm. The two forms of diabetes mellitus, type 1 and type 2, affect approximately 1 percent of the U.S. population, and is increasing as obesity becomes more common. Even with treatment, patients with diabetes mellitus often develop chronic medical problems. In general, these problems are related to cardiovascular abnormalities. The most common examples include the following:

1. Vascular changes occur at the retina, including proliferation of capillaries and hemorrhaging, often causing partial or total blindness. This condition is called **diabetic retinopathy**.

2. Changes occur in the clarity of the lens, producing cataracts.

3. Small hemorrhages and inflammation at the microvasculature of the kidneys cause degenerative changes that can lead to kidney failure. This condition is called **diabetic nephropathy**.

4. A variety of neural problems appear, including nerve palsies, paresthesias, and autonomic dysfunction, including male erectile dysfunction. These disorders, collectively termed **diabetic neuropathy,** are probably related to disturbances in the blood supply to neural tissues.

5. Degenerative changes in cardiac circulation can lead to accelerated coronary artery disease and early heart attacks. For a given age group, heart attacks are three to five times more likely in diabetic individuals.

6. Other peripheral changes in the vascular system can disrupt normal circulation to the limbs. For example, a reduction in blood flow to the feet can lead to tissue death, ulceration, infection, and loss of toes or a major portion of one or both feet.

■ INSULIN-DEPENDENT DIABETES MELLITUS

The primary cause of **insulin-dependent diabetes mellitus (IDDM),** or **type 1 diabetes,** is either loss of, or inadequate insulin production by, the beta cells of the pancreatic islets. In most cells, glucose transport cannot occur in the absence of insulin. When insulin concentrations decline, cells can no longer absorb glucose; tissues remain glucose-starved, despite the presence of adequate or even excessive amounts of glucose in the bloodstream. After a meal rich in glucose, blood glucose concentrations may become so elevated that the kidney cells cannot reclaim all the glucose molecules that enter the urine. The high urinary concentration of glucose limits the ability of the kidneys to conserve water, so the individual urinates frequently and may become dehydrated. The chronic hyperglycemia and dehydration leads to disturbances of neural function (blurred vision, tingling sensations, disorientation, and fatigue) and muscle weakness.

Despite high blood concentrations, glucose cannot enter endocrine tissues, and the endocrine system responds as if glucose were in short supply. Alpha cells release glucagon, and glucocorticoid production accelerates. Peripheral tissues then break down lipids and proteins to obtain the energy needed to continue functioning. The breakdown of large numbers of fatty acids promotes the generation of molecules called **ketone bodies**. These small molecules are metabolic acids whose accumulation in large numbers can cause a dangerous reduction in blood pH. This condition, called **ketoacidosis,** may trigger vomiting. In severe cases, it can progress to coma and death.

If the individual survives (an impossibility without insulin therapy), long-term treatment involves a combination of dietary control, monitoring of blood glucose levels several times a day, and the administration of insulin, either by injection or by infusion, using an **insulin pump** connected by tubing and a small catheter and needle to the subcutaneous tissue. The treatment is complicated by the fact that tissue glucose demands vary with food eaten, physical activity, emotional state, stress, and other factors that are hard to assess or predict. Dietary control, including the regulation of the type of food, time of meals, and amount consumed, can help reduce oscillations in blood glucose levels.

Modern insulin comes in many forms, with varying durations of activities. Single or, more commonly, multiple injections throughout the day are guided by blood glucose measurements in an attempt to approach normal homeostatic glucose regulation. At present this testing involves needle sticks and finger pricks, but visual glucose monitors are now available and insulin injection can be avoided by using an insulin pump or nasal spray. However, it remains difficult to maintain stable and normal blood glucose levels over long periods, even with an insulin pump.

Precise glucose control has been shown to reduce and delay the onset of the serious chronic complications of diabetes. These complications include accelerated coronary artery disease, kidney failure, and microvascular complications that are the leading cause of blindness and foot amputations in the United States.

Since 1990, pancreas transplants have been used to treat diabetes in the United States. The procedure is generally limited to gravely ill patients already undergoing kidney transplantation. The graft success rate over five years is roughly 50 percent, and the procedure is controversial. Pancreatic islet transplantation has recently shown promise, but it requires a large number of islets (two cadaver organs are needed to serve as source), and, as with all transplants, suppression of the immune system to prevent rejection is necessary.

Another approach is the use of a **biohybrid artificial pancreas**. This procedure has been used to treat type 1 diabetes in dogs. Islet cells can be cultured in the laboratory and inserted within an artificial membrane. The membrane contains pores that allow movement of fluid, but prevent interactions between the islet cells and white blood cells that would reject them as foreign. The islet cells monitor the blood glucose concentration and secrete insulin or glucagon as needed. The biohybrid artificial pancreas can be located almost

19

anywhere that has an adequate blood supply; in human trials, it is inserted under the skin of the abdomen.

Type 1 diabetes most commonly occurs in individuals under age 40. Because it typically appears in childhood, it has been called **juvenile-onset diabetes**. Roughly 80 percent of people with this type of diabetes have circulating antibodies that target the surfaces of beta cells. The disease may therefore be an *autoimmune disorder*—a condition that results when the immune system attacks normal body cells. Consequently, attempts have been made to prevent the appearance of type 1 diabetes with *azathioprine (Imuran)*, a drug that suppresses the immune system. This treatment is potentially dangerous, however, because compromising immune function indefinitely increases the risk of acquiring serious infections or of developing cancer. Type 1 diabetes is complex and probably reflects a combination of genetic and environmental factors. To date, genes associated with the development of type 1 diabetes have been localized to chromosomes 6, 11, and 18.

■ NON-INSULIN-DEPENDENT DIABETES MELLITUS

Non-insulin-dependent diabetes mellitus (NIDDM), or **type 2 diabetes,** typically affects obese individuals over age 40. Because of the age factor, this condition is also called **maturity-onset diabetes,** although with more childhood obesity, significant numbers of adolescents are developing the disease. Type 2 diabetes is far more common than type 1, accounting for more than 90 percent of all cases and occurring in an estimated 6.6 percent of the U.S. population. Some 500,000 new cases are diagnosed each year in the United States alone.

In type 2 diabetes, insulin levels are normal or elevated, but peripheral tissues no longer respond normally. Because most patients are overweight, treatment consists of weight loss and dietary restrictions that may elevate insulin production and tissue response. The drug *metformin (Glucophage)* lowers plasma glucose concentrations, primarily by reducing glucose synthesis and release at the liver. The use of metformin in combination with other drugs that affect glucose metabolism promises to improve the quality of life for many type 2 diabetes patients.

A diagnosis of diabetes mellitus is based on two observations: a high fasting blood glucose and the persistence of an elevated blood glucose level two hours after drinking a fixed amount of glucose. These criteria have largely replaced the six-hour glucose tolerance test, which involved having the patient drink 75–100 g of glucose and then testing the blood glucose level multiple times over four to six hours. Careful control to avoid high glucose levels and to keep the long-term marker (hemoglobin A1c) at a low level has been shown to reduce the risk of chronic kidney, eye, and cardiovascular complications.

■ LIGHT AND BEHAVIOR   *HA p. 517*

Exposure to sunlight can do more than induce a tan or promote the formation of vitamin D_3. Evidence indicates that daily light-dark cycles have widespread effects on the central nervous system, with melatonin playing a key role. Several studies have indicated that residents of temperate and higher latitudes in the Northern Hemisphere undergo seasonal changes in mood and activity patterns. These people feel most energetic from June through September, and they experience relatively low spirits from December through March. (The opposite situation occurs in the Southern Hemisphere, where winter and summer are reversed relative to the Northern Hemisphere.) The degree of seasonal variation differs from individual to individual: Some people display no symptoms; other people are affected so severely that they seek medical attention. The observed symptoms are called **seasonal affective disorder (SAD)**. Individuals with SAD experience depression and lethargy and have difficulty concentrating. They tend to sleep for long periods, perhaps 10 hours or more a day. They may also go on eating binges and crave carbohydrates.

Melatonin secretion appears to be regulated by exposure to sunlight, not simply by exposure to light. Normal interior lights are apparently not strong enough or do not release the right mixture of light wavelengths to depress melatonin production. Because many people spend very little time outdoors in the winter, melatonin production increases then; the depression, lethargy, and concentration problems appear to be linked to elevated melatonin levels in blood. In experiments, comparable symptoms can be produced in a healthy individual by an injection of melatonin.

Many individuals with SAD are successfully treated by exposure to sunlamps that produce full-spectrum light. Experiments are underway to define exactly how intense the light must be and to determine the minimal effective time of exposure. Some people have been using melatonin obtained (in varying doses and purity) from health-food stores to treat insomnia and jet lag. Because the health-food market is unregulated and few, if any, controlled studies have been done, it remains unclear whether melatonin is truly an effective therapy for sleep disorders or whether it aggravates SAD and depression.

ENDOCRINE DISORDERS   *HA p. 517*

Endocrine disorders fall into two basic categories: They reflect either abnormal hormone production (too much or too little) or abnormal cellular sensitivity (extremely sensitive or insensitive). The symptoms are interesting because they highlight the significance of normally "silent" hormonal contributions. An abbreviated summary is presented in Table 21.

TABLE 21   Clinical Implications of Endocrine Malfunctions

Hormone	Underproduction Syndrome	Principal Symptoms	Overproduction Syndrome	Principal Symptoms
Growth hormone (GH)	Pituitary growth failure (children) (p. 87)	Retarded growth, abnormal fat distribution, low blood glucose hours after a meal	Gigantism (children), acromegaly (adults) (p. 87)	Excessive growth in stature of a child or in face and hand in an adult
Antidiuretic hormone (ADH)	Diabetes insipidus (p. 87)	Polyuria	SIADH (syndrome of inappropriate ADH secretion)	Increased body water content and hyponatremia
Thyroxine (TX, T_4)	Myxedema, cretinism (p. 89)	Low metabolic rate, body temperature; impaired physical and mental development	Graves' disease (p. 89)	High metabolic rate, body temperature; tachycardia; weight loss
Parathyroid hormone (PTH)	Hypoparathyroidism (p. 89)	Muscular weakness, neurological problems, tetany due to low blood calcium concentrations	Hyperparathyroidism	Neurological, mental, muscular problems due to blood calcium concentrations; weak and brittle bones
Insulin	Diabetes mellitus (Type 1) (p. 87)	High blood glucose, impaired glucose utilization, dependence on lipids for energy, glycosuria, ketosis	Excess insulin production or administration	Low blood glucose levels, possibly causing coma
Mineralocorticoids (MC)	Hypoaldosteronism (p. 90)	Polyuria, low blood volume, high blood potassium concentrations	Aldosteronism (p. 90)	Increased body weight due to water retention, low blood potassium concentration
Glucocorticoids (GC)	Addison's disease (p. 90)	Inability to tolerate stress, mobilize energy reserves, maintain normal blood glucose concentrations	Cushing's disease (p. 87)	Excessive breakdown of tissue proteins and lipid reserves, impaired glucose metabolism
Epinephrine (E), norepinephrine (NE)	None identified		Pheochromocytoma (p. 91)	High metabolic rate, body temperature, and heart rate; elevated blood glucose levels; other symptoms comparable to those of excessive autonomic stimulation
Estrogens (female)	Hypogonadism	Sterility, lack of secondary sexual characteristics	Androgenital syndrome (p. 90)	Overproduction of androgens by zona reticularis of adrenal leads to masculinization
	Menopause	Cessation of ovulation	Precocious puberty	Early production of developing follicles and estrogen secretion
Androgens (male)	Hypogonadism, eunuchoidism	Sterility, lack of secondary sexual characteristics	Gynecomastia (p. 90)	Abnormal production of estrogens, sometimes due to adrenal or interstitial cell tumors, leads to breast enlargement
			Precocious puberty, acne (p. 26)	Early production of androgens, leading to premature physical development and behavioral changes

19

CHAPTER 20
BLOOD

AN INTRODUCTION TO THE CARDIOVASCULAR SYSTEM AND ITS DISORDERS *HA pp. 525, 542*

The components of the cardiovascular system include the blood, heart, and blood vessels. Blood flows through a network of thousands of miles of vessels in the body, transporting nutrients, gases, wastes, hormones, and ions and redistributing the heat generated by active tissues. The exchange of materials between blood and peripheral tissues occurs across the thin walls of tiny capillaries that are situated between the arterial and venous systems. The total capillary surface area for exchange is truly enormous, averaging about 6300 square meters—about 50 percent larger than a football field.

Because the cardiovascular system plays a key role in supporting all other systems, disorders of this system affect every cell in the body. One method of organizing the many potential disorders involving the cardiovascular system is by the nature of the primary problem—whether it affects the blood, the heart, or the vascular network. Figure 36a,b● provides an introductory overview of major blood disorders and cardiovascular disorders that are discussed in the text and in later sections of the *Clinical Issues* supplement.

■ THE PHYSICAL EXAMINATION AND THE CARDIOVASCULAR SYSTEM *HA pp. 525, 542*

Individuals with cardiovascular problems commonly seek medical attention. In many cases, they describe one or more of the following as chief complaints:

- **Weakness and fatigue.** These symptoms develop when the cardiovascular system can no longer meet tissue demands for oxygen and nutrients. They may occur because cardiac pumping function is impaired, as in *heart failure* (p. 109) or *cardiomyopathy* (p. 107), or because the blood is unable to carry normal amounts of oxygen, as in the various forms of *anemia* (p. 91). In the early stages of these conditions, the individual feels healthy at rest, but becomes weak and fatigued with any significant degree of exertion because the cardiovascular system cannot keep pace with the rising tissue oxygen demands. In more advanced stages of the disorders, weakness and fatigue persist even at rest.

- **Cardiac pain.** This pain is usually perceived as a deep pressure pain felt in the substernal region; it typically radiates down the left arm or up into the shoulder and neck. Cardiac pain has two major causes:

  1. Constant severe pain can result from inflammation of the pericardial sac, a condition known as *pericarditis*. This *pericardial pain* can superficially resemble the pain experienced in a *myocardial infarction (MI),* or heart attack. Pericardial pain differs from the pain of an MI in that (a) it may change with breathing or with changes in position, and is often less severe when sitting; (b) a fever may be present; and (c) the pain does not respond to the

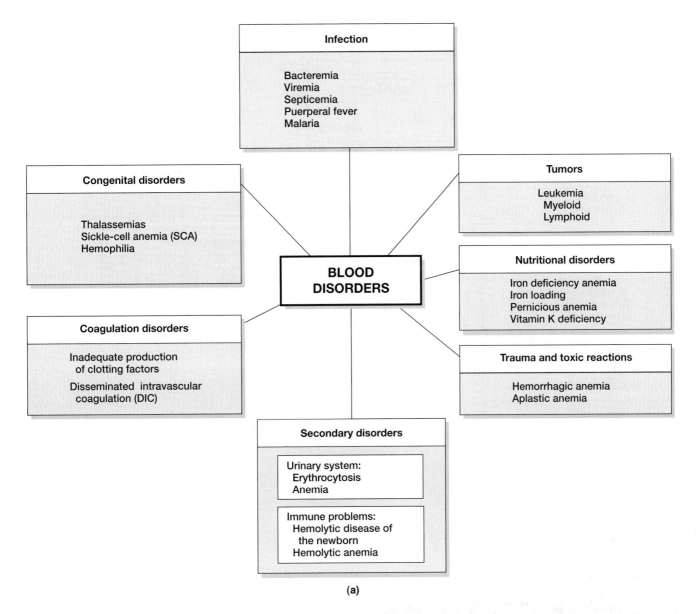

● FIGURE 36a
Disorders of the Cardiovascular System. (a) Blood disorders.

administration of drugs, such as *nitroglycerin*, that dilate coronary blood vessels.

2. Cardiac pain also results from inadequate blood flow to the myocardium. This type of pain is called *myocardial ischemic pain*. Ischemic pain occurs in *angina pectoris* and in an MI. Angina pectoris, discussed in the text, most commonly results from the narrowing of coronary blood vessels by atherosclerosis. The associated pain first appears during physical exertion, when myocardial oxygen demands increase, and the pain is relieved by rest and drugs such as nitroglycerin, which dilate coronary vessels and improve coronary blood flow. The pain associated with an MI is commonly felt as a heavy weight or a constriction of the chest. The pain of an MI is also distinctive because (a) it is not necessarily linked to exertion; (b) it is persistent, lasting longer than 15–20 minutes, and is not relieved by rest, nitroglycerin, or other coronary vasodilators; and (c) nausea, vomiting, and sweating may occur during the attack.

- **Palpitations.** Palpitations are a person's perception of an altered heart rate. The individual may complain of the heart "skipping a beat" or "racing." The most likely cause of palpitations is an abnormal pattern of cardiac activity known as an *arrhythmia*. The detection and analysis of arrhythmias are considered in a later section (p. 109).

- **Pain on movement.** Individuals with advanced peripheral vascular disease have atherosclerosis of peripheral arteries and may experience pain in the extremities during exercise. The pain, called claudication, may be so severe that the person is unwilling or unable to walk or perform other common activities. The underlying problem is the constriction or

20

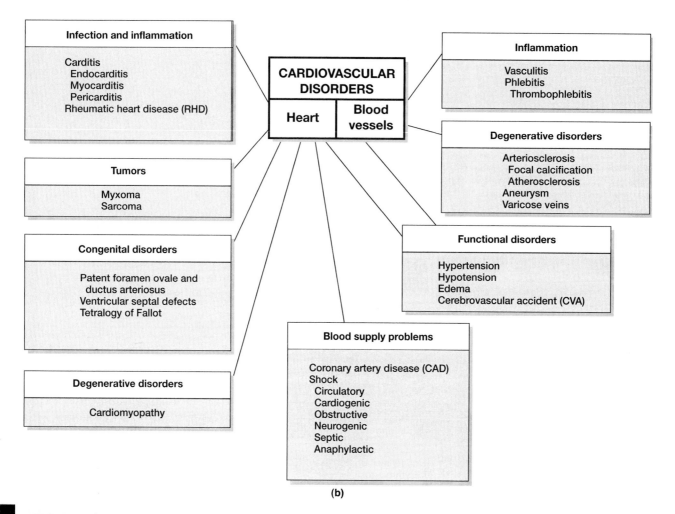

● **FIGURE 36b**
Disorders of the Cardiovascular System (continued). (b) Cardiovascular disorders.

partial occlusion of major arteries, such as the external iliac arteries to the lower limb muscles, by plaque formation.

These are only a few of the many symptoms that can be caused by cardiovascular disorders. In addition, the individual may have characteristic observable signs of underlying cardiovascular problems. A partial listing of important signs includes the following:

- **Edema** is an increase of fluid in tissues that can occur when (a) the pumping efficiency of the heart is decreased, (b) the plasma protein content of the blood is reduced, or (c) venous pressures are abnormally high. If the person can stand or sit, the tissues of the limbs are most commonly affected, and swollen feet, ankles, and legs occur. If the person is bedridden, swelling of the eyelids, the hands, and the area over the sacrum may occur. When edema is so severe that pressing on the affected area leaves an indentation, the sign is called *pitting edema.* (Edema is also discussed in the text.)

- Breathlessness, or *dyspnea,* occurs when cardiac output is inadequate for tissue oxygen demands. Dyspnea may also occur with *pulmonary edema,* a buildup of fluid within the alveoli of the lungs. Pulmonary edema and dyspnea are typically associated with *congestive heart failure (CHF)* (p. 109).

- **Varicose veins** are dilated superficial veins that are visible at the skin surface. This condition, which develops when venous valves malfunction, can be caused or exaggerated by increased systemic venous pressures. Varicose veins are considered further on p. 114.

- There may be characteristic and distinctive changes in skin coloration. For example, *pallor* is the lack of normal red or pinkish color in the skin of a light-skinned person or in the conjunctiva, nail beds, and oral mucosa of darker-skinned people. Pallor accompanies many forms of anemia, but can also be the result of inadequate cardiac output, shock, or circulatory collapse. *Cyanosis* is the bluish color of the same areas that occurs when tissues are deficient in oxygen. Cyanosis generally results from cardiovascular or respiratory disorders.

- **Vascular skin lesions** were introduced in the discussion of skin disorders in Chapter 4. Characteristic vascular lesions can occur in clotting disorders (p. 105) and in *leukemia* (p. 104).

For example, abnormal bruising may be the result of a disorder that affects the clotting system, platelet production, or the structure of blood vessels. *Petechiae,* which appear as purple spots on the skin, are typically seen in individuals with certain types of leukemia or other diseases associated with low platelet counts.

DIAGNOSTIC PROCEDURES FOR CARDIOVASCULAR DISORDERS

Functional abnormalities of the heart and blood vessels can be detected through physical assessment and the recognition of characteristic signs and symptoms In some cases, the initial detection of a cardiovascular disorder occurs during the physical examination:

1. When the vital signs are taken, the pulse is checked for strength, rate, and rhythm. Weak or irregular heartbeats will commonly be noticed at this time.

2. The blood pressure is monitored with a stethoscope and blood pressure cuff (sphygmomanometer). Unusually high or low readings can alert the examiner to potential problems with cardiac or vascular function. However, a diagnosis of chronic hypertension (p. 87) is usually not made on the basis of a single reading but after several readings over a period of time.

3. The heart sounds are assessed by auscultation with a stethoscope:

   • Cardiac rate and rhythm can be checked and arrhythmias detected.

   • Abnormal heart sounds, or *murmurs,* may indicate problems with atrioventricular or semilunar valves. Murmurs are noted in relation to their location in the heart (as determined by the position of the stethoscope on the chest wall), the time of occurrence in the cardiac cycle, whether the sound is low pitched or high pitched, and whether variations in intensity are present.

   • Usually nothing is heard during auscultation of normal blood vessels. *Bruits* are the sounds that result from turbulent blood flow, in many cases around an obstruction within an artery. Bruits are typically heard where large atherosclerotic plaques have formed.

The structural basis of heart and blood vessel problems is often determined by the use of scans and X-rays and by monitoring electrical activity in the heart. For problems with the various components of blood, laboratory tests performed on blood samples normally provide the information necessary to reach a diagnosis. Table 22a summarizes information about representative diagnostic procedures used to evaluate the health of the cardiovascular system.

Changes in blood volume have direct effects on other body systems. Clinicians use the terms *hypovolemia* (hī-pō-vō-LĒ-mē-ah), *normovolemia* (nor-mō-vō-LĒ-mē-ah), and *hypervolemia* (hī-per-vō-LĒ-mē-ah) to refer to low, normal, and excessive blood volumes, respectively. Hypovolemia and

hypervolemia are potentially dangerous, because variations in blood volume affect other components of the cardiovascular system. For example, an abnormally large blood volume can place a severe stress on the heart, which must keep the extra fluid circulating through the lungs and throughout the body. Hypovolemia, such as dehydration secondary to gastrointestinal problems or blood loss from trauma, is clinically much more common than hypervolemia. Short-term therapies to treat hypovolemia may include the intravenous use of an electrolyte solution or a *transfusion*—the administration of blood components to restore blood volume or to remedy a deficiency in blood composition. Whole blood, packed red blood cells, plasma, platelets, extracted proteins, or clotting factors may be administered; these therapies will be considered in a later section.

■ BLOOD TESTS AND RBCs   *HA p. 529*

Several common blood tests are used to assess circulating RBCs (Table 22b):

■ RETICULOCYTE COUNT

Reticulocytes are immature RBCs that are still synthesizing hemoglobin. Most reticulocytes remain in the bone marrow until they complete their maturation, but some enter the bloodstream. Reticulocytes normally account for about 0.8 percent of the erythrocyte population. Values above 1.5 percent or below 0.5 percent indicate that something is wrong with the rates of RBC survival or maturation.

■ HEMATOCRIT (Hct)

The hematocrit value is the percentage of whole blood occupied by red blood cells. The hematocrit of a normal adult averages 46 for men and 42 for women, with ranges of 42–52 for men and 37–47 for women.

■ HEMOGLOBIN CONCENTRATION (Hb)

This test determines the amount of hemoglobin in the blood, expressed in grams per deciliter (g/dl). Normal ranges are 14–18 g/dl for males and 12–16 g/dl for females. The differences in hemoglobin concentration reflect the differences in hematocrit. For both genders, a normal, single RBC contains 27–33 picograms (pg) of hemoglobin.

■ RBC COUNT

Calculations of the RBC count—the number of RBCs per microliter of blood—are based on the hematocrit and hemoglobin content and can be used to get more information about the condition of the RBCs. Values typically reported in blood tests include the following:

• **Mean corpuscular volume (MCV),** the average volume of an individual RBC, in cubic micrometers [μm^3]. The MCV is

20

TABLE 22a   **Examples of Tests Used in Diagnosing Cardiovascular Disorders**

Diagnostic Procedure	Method and Result	Representative Uses
Electrocardiography (ECG/EKG)	Electrodes placed on the chest detect electrical activity of cardiac muscle during a cardiac cycle; information is transmitted to a monitor for graphic recording. Also, a paper record is produced.	Heart rate can be determined through study of the ECG; abnormal wave patterns may occur with cardiac irregularities such as myocardial infarction, chamber hypertrophy, or arrhythmias, such as premature ventricular contractions (PVCs) and defects in the conduction system.
Echocardiography	Standard ultrasound examination of the heart	Detects structural abnormalities of the heart; useful in determination of valve function, chamber size, vessel size, and ejection fraction
Transesophageal echocardiography	Use of an ultrasound probe attached to an endoscope and inserted into the esophagus	Provides enhanced view of posterior and inferior heart chambers
Phonocardiography	Heart sounds are monitored and graphically recorded	Detects murmurs and abnormal heart sounds (a useful aid for teaching cardiac auscultation)
Exercise stress test	ECG, blood pressure, and heart rate are monitored during exercise on treadmill; often teamed with before-and-after echocardiograms	Coronary artery narrowing is suspected if ECG patterns or echocardiography results change
Chest x-ray	X-rays penetrate body, recorded on film sheet with radiodense tissues (white-on-black negative image)	Determines abnormal shape or size of the heart and abnormalities of the aorta and great vessels; detects masses or fluid in lungs
Cardiac catheterization (coronary arteriography)	For the study of the left side of the heart, a catheter is inserted into the femoral or brachial artery; for the study of the right side, the catheter is inserted into the femoral or subclavian vein; the catheter passes along the vessels to the heart; contrast dye is injected to visualize vessels and chambers as x-rays are taken	Detects blockages and spasms in coronary arteries; also helps evaluate congenital defects and ventricular hypertrophy and determine the severity of a valvular defect; used to monitor volumes and pressures within the chambers
MRI of heart	Noninvasive images of coronary arteries; limited by heart motion and imaging speed	Detects coronary artery calcification, indicating atherosclerotic plaques and blockages
Technesium scan	Radionuclides are injected into the bloodstream; accumulation occurs in perfused areas of the heart; a computer image is produced for examination	Detects tissue areas with a reduced perfusion; determines blood flow through the coronary arteries; after an MI, it helps determine extent of damage from infarction; may detect exercise ischemia when used with stress test
Pericardiocentesis	Needle aspiration of fluid from the pericardial sac for analysis; therapeutic for cardiac tamponade	Fluid is analyzed for appearance, number and types of blood cells, protein and glucose levels, and presence of pathogens.
Doppler ultrasound	Transducer is placed over the vessel to be examined, and the echoes are analyzed by computer to provide information on blood velocity and flow direction	Detects venous occlusion or venous valve insufficiency; determines cardiac efficiency; monitors fetal circulation and blood flow in umbilical vessels
Venography	Radiopaque dye is injected into a peripheral vein of a limb, and an x-ray study is performed to detect venous occlusions	Detects venous thrombosis (phlebitis)

calculated by dividing the hematocrit by the RBC count, using the formula

$$MCV = \frac{Hct}{RBC\ count\ (in\ millions)} \times 10$$

This is a "cookbook" method that takes advantage of the fact that the hematocrit closely approximates the relative volume of RBCs in any unit sample of whole blood. Normal values for the MCV range from 80 to 98. For a representative hematocrit of 46 and an RBC count of 5.2 million, the mean corpuscular volume is

$$MCV = \frac{46}{5.2} \times 10 = 88.5\ \mu m^3$$

Cells of normal size are said to be **normocytic**, whereas larger-than-normal or smaller-than-normal RBCs are called **macrocytic** and **microcytic**, respectively.

TABLE 22b   Representative Hematology Studies for Diagnosing Blood Disorders

Laboratory Test	Normal Values in Blood Plasma or Serum	Significance of Abnormal Values
Complete Blood Count		
RBC count	Adult males: 4.7–6.1 million/mm^3 Adult females: 4.0–5.0 million/mm^3	Increased RBC count, Hb, and Hct occur in polycythemia vera, congenital heart disease, and events that induce chronic hypoxia, such as moving to a high altitude, or lung disease
Hemoglobin (Hb, Hgb)	Adult males: 14–18 g/dl Adult females: 12–16 g/dl	Decreased RBC count, Hb, and Hct occur with hemorrhage and the various forms of anemia, including hemolytic anemia, iron-deficiency anemia, vitamin B$_{12}$ deficiency, sickle cell anemia, and thalassemia, or as a result of bone marrow failure or leukemia
Hematocrit (Hct)	Adult males: 42–52 Adult females: 37–47	See sections on RBC count and hemoglobin
RBC indices Mean corpuscular volume (MCV; measure of average volume of a single RBC)	Adults: 80–98 μl^3	Increased MCV and MCH occur in types of macrocytic anemia, including vitamin B$_{12}$ deficiency
Mean corpuscular hemoglobin (MCH; measure of average amount of Hb per RBC)	Adults: 27–31 pg	Decreased MCV and MCH occur in types of microcytic anemia, such as iron-deficiency anemia and thalassemia
Mean corpuscular hemoglobin concentration (MCHC)	Adults: 32–36% (derived by dividing the total Hb concentration number by the Hct value)	Decreased levels (hypochromic erythrocytes) suggest iron-deficiency anemia or thalassemia

• **Mean corpuscular hemoglobin concentration (MCHC)**, the average amount of hemoglobin in a single RBC, in picograms. Normal values range from 27 to 33 pg. The MCHC is calculated as

$$MCHC = \frac{Hb}{RBC\ count\ (in\ millions)} \times 10$$

RBCs containing normal amounts of hemoglobin are said to be **normochromic**, whereas the names **hyperchromic** and **hypochromic** indicate higher-than-normal and lower-than-normal hemoglobin content, respectively.

Anemia exists whenever the oxygen-carrying capacity of the blood is reduced, diminishing the delivery of oxygen to peripheral tissues. Such a reduction causes a variety of symptoms, including premature muscle fatigue, weakness, lethargy, and a lack of energy. Anemia may exist because the hematocrit is abnormally low or because the amount of hemoglobin in the RBCs is reduced. Standard laboratory tests can be used to differentiate among the various forms of anemia on the basis of the number, size, shape, and hemoglobin content of RBCs. As an example, Table 23 shows how this information can be used to distinguish among four major types of anemia:

1. **Hemorrhagic anemia** results from severe blood loss. Erythrocytes are of normal size; each contains a normal amount of hemoglobin, and reticulocytes are present in normal concentrations, at least initially before the homeostatic mechanisms increase blood production. Blood tests would therefore show a low hematocrit and low hemoglobin, but the MCV, MCHC, and reticulocyte counts would be normal.

2. In **aplastic** (ā-PLAS-tik) **anemia**, the bone marrow fails to produce new RBCs. Presumed causes include radiation, toxic chemicals, and immunologic or infectious diseases, but in most cases the precise cause is unknown. The 1986 nuclear accident in Chernobyl (in the former USSR) caused a number of cases of aplastic anemia. The condition is fatal unless surviving stem cells repopulate the marrow or a transplant of hematologic stem cells is performed. In aplastic anemia, the circulating RBCs are normal in all respects, but because new RBCs are not being produced, the RBC count, Hct, Hb, and reticulocyte count are low.

3. In **iron-deficiency anemia**, normal hemoglobin synthesis cannot occur because iron reserves are inadequate (p. 101). Because developing RBCs cannot make functional hemoglobin, they are unusually small. A blood test shows a low hematocrit, low hemoglobin content, low MCV, and low MCHC, but generally a normal reticulocyte count. An estimated 60 million women worldwide have iron-deficiency anemia.

4. In **pernicious** (per-NISH-us) **anemia**, normal RBC maturation ceases because the supply of vitamin B$_{12}$ is inadequate. Erythrocyte production declines, and the RBCs are abnormally large and may develop a variety of bizarre shapes.

TABLE 22c Representative Hematology Studies for Diagnosing Blood Disorders (Continued)

Laboratory Test	Normal Values in Blood Plasma or Serum	Significance of Abnormal Values
WBC count	Adults: 5000–10,000/mm^3	Increased levels occur with chronic and acute infections, tissue death (MI, burns), leukemia, parasitic diseases, and stress; decreased levels occur in aplastic and pernicious anemias, overwhelming bacterial infection (sepsis), and viral infections, including AIDS
Differential WBC count	Neutrophils: 50–70%	Increased levels occur in acute bacterial infection, myelocytic leukemia, rheumatoid arthritis, and stress; decreased levels occur in aplastic and pernicious anemia, viral infections, radiation treatment, and with some medications
	Lymphocytes: 20–30%	Increased levels occur in lymphocytic leukemia, infectious mononucleosis, and viral infections; decreased levels occur in radiation treatment, AIDS, and corticosteroid therapy
	Monocytes: 2–8%	Increased levels occur in chronic inflammation, viral infections, and tuberculosis; decreased levels occur in aplastic anemia and corticosteroid therapy
	Eosinophils: 2–4%	Increased levels occur in allergies, parasitic infections, and some immune disorders; decreased levels occur with steroid therapy
	Basophils: 0.5–1%	Increased levels occur in inflammatory processes and during healing; decreased levels occur in hypersensitivity reactions
Platelet count	Adults: 150,000–400,000/mm^3	Increased count can cause vascular thrombosis and occurs in polycythemia vera; decreased levels can result in spontaneous bleeding and occurs in different types of anemia and in some leukemias

Clotting System Tests

Bleeding time (amount of time for bleeding to stop after a small incision is made in skin)	3–7 minutes	Prolongation occurs in patients with decreased platelet count, anticoagulant therapy, aspirin ingestion, leukemia, or clotting factor deficiencies.
Factors assay (coagulation factors I, II, V, VIII, IX, X, XI, XII)	Measured for their hemostatic activity	Decreased activity of the coagulation factors will result in defective clot formation; deficiencies can be caused by liver disease or vitamin K deficiencies
Plasma prothrombin time (PT)	Clotting within 2 seconds of control (control should be 11–15 seconds)	Prolonged time to clotting can occur after trauma, MI, or infection. Decreased levels occur in DIC and liver disease. Therapeutic use of the anticoagulant coumadin is monitored by the PT ratio of tested blood to control results, or INR (international normalized ratio).

Other Tests

Erythrocyte sedimentation rate (ESR) (measure of rate at which erythrocytes settle in a column of whole blood collected in a non-clotting tube)	Adult males: 1–15 mm/h Adult females: 1–20 mm/h	Increased by disease processes that increase the protein concentration in the plasma, including inflammatory conditions, infections, and cancer; not useful in diagnosis of specific disorders but a rough gauge severity of active inflammation
Bone marrow aspiration biopsy (involves laboratory examination of shape and cell size of erythrocytes, leukocytes, and megakaryocytes)	For the evaluation of hematopoiesis or the presence of tumor cells or infection	Increased RBC precursors with polycythemia vera; increased WBC precursors with leukemia; radiation therapy or chemotherapy can cause a decrease in all cell populations

TABLE 23   RBC Tests and Anemias

Type of Anemia	Hct	Hb	Reticulocyte Count	MCV	MCHC
Hemorrhagic (acute)	Low	Low	Normal or high	Normal	Normal
Aplastic	Low	Low	Very low	Normal	Normal
Iron deficiency	Low	Low	Normal or low	Low	Low
Pernicious (B_{12} deficiency)	Low	Low	Very low	High	Normal or low
Hemolytic	Low	Low	Very high (three times normal)	Normal or low	Normal

Blood tests from a person with pernicious anemia indicate a low hematocrit with a high MCV and a normal or low reticulocyte count. A similar macrocytic anemia occurs with deficiency of another B vitamin, folate.

5. In **hemolytic anemia**, RBCs are breaking down in the bloodstream. The individual RBCs are generally normal in size and hemoglobin content, although in some cases RBC fragments are present. The hematocrit and hemoglobin concentration are low. In chronic cases, reticulocyte counts are high as reticulocytes enter the bloodstream prematurely in response to the anemia.

■ ABNORMAL HEMOGLOBIN

HA p. 530

Several inherited disorders are characterized by the production of abnormal hemoglobin. Two of the best known are thalassemia and sickle cell anemia.

■ THALASSEMIA

The various forms of **thalassemia** (thal-ah-SĒ-mē-uh) result from an inability to produce adequate amounts of alpha or beta chains of hemoglobin. As a result, the rate of RBC production is slowed, and the mature RBCs are fragile and short lived. The scarcity of healthy RBCs reduces the oxygen-carrying capacity of the blood and leads to problems with the development and growth of systems throughout the body. Individuals with severe thalassemia must periodically undergo *transfusions*—the administration of blood components—to keep adequate numbers of RBCs in the bloodstream.

The thalassemias are categorized as an **alpha-thalassemia** or a **beta-thalassemia**, depending, respectively, on whether the alpha or beta hemoglobin chains are affected. Normal individuals inherit two copies of alpha-chain genes from each parent, and alpha-thalassemia develops when one or more of these genes are missing or inactive. The severity of the symptoms varies with the number of normal alpha-chain genes that remain functional. For example, an individual with three normal alpha-chain genes will not develop symptoms, but can be a carrier, passing the defect to the next generation. A child whose parents are both carriers is at risk to develop a more severe form of the disease:

- Individuals with two copies, rather than four, of the normal alpha-chain gene (one from each parent) have somewhat impaired hemoglobin synthesis. This condition is known as *alpha-thalassemia trait*. The RBCs are small and contain less than the normal quantity of hemoglobin. About 2 percent of African Americans and many Southeast Asians have alpha-thalassemia trait.

- Individuals with only one copy of the normal alpha-chain gene have very small (*microcytic*) RBCs that are relatively fragile.

- Most individuals with no functional copies of the normal alpha-chain gene die before birth or shortly after, because the hemoglobin that is synthesized cannot bind and transport oxygen normally. The incidence of fatal alpha-thalassemia is highest among Southeast Asians.

Each person inherits only one gene for the beta hemoglobin chain from each parent. These genes contain several possible mutations, and beta-thalassemia can take a variety of forms. If an individual does not receive a copy of a normal, functioning gene from either parent, the condition of **beta-thalassemia major** develops. Symptoms of this condition include severe anemia with microcytosis and a low hematocrit (below 20) and enlargement of the spleen, the liver, the heart, and areas of red bone marrow. Treatments for people with severe symptoms include transfusions, splenectomy (to slow the rate of RBC recycling), and bone marrow transplantation. Beta-thalassemia minor, or beta-thalassemia trait, seldom produces clinical symptoms. An individual with the condition has one normal gene for the beta hemoglobin chain, and the rate of hemoglobin synthesis is depressed by roughly 15 percent. This decrease does not affect the RBCs' functional abilities, however, so no treatment is necessary. Blood counts from individuals with alpha-thalassemia trait or beta-thalassemia trait are similar to those from individuals who have iron-deficiency anemia, and there is a risk of inappropriate treatment with iron supplements.

■ SICKLE CELL ANEMIA

Sickle cell anemia (SCA) results from a mutation affecting the amino acid sequence of the beta chains of the Hb molecule. The abnormal subunit is called *hemoglobin S*. When

20

blood contains abundant oxygen, the Hb molecules and the RBCs that carry them appear normal (Figure 37a●). But when the defective hemoglobin gives up enough of its bound oxygen, the adjacent Hb molecules cluster into rods, and the cells become stiff and curved (Figure 37b●). This "sickling" makes the RBCs fragile and easily damaged, Moreover, even though RBCs that have folded to squeeze into a narrow capillary deliver their oxygen to the surrounding tissue, the cells can become stuck as sickling occurs. A circulatory blockage results, and nearby tissues become starved for oxygen. Today, sickle cell anemia affects 60,000–80,000 African Americans, or approximately 0.2 percent of the African American population, and from 0.07 to 0.1 percent of Hispanic Americans.

An individual with sickle cell anemia carries two copies of the abnormal gene—one from each parent. If only one copy is present, the individual has a *sickling trait.* One African American in 12 carries the sickling trait. These genes are also present in some people of Mediterranean, Middle Eastern, and East Indian ancestry.

In an individual with the sickling trait, most of the hemoglobin is of the normal form, and the RBCs function normally. But the presence of the abnormal hemoglobin gives the individual some ability to resist the parasitic infection that causes **malaria**, a mosquito-borne illness. Malaria parasites enter the bloodstream when an individual is bitten by an infected mosquito. The microorganisms invade, and reproduce within, the RBCs. But when they enter the RBC of a person with the sickling trait, the cell responds by sickling. Either the sickling itself kills the parasite, or the sickling attracts the attention of a phagocyte that engulfs the RBC and kills the parasite. In either event, the individual better tolerates the parasitic infection, whereas individuals without the sickling trait are more likely to sicken and die of malaria. Genetic studies indicate that the sickling mutation has evolved at least five times at different locations in Africa and India—regions where malaria poses a serious health problem.

When sickled RBCs get stuck in small capillaries and obstruct blood flow, they cause pain and eventually damage to a variety of organs and systems, depending on the location and duration of the obstructions. In addition, the trapped RBCs die and break down, producing a characteristic hemolytic anemia.

A normal RBC stays in circulation 120 days; sickled cells are gone in 10–20 days. This rapid RBC loss damages the spleen and exceeds the maximum rate of RBC production. Transfusions of normal blood can temporarily prevent additional complications, and treatment of affected infants with antibiotics reduces deaths due to infections. *Hydroxyurea* is an anticancer drug that stimulates the production of fetal hemoglobin, a slightly different form of hemoglobin normally produced during development of the fetus. The drug is effective, but has toxic side effects (not surprising in an anticancer drug). The food additive **butyrate**, found in butter and other foods, appears to be even more effective in promoting the synthesis of fetal hemoglobin. In clinical trials, it has been effective in treating sickle cell anemia and other conditions caused by abnormal hemoglobin structure, such as beta-thalassemia. Gene therapy to increase levels of fetal hemoglobin has helped mice with sickle cell anemia and may soon be tested in human patients.

■ ERYTHROCYTOSIS   *HA p. 531*

In erythrocytosis the blood contains abnormally large numbers of red blood cells. Erythrocytosis usually results from the massive release of erythropoietin by tissues deprived of oxygen. People moving to high altitudes usually experience erythrocytosis, because the air contains less oxygen than it does at sea level. The increased number of red blood cells compensates for the fact that individually each RBC is carrying less oxygen than it would at sea level. Mountaineers and those living at altitudes of 10,000–12,000 feet may have hematocrits as high as 65. Individuals whose hearts or lungs are functioning inadequately may also develop erythrocytosis. For example, this condition is often seen in heart failure and emphysema, two conditions discussed in later chapters. Whether the blood fails to circulate efficiently or the lungs do not deliver enough oxygen to the blood, peripheral tissues remain oxygen-poor despite the rising hematocrit. Having a higher concentration of red blood cells increases the oxygen-carrying capacity of the blood, but it also makes the blood thicker and harder to push around the circulatory system. This change increases the workload on the heart, making a bad situation even worse.

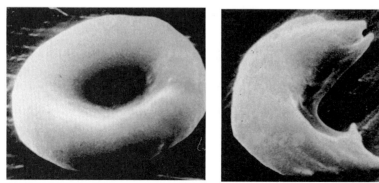

(a) Normal RBC   (b) Sickled RBC

● **FIGURE 37**
"Sickling" in Red Blood Cells. (a) When fully oxygenated, the cells of an individual with the sickling trait appear relatively normal. **(b)** At lower oxygen concentrations, the RBCs change shape, becoming more rigid and sharply curved.

■ HEMOLYTIC DISEASE OF THE NEWBORN *HA p. 531*

Genes controlling the presence or absence of any surface antigen in the membrane of a red blood cell are provided by both parents, so a child can have a blood type different from that of either parent. During pregnancy, when fetal and maternal circulatory systems are closely intertwined, the mother's antibodies may cross the placenta, attacking and destroying fetal RBCs. The resulting condition is called **hemolytic disease of the newborn (HDN)**.

This disease has many forms, some so mild as to remain undetected. Those involving the Rh surface antigen are quite dangerous, because unlike anti-A and anti-B antibodies, anti-Rh antibodies are able to cross the placenta and enter the fetal

bloodstream. An Rh-positive mother (who lacks anti-Rh antibodies) can carry an Rh-negative fetus without difficulty. Because maternal antibodies can cross the placenta, problems may appear when an Rh-negative woman carries an Rh-positive fetus. Sensitization generally occurs at delivery, when bleeding takes place at the placenta and uterus. Such mixing of fetal and maternal blood can stimulate the mother's immune system to produce anti-Rh antibodies. Roughly 20 percent of Rh-negative mothers who carried Rh-positive children become sensitized within six months of delivery.

Because the anti-Rh antibodies are not produced in significant amounts until after delivery, a woman's first infant is not affected (Figure 38●). (Some fetal RBCs cross into the maternal bloodstream during pregnancy, but generally not in

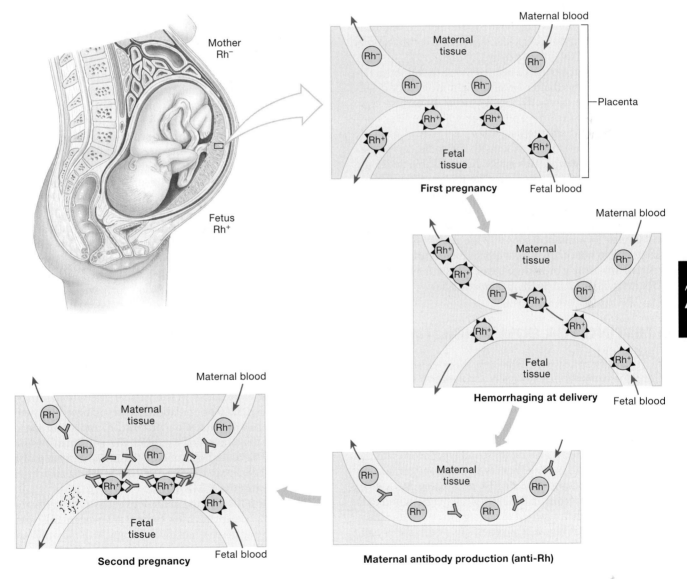

● FIGURE 38
Rh Factors and Pregnancy. When an Rh-negative woman delivers her first Rh-positive child, fetal and maternal blood mix when the placenta breaks down. The presence of Rh-positive blood cells in the maternal bloodstream sensitizes the mother, stimulating the production of anti-Rh antibodies. If another pregnancy involves an Rh-positive fetus, maternal anti-Rh antibodies can cross the placenta and attack fetal blood cells, producing hemolytic disease of the newborn.

numbers sufficient to stimulate antibody production.) But if a subsequent pregnancy involves an Rh-positive fetus, maternal anti-Rh antibodies produced after the first delivery cross the placenta and enter the fetal bloodstream. These antibodies destroy fetal RBCs and produce a dangerous anemia. The fetal demand for blood cells increases, and they begin leaving the bone marrow and entering the bloodstream before completing their development. Because these immature RBCs are erythroblasts, HDN is also known as **erythroblastosis fetalis** (ē-rith-rō-blas-TŌ-sis fē-TAL-is).

Without treatment, the fetus will probably die before delivery or shortly thereafter. A newborn with severe HDN is anemic, and the high concentration of circulating bilirubin produces jaundice. Because the maternal antibodies will remain active for one to two months after delivery, the entire blood volume of the infant may have to be replaced. Replacing the blood removes most of the maternal anti-Rh antibodies, as well as the affected RBCs, reducing complications and the chance of the infant's dying.

If the fetus is in danger of not surviving to full term, delivery may be induced after seven to eight months of development. In severe cases affecting a fetus at an earlier stage, one or more transfusions can be given while the fetus continues to develop in the uterus.

The maternal production of anti-Rh antibodies can be prevented by administering such antibodies (available under the name *RhoGam*) to the mother in the last three months of pregnancy and during and after delivery. (Anti-Rh antibodies are also given after a miscarriage or an abortion.) These antibodies will destroy any fetal RBCs that cross the placenta before they can stimulate an immune response in the mother. Because sensitization does not occur, no anti-Rh antibodies are produced. This relatively simple procedure has almost entirely prevented HDN mortality caused by Rh incompatibilities.

■ TESTING FOR COMPATIBILITY

HA p. 531

Testing for compatibility normally involves two steps: a determination of blood type and a cross-match test. At least 50 surface antigens have been identified on red blood cell surfaces, but the standard test for blood type categorizes a blood sample on the basis of the three most likely to produce dangerous cross-reactions. The test for blood type involves taking drops of blood and mixing them with solutions containing anti-A, anti-B, and anti-Rh (anti-D) antibodies. Any cross-reactions are then recorded. For example, if the red blood cells clump together when exposed to anti-A and anti-B, the individual has Type AB blood. If no reactions occur, the person must be Type O. The presence or absence of the Rh surface antigen is also noted, and the individual is classified as Rh-positive or Rh-negative on that basis. Type O-positive is the most common blood type. The red blood cells of these individuals do not have surface antigens A and B, but they do have surface antigen D. Standard blood typing can be completed in a matter of minutes, and Type O-negative blood can be safely administered to anyone of any blood type in an emergency. However, with at least 48 other possible surface antigens on the cell surface, cross-reactions can occur, even to Type O blood. Whenever time and facilities permit, further testing is performed to ensure complete compatibility. Cross-match testing involves exposing the donor's red blood cells to a sample of the recipient's plasma under controlled conditions. This procedure reveals the presence of significant cross-reactions involving other surface antigens and antibodies.

■ THE LEUKEMIAS *HA p. 533*

Leukemias characterized by the presence of abnormal granulocytes or other cells of the bone marrow are called **myeloid**; leukemias that involve abnormal lymphocytes are termed **lymphoid**. The first symptoms appear as immature and abnormal white blood cells (WBCs) enter the bloodstream. As the number of white blood cells increases, they travel through the circulation, invading tissues and organs throughout the body.

These cells are extremely active, and they require abnormally large amounts of energy. As in other cancers, described in the text and elsewhere in this manual (p. 17), leukemic cells replace normal cells, especially in the bone marrow. Red blood cell, normal WBC, and platelet formation declines, with resulting anemia, infection, and impaired blood clotting. Untreated leukemias are invariably fatal.

Leukemias are classified as *acute* (short and severe) or *chronic* (prolonged). Acute leukemias are linked to radiation exposure, hereditary susceptibility, viral infections, or unknown causes. Chronic leukemias are related to chromosomal abnormalities or immune system malfunctions. Survival in untreated acute leukemia averages about three months; individuals with chronic leukemia may survive for years.

Effective treatments exist for some forms of leukemia, but not for others. For example, when acute lymphoid leukemia in children is detected early, 85–90 percent of patients can be held in remission for five years or longer, but only 10–15 percent of patients with acute myeloid leukemia (which is more common in adults) survive five or more years. The yearly mortality rate for all types of leukemia in the United States has not declined appreciably in the past 30 years, remaining at about 6.8 per 100,000 population. However, new treatments are being developed that show promise in combating specific forms of leukemia. For example, the administration of *gamma-interferon,* a hormone of the immune system, has been highly effective in treating hairy cell leukemia and chronic myeloid leukemia (CML).

One option for treating acute leukemias is to perform a bone marrow transplant. Most bone marrow/stem cell transplants involve aggressive "ablative" therapy with radiation and chemotherapy, which is meant to destroy the patient's cancer and also destroys the patient's bone marrow stem cells. This result then requires the replacement of functioning stem cells for the patient's survival. Following the ablative therapy, the individual receives an infusion of healthy bone marrow with its stem cells, which repopulate the blood and marrow tissues. Recently, transplants using stem cells collected by filtering the donor's blood have been successful, reducing the need for bone marrow collection.

If the hematologic stem cells are extracted from another person (a **heterologous transplant**), care must be taken to ensure that the blood types and tissue types are compatible. If they are not, the new lymphocytes may attack the patient's tissues (graft-versus-host disease, or GVHD), with potentially fatal results. The best results are obtained when the donor is a close relative. Fetal blood with fetal hematologic stem cells obtained from the umbilical cord after birth appears to be more compatible with unrelated recipients, and stored cord-blood banks exist in Europe and North America. While the number of stem cells required for successful transplantation increases with body size, cord-blood transplantation has been successful in both children and adults. In older patients too fragile to tolerate aggressive therapy, nonablative transplant therapy has been tried. Reduced amounts of chemo and radiation therapy are used, and an allograft of stem cells is performed. Any graft-versus-host disease that occurs is hoped to attack any remaining leukemia cells, with serious GVHD avoided by immunosuppressive treatment.

In an **autologous marrow transplant**, bone marrow is removed from the patient, cleansed of cancer cells, and reintroduced after radiation or chemotherapy treatment. Although this method produces fewer complications, the preparation and cleansing of the marrow are technically difficult, and a recurrence of leukemia is more likely.

Whatever the source (bone marrow, peripheral blood, or umbilical cord blood), successful transplants require a large number of hematologic stem cells, and may require ongoing immunosuppression for less well matched recipient/donor pairs.

■ TESTING THE CLOTTING SYSTEM

HA p. 533

Several clinical tests check the efficiency of the clotting system (Table 22c).

▫ BLEEDING TIME

This test measures the time it takes a small skin wound to seal. There are variations on the procedure, with normal values ranging from three to seven minutes. Aspirin prolongs bleeding time by affecting platelet function and suppressing the extrinsic pathway.

▫ COAGULATION TIME

In this test, a sample of whole blood stands under controlled conditions until a visible clot has formed. Normal values range from 3 to 15 minutes. The test has several potential sources of error and so is not very accurate. Nevertheless, it is the simplest test that can be performed on a blood sample.

▫ PARTIAL THROMBOPLASTIN TIME (PTT)

In this test, a plasma sample is mixed with chemicals that mimic the effects of activated platelets. Plasma is obtained by adding citrate ions to a blood sample. This ties up calcium ions and prevents clotting. Calcium ions are then added to the plasma, and the clotting time is recorded. Clotting normally occurs in 35–50 seconds if the enzymes and clotting factors of the intrinsic pathway are present in normal concentrations. The PTT is prolonged by heparin and is used to monitor its therapeutic use.

▫ PROTHROMBIN TIME

This test checks the performance of the extrinsic pathway. The procedure is similar to that used in the PTT test, but the clotting process is triggered by exposure to a combination of tissue thromboplastin and calcium ions. Clotting normally occurs in 13–17 seconds. The **prothrombin time** (PT) is prolonged by *coumadin* and also used to monitor its use.

■ ABNORMAL HEMOSTASIS   *HA p. 533*

▫ EXCESSIVE OR ABNORMAL BLOOD CLOTTING

If the clotting response is inadequately controlled, blood clots will begin to form in the bloodstream rather than at the site of an injury. These blood clots may not stick to the wall of the vessel, but continue to drift around until either plasmin digests them or they become lodged in a small blood vessel. A drifting blood clot is a type of **embolus** (EM-bō-lus; *embolos,* plug), an abnormal mass within the bloodstream. An embolus that becomes stuck in a blood vessel blocks circulation to the area downstream, killing the affected tissues. The sudden blockage is called an **embolism**, and the area of tissue damage caused by the circulatory interruption is one form of an **infarct**. Infarctions in the brain are known as *strokes;* infarctions in the heart are called *myocardial infarctions,* or *heart attacks.*

An embolus in the arterial system can get stuck in capillaries in the brain, causing an embolic stroke. An embolus in the venous system will probably become lodged in one of the capillaries of the lungs, resulting in *pulmonary embolism.*

A **thrombus** (*thrombos,* clot), or blood clot attached to a vessel wall, begins to form when platelets stick to the wall of an intact blood vessel. Often, the platelets are attracted to areas called *plaques,* where endothelial and smooth muscle cells contain large quantities of lipids and may be inflamed. (The mechanism of plaque formation is discussed in Chapter 22 of the text.) The thrombus gradually enlarges, projecting into the lumen of the vessel and reducing its diameter. Eventually, smaller vessels (usually arteries) may be completely blocked, creating an infarct, or a large chunk of the clot may break off, creating an equally dangerous embolus.

To treat these circulatory blockages, clinicians may attempt surgery or introduce catheters into the affected vessels to relieve the obstruction. Locally or systemically administered enzymes that attack blood clots and prevent further clot formation are also used in selected patients. Important anticoagulant drugs include the following:

20

- Heparin, which activates antithrombin-III.
- **Coumadin** (COO-ma-din) or *warfarin* (WAR-fa-rin), and **dicumarol** (dī-KOO-ma-rol), which depress the synthesis of several clotting factors by blocking the action of vitamin K.
- Recombinant DNA–synthesized **tissue plasminogen activator (t-PA)**, which stimulates plasmin formation.
- **Streptokinase** (strep-tō-KĪ-nās) and **urokinase** (ū-rō-KĪ-nās), enzymes that convert plasminogen to plasmin.
- Aspirin, which (1) inactivates platelet enzymes involved in the production of thromboxanes and prostaglandins and (2) inhibits the production of prostacyclin by endothelial cells. Daily ingestion of small quantities of aspirin reduces the sensitivity of the clotting process. This method has been proven effective in preventing heart attacks in people with significant coronary artery disease.

For some blood tests, it is necessary to prevent clotting within a collected blood sample to avoid changes in plasma composition. Blood samples can be stabilized temporarily by the addition of heparin or *EDTA* (*e*thylene*d*iamine*t*etroacetic *a*cid) to the collection tube. EDTA removes Ca^{2+} from plasma, effectively preventing clotting. In units of whole blood held for extended periods in a blood bank, *citratephosphate dextrose* (CPD) is typically added. Like EDTA, CPD ties up plasma Ca^{2+}.

■ INADEQUATE BLOOD CLOTTING

Hemophilia (hē-mō-FĒL-ē-a) is one of many inherited disorders characterized by the inadequate production of clotting factors. The condition affects about one in 10,000 people, 80–90 percent of whom are males. In hemophilia, the production of a single clotting factor (most commonly, Factor VIII) is inadequate; the severity of the condition depends on the degree of underproduction. In severe cases, extensive bleeding accompanies the slightest mechanical stress; hemorrhages occur spontaneously at joints and around muscles.

In many cases, transfusions of clotting factors can reduce or control the symptoms of hemophilia, but plasma from many individuals must be pooled (combined) to obtain adequate amounts of clotting factors. This procedure makes the treatment very expensive and increases the risk of blood-borne infections such as hepatitis or AIDS. Gene-splicing techniques have been used to manufacture clotting Factor VIII, an essential component of the intrinsic clotting pathway. As methods are developed to synthesize other clotting factors, treatment of the various forms of hemophilia will become safer and cheaper.

The condition known as **von Willebrand disease** is the most common inherited coagulation disorder. The *von Willebrand factor* (vWF) is a plasma protein that binds and stabilizes Factor VIII. The symptoms and severity of the bleeding vary widely. Many individuals with mild forms of von Willebrand disease remain unaware of any bleeding problems until they have an accident or undergo surgery. Because a manufactured form of vWF is unavailable, treatment consists of the administration of pooled serum Factor VIII, because pooled Factor VIII contains normal vWF. In some forms of this disease in which normal vWF is produced but plasma levels are abnormally low, bleeding can be controlled by the use of nasal sprays containing a synthetic form of antidiuretic hormone (ADH). The absorbed ADH appears to stimulate the release of vWF from endothelial cells.

In *disseminated intravascular coagulation* (DIC), bacterial toxins or factors released by damaged tissues activate thrombin, which then converts fibrinogen to fibrin within the circulating blood. Much of the fibrin is removed by phagocytes or dissolved by plasmin, but tiny clots may block small vessels and damage the associated tissues. If the liver cannot keep pace with the increased demand for fibrinogen, clotting abilities decline and uncontrolled bleeding may occur. DIC is one of the complicating factors of septicemia, a dangerous infection of the bloodstream that spreads bacteria and bacterial toxins throughout the body.

20

DISORDERS OF THE HEART

■ INFECTION AND INFLAMMATION OF THE HEART   *HA p. 546*

Many different microorganisms may infect heart tissue, leading to serious cardiac abnormalities. **Carditis** (kar-DĪ-tis) is a general term for inflammation of the heart. Clinical conditions resulting from cardiac infection are usually identified by the primary site of the infection. For example, infections that affect the endocardium produce symptoms of **endocarditis**, a condition that damages primarily the chordae tendineae and heart valves; the mortality rate may reach 21–35 percent. The most severe complications of endocarditis result from the formation of blood clots on the damaged surfaces. These clots subsequently break free, entering the bloodstream as drifting emboli that may cause strokes, heart attacks, or kidney failure. The destruction of heart valves by the infection may lead to valve leakage, heart failure, and death.

Myocarditis, inflammation of the heart muscle, can be caused by bacteria, viruses, protozoans, or fungal pathogens that either attack the myocardium directly or produce toxins that damage the myocardium. The microorganisms implicated include those responsible for many of the conditions discussed in previous sections, such as diphtheria, syphilis, polio, tuberculosis, and malaria. The membranes of infected heart muscle cells become facilitated, and the heart rate may rise dramatically. Over time, abnormal contractions may appear and the heart muscle weakens; these problems may eventually prove fatal.

If the pericardium becomes inflamed or infected, fluid may accumulate around the heart *(cardiac tamponade)* or the elasticity of the pericardium may be reduced *(constrictive pericarditis)*. In both conditions, the expansion and filling of the heart is restricted and cardiac output is reduced. Treatment includes draining the excess fluid or cutting a window in the pericardial sac.

■ THE CARDIOMYOPATHIES   *HA p. 550*

The **cardiomyopathies** (kar-dē-ō-mī-OP-a-thēz) are an assortment of diseases with a common symptom: the progressive, irreversible degeneration of the myocardium. Cardiac muscle cells are damaged and replaced by fibrous tissue, and the muscular walls of the heart become thin and weak. As muscle tone declines, the ventricular chambers enlarge greatly. When the remaining cells cannot develop enough force to maintain cardiac output, symptoms of heart failure develop.

Chronic alcoholism and coronary artery disease are probably the most common causes of cardiomyopathy in the United States. Infectious agents, especially viruses, also cause cardiomyopathy. In addition, diseases affecting neuromuscular performance, such as muscular dystrophy (p. 53), damage cardiac muscle cells, as can starvation or chronic variations in the extracellular concentrations of calcium or potassium ions.

Several forms of cardiomyopathy are inherited. **Hypertrophic cardiomyopathy** *(HCM)* is an inherited disorder that makes the wall of the left ventricle thicken to the point at which it has difficulty pumping blood. Most people with HCM do not become aware of it until relatively late in life. However, HCM can also cause a fatal arrhythmia and has been implicated in the sudden deaths of several young athletes. The implantation of an electronic cardiac pacemaker has proved to be beneficial in controlling HCM-related arrhythmias.

Finally, there are a significant number of cases of *idiopathic cardiomyopathy,* a term used when the primary cause cannot be determined.

◼ HEART TRANSPLANTS AND ASSIST DEVICES

Individuals with severe cardiomyopathy may be considered as candidates for a heart transplant. This surgery involves the removal of the weakened heart and its replacement with a heart taken from a suitable donor. To survive the surgery, the recipient must be in otherwise satisfactory health. In the United States in 2002, more than 4000 patients, ranging in age from infancy to over 65, were waiting for a heart transplant. In 2001, 622 patients died while on the waiting list, and 2202 heart transplants were performed. After successful transplantation, the one-year survival rate is 80–85 percent and the five-year survival rate is 50–70 percent. These rates are quite good, especially considering that most of these patients were expected to die if the transplant had not been performed.

As noted, many individuals with cardiomyopathy who are initially selected for heart transplant surgery succumb to the disease before a suitable donor becomes available. For that reason, there continues to be considerable interest in the development of an artificial heart. One model, the *Jarvik-7,* had limited clinical use in the 1980s. Attempts to implant it on a permanent basis were unsuccessful. One problem was that blood clots tended to form on the mechanical valves. When the clots broke free, they became drifting emboli that plugged peripheral vessels, producing strokes, kidney failure, and other complications. In addition, infections involving the connections of the heart to its external power source sometimes led to sepsis and death. Since 2001, the AbioCor artificial heart (which requires no external power source) has been implanted in 12 near-death patients. The first recipient survived for several months, and one lived 17 months after surgery. However, the average survival has been five months, and it is still an experimental treatment of last resort.

Modified versions of these units and others now under development are used to maintain transplant candidates who are awaiting the arrival of a donor organ. The new machines are called *left ventricular assist devices* (LVADs). As the name implies, they assist, rather than replace, the damaged heart. In one study, the survival rate was 75 percent for patients relying on such a device for three and a half months. The devices are also being called on to reduce the workload on weak hearts, giving them time to heal, or even to supplement the heart indefinitely, rather than acting solely as a bridge to heart transplantation.

In 1996, attention focused on a Brazilian surgeon who developed a surgical procedure to improve cardiac function in patients with cardiomyopathy. The surgeon removes a largely nonfunctional portion of the weakened left ventricle. The ventricular muscle that remains is sewn together, forming a smaller chamber. In part because the muscle cells in the dilated heart were excessively stretched, the smaller, remodeled ventricle pumps blood more efficiently. Heart surgeons in the United States have confirmed the effectiveness of this therapy, which may reduce the demand for heart transplants in the years to come.

As public-health knowledge and education about the risk factors for heart disease have improved, and more effective treatments for hypertension, high cholesterol, and heart failure have been developed, death rates from coronary artery disease have declined significantly. Between 1963 and 1995, the rates fell from over 200 to 134 per 100,000 population. It has been estimated that during this period there were 621,000 fewer deaths from coronary heart disease than would have been expected had mortality rates stayed at their mid-1960s levels.

◼ RHD AND VALVULAR STENOSIS
HA p. 550

Rheumatic (roo-MA-tik) **fever** is an inflammatory condition that can develop after untreated infection by streptococcal bacteria ("strep throat"). Rheumatic fever typically affects children of 5–15 years of age; symptoms include high fever, joint pain and stiffness, and a distinctive full-body rash. Obvious symptoms generally persist for less than six weeks, although severe cases may linger for six months or more. The longer the duration of the inflammation, the more likely it is that carditis will develop. The carditis that does develop in 50–60 percent of individuals may escape detection, and scar tissue gradually forms in the myocardium and the heart valves. Valve condition deteriorates over time, and valve problems serious enough to affect cardiac function may not appear until 10–20 years after the initial infection.

During the interim, the affected valves become thickened and may also calcify to some degree. This thickening narrows the opening guarded by the valves, producing a condition called **valvular stenosis** (ste-NŌ-sis; *stenos,* narrow). The resulting clinical disorder is known as **rheumatic heart disease (RHD)**. The thickened cusps stiffen in a partially closed position, but the valves do not completely block the blood flow, because the edges of the cusps are rough and irregular. Regurgitation may occur, and much of the blood pumped out of the heart may flow back in. The abnormal valves are also much more susceptible to bacterial infection, a type of endocarditis (p. 119). Fortunately, with the detection and prompt antibiotic treatment of strep infections, the number of cases of RHD has declined dramatically since the 1940s.

Mitral stenosis and **aortic stenosis** are the most common forms of RHD. About 40 percent of patients with RHD develop mitral stenosis, and two-thirds of them are women. The

21

reason for the correlation between gender and mitral stenosis is unknown. In mitral stenosis, blood enters the left ventricle at a slower-than-normal rate; when the ventricle contracts, blood flows back into the left atrium, as well as into the aortic trunk. As a result, the left ventricle has to work much harder to maintain adequate systemic circulation. The right and left ventricles discharge identical amounts of blood with each beat, so, as the output of the left ventricle declines, blood "backs up" in the pulmonary circuit. Venous pressures then rise in the pulmonary circuit, and the right ventricle must develop greater pressures to force blood into the pulmonary trunk. In severe cases of mitral stenosis, the right ventricular musculature is not up to the task. The heart weakens and peripheral tissues begin to suffer from oxygen and nutrient deprivation. (This condition is called *congestive heart failure* (*CHF*).)

Symptoms of aortic stenosis develop in roughly 25 percent of patients with valvular heart disease. Stenosis frequently develops in people born with bicuspid aortic valves, 80 percent of whom are males. If rheumatic fever is the cause, the mitral valve is usually affected as well. Symptoms of aortic stenosis are initially less severe than those of mitral stenosis. Although the left ventricle enlarges and works harder, normal circulatory function can typically be maintained for years. Clinical problems develop only when the opening narrows enough to prevent adequate blood flow. Symptoms then resemble those of mitral stenosis.

One reasonably successful treatment for severe stenosis involves the replacement of the damaged valve. Figure 39a● shows a stenotic heart valve; two possible replacements are a valve from a pig (Figure 39b●) and a synthetic valve (Figure 39c●), one of a number of designs that have been employed. Pig valves do not require anticoagulant therapy, but may wear out and begin leaking after roughly 10 years of service. The plastic or stainless-steel components of the artificial valve are more durable, but they activate the clotting system of the recipient, leading to inflammation, clot formation, and other complications. Synthetic-valve recipients must take anticoagulant drugs to prevent strokes and other disorders caused by

the formation of emboli. Valve replacement operations are quite successful, with about 95 percent of the surgical patients surviving for three years or more and 70 percent surviving more than five years.

■ DIAGNOSING ABNORMAL HEARTBEATS *HA p. 557*

Damage to the conduction pathways caused by mechanical distortion, ischemia, infection, or inflammation can affect the normal rhythm of the heart. The resulting condition is called a **heart block**, or **conduction deficit**. Figure 40● shows the electrocardiogram of a normal heart and heart blocks of varying severity. In a **first-degree heart block** (Figure 40b●), the AV node and the proximal portion of the AV bundle slow the passage of impulses that are heading for the ventricular myocardium. As a result, a pause appears between the atrial and ventricular contractions. Although a delay exists, the regular rhythm of the heart continues, and each atrial beat is followed by a ventricular contraction.

If the delay lasts long enough, the nodal cells will still be repolarizing from the previous beat when the next impulse arrives from the SA node. The arriving impulse will then be ignored, the ventricles will not be stimulated, and the normal "atria–ventricles, atria–ventricles" pattern will disappear. This condition is a **second-degree heart block** (Figure 40c●). A mild second-degree block may produce only an occasional skipped beat, but with more substantial delays, the ventricles will follow every second atrial beat. The resulting pattern of "atria, atria–ventricles, atria, atria–ventricles" is known as a **two-to-one (2:1) block**. Three-to-one or even four-to-one blocks are also encountered.

In a **third-degree heart block**, or **complete heart block**, the conducting pathway stops functioning (Figure 40d●). The atria and ventricles continue to beat, but their activities are no longer synchronized. The atria follow the pace set by the SA node, beating 70–80 times per minute, and the ventricles follow the commands of the AV node, beating at a rate of 40–60 beats per minute. A temporary heart block can be

21

(a) (b) (c)

● **FIGURE 39**
Artificial Heart Valves. (a) A stenotic semilunar valve; note the irregular, stiff cusps. (b) Intact Bioprosthetic heart valve, which uses the valve from a pig's heart. (c) Medtronic Hall prosthetic heart valve.

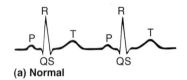

(a) Normal

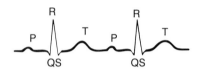

(b) First-degree heart block (long P–R interval)

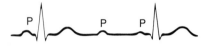

Skipped ventricular beat

2:1 block (ventricles follow every other atrial beat)

3:1 block (ventricles follow every third atrial beat)

(c) Second-degree blocks

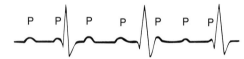

(d) Complete block (third-degree block)
(atrial beats occur regularly, ventricular beats occur at slower, unrelated pace)

● **FIGURE 40**
Heart Blocks

41b●, often occur in healthy individuals. In a PAC, the normal atrial rhythm is momentarily interrupted by a "surprise" atrial contraction. Stress, caffeine, and various drugs may increase the incidence of PAC, presumably by increasing the permeabilities of the SA pacemakers. The impulse spreads along the conduction pathway, and a normal ventricular contraction follows the atrial beat.

In **paroxysmal** (par-ok-SIZ-mal) **atrial tachycardia,** or **PAT,** a premature atrial contraction triggers a flurry of atrial activity (Figure 41c●). The ventricles are still able to keep pace, and the heart rate jumps to about 180 beats per minute. In **atrial flutter,** the atria contract in a coordinated manner, but the contractions occur very frequently. During a bout of **atrial fibrillation** (fi-bri-LA-shun), the impulses move over the atrial surface at rates of perhaps 500 beats per minute (Figure 41d●). The atrial wall quivers instead of producing an organized contraction. The ventricular rate in atrial flutter or atrial fibrillation cannot follow the atrial rate and may remain within normal limits. Despite the fact that the atria are now essentially nonfunctional, the condition

(a) Normal

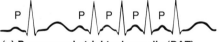

(b) Premature atrial contraction (PAC)

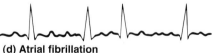

(c) Paroxysmal atrial tachycardia (PAT)

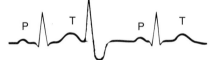

(d) Atrial fibrillation

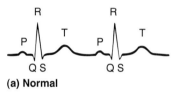

(e) Premature ventricular contraction (PVC)

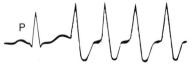

(f) Ventricular tachycardia (VT)

(g) Ventricular fibrillation (VF)

● **FIGURE 41**
Cardiac Arrhythmias

induced by stimulating the vagus nerve. In addition to slowing the rate of impulse generation by the SA node, such stimulation inhibits the AV nodal cells to the point at which they cannot respond to normal stimulation. Comments such as "my heart stopped" or "my heart skipped a beat" generally refer to this phenomenon. The pause typically lasts just a few seconds. Longer delays end when a conducting cell, normally one of the Purkinje fibers, depolarizes to threshold. This phenomenon is called **ventricular escape,** because the ventricles are escaping from the control of the SA node. Ventricular escape can be a lifesaving event if the conduction system is damaged. Even without instructions from the SA or AV nodes, the ventricles will continue to pump blood at a slow (40–50 beats per minute), but steady, rate.

■ TACHYCARDIA AND FIBRILLATION

Additional important examples of arrhythmias are shown in Figure 41●. Figure 41a● shows a normal heart rhythm. **Premature atrial contractions (PACs),** indicated in Figure

may go unnoticed, especially in older individuals who lead sedentary lives. In chronic atrial fibrillation, blood clots may form near the atrial walls. Pieces of the clot may break off, creating emboli and increasing the risk of stroke. As a result, most people diagnosed with this condition are placed on anticoagulant therapy. PACs, PAT, atrial flutter, and even atrial fibrillation are not considered very dangerous, unless they are prolonged or associated with some more serious indications of cardiac damage, such as coronary artery disease or valve problems.

In contrast, ventricular arrhythmias can be fatal. Because the conduction system functions in one direction only, from atria to ventricle, a ventricular arrhythmia is not linked to atrial activities. **Premature ventricular contractions (PVCs)** occur when a Purkinje cell or ventricular myocardial cell depolarizes to threshold and triggers a premature contraction (Figure 41e●). Single PVCs are common and not dangerous. The cell responsible is called an *ectopic pacemaker*. The frequency of PVCs can be increased by exposure to epinephrine, to other stimulatory drugs, or to ionic changes that depolarize cardiac muscle cell membranes. Similar factors may be responsible for periods of **ventricular tachycardia** (defined as four or more PVCs without intervening normal beats), also known as **VT** or *V-tach* (Figure 41f●). Multiple PVCs and VT often precede the most serious arrhythmia, **ventricular fibrillation (VF)** (Figure 41g●). The resulting condition, also known as **cardiac arrest**, is rapidly fatal, because the heart quivers and stops pumping blood. During ventricular fibrillation, the cardiac muscle cells are overly sensitive to stimulation, and the impulses are traveling from cell to cell, around and around the ventricular walls. A normal rhythm cannot become established, because the ventricular muscle cells are stimulating one another at such a rapid rate. The problem is exaggerated by a sustained rise in free intracellular calcium ion concentrations, due to a massive stimulation of alpha and beta receptors following sympathetic activation.

■ TREATING PROBLEMS WITH PACEMAKER FUNCTION *HA p. 559*

Symptoms of severe **bradycardia** (less than 50 beats per minute) include weakness, fatigue, fainting, and confusion. Drug therapies are seldom helpful, but artificial pacemakers can be used with considerable success. Wires are run to the atria, the ventricles, or both, depending on the nature of the problem, and the unit delivers small electrical pulses to stimulate the myocardium. Internal pacemakers are surgically implanted, batteries and all. These units last seven to eight years or more before another operation is required to change the battery. External pacemakers are used for temporary emergencies, such as immediately after cardiac surgery. Only the wires are implanted, and an external control box is worn on a belt.

More than 50,000 artificial pacemakers are in use at present (Figure 42●). The simplest provide constant stimulation to the ventricles at rates of 70–80 impulses per minute. More

sophisticated pacemakers stimulate the atria and ventricles in sequence and may vary their rates to adjust to changing circulatory demands, such as during exercise. Others are able to monitor cardiac activity and respond whenever the heart begins to function abnormally.

A **defibrillator** is a device that attempts to eliminate ventricular fibrillation and restore normal cardiac rhythm. An external defibrillator has two electrodes that are placed in contact with the chest, and a powerful electrical shock is administered. The electrical stimulus depolarizes the entire myocardium simultaneously. With luck, after repolarization, the SA node will be the first area of the heart to reach threshold. Thus, the primary goal of defibrillation is not just to stop the fibrillation, but to give the ventricles a chance to respond to normal SA commands.

Early defibrillation can result in dramatic recovery of an unconscious cardiac-arrest victim. **Automatic external defibrillators (AEDs)** are easily used portable machines that can detect lethal ventricular rhythms in people who have collapsed and administer a defibrillating shock. These devices are increasingly being placed on planes and in airports. Implantable pacemakers that are able to sense ventricular fibrillation and deliver an immediate defibrillating shock have been successful in preventing sudden death in patients with previous episodes of ventricular tachycardia and/or ventricular fibrillation.

Tachycardia, usually defined as a heart rate of more than 100 beats per minute, increases the workload on the heart. Cardiac performance suffers at very high heart rates, because the ventricles do not have enough time to refill with blood before the next contraction occurs. Chronic or acute incidents of tachycardia may be controlled by drugs that affect the permeability of pacemaker membranes or block the effects of sympathetic stimulation.

21

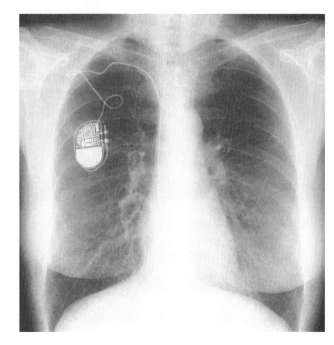

● **FIGURE 42**
An Artificial Pacemaker

EXAMINING THE HEART *HA p. 560*

Many techniques can be used to examine the structure and performance of the heart. No single diagnostic procedure provides the complete picture, so the tests used will vary with the suspected nature of the problem. A standard chest x-ray will show the basic size, shape, and orientation of the heart. Additional details require more specialized procedures to enhance the clarity of the images.

Coronary arteriography (ar-tē-rē-OG-ra-fē) is often used to look for abnormalities in the coronary circulation. In this procedure, a catheter is inserted into the femoral artery and is maneuvered back along the arterial passageways until its tip reaches the heart. A radiopaque dye can then be released at the openings of the coronary arteries, and its distribution can be followed in a series of high-speed x-rays. The images obtained are called **coronary angiograms** (Figure 43a●). For direct analyses of cardiac performance, the monitoring of chamber pressures, or the collection of blood samples, a catheter may be introduced past the aortic or tricuspid valves and into the ventricles and atria. The catheters enter via a femoral artery and the aorta, or the venous system by way of a femoral vein and the inferior vena cava.

Because the heart is constantly moving, ordinary ultrasound, computerized tomography (CT), and magnetic resonance imaging (MRI) scans create blurred images. Special instruments and computers, however, that can generate images at high speeds can be used with these techniques to develop three-dimensional still or moving pictures of the heart as it beats (Figure 43a,b,c,d●). The resulting images are dramatic, but the cost and complexity of the equipment have so far limited their use to major research institutions. MRI scans can detect calcifications in coronary arteries, which is associated with coronary artery disease and may be detectable before symptoms appear. The value of scanning for early coronary heart disease versus the cost and complications of treatment in asymptomatic people diagnosed using these scans is controversial.

Ultrasound analysis, called **echocardiography** (ek-ō-kar-dē-OG-ra-fē) (Figure 43b●), provides images that lack the clarity of CT or MRI scans, but the equipment is relatively inexpensive and portable. Recent advances in data processing have made the images suitable for following details of cardiac contractions, including valve function and blood flow dynamics. Echocardiography is now an important diagnostic tool.

21

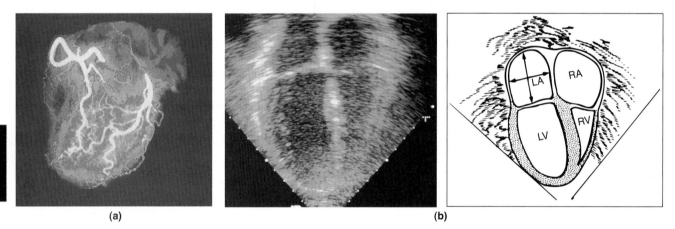

(a)

(b)

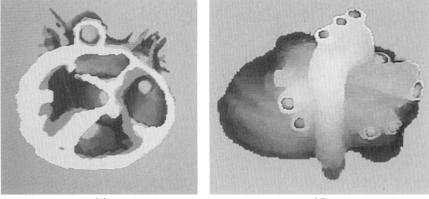

(c)

(d)

● **FIGURE 43**
Monitoring the Heart. (a) A coronary angiogram. (b) An echocardiogram (left) with interpretive drawing (right). (c) A three-dimensional CT scan of an oblique section and (d) of a posterior–superior view of the heart and great vessels.

DISORDERS OF THE VASCULAR SYSTEM

■ ANEURYSMS   *HA p. 566*

An **aneurysm** (AN-ū-rizm) is a bulge in the weakened wall of a blood vessel (generally an artery) or a heart chamber (usually in the left ventricle). The bulge resembles a bubble in the wall of a tire—and like a bad tire, the affected artery may suffer a catastrophic "blowout," or rupture. The most dangerous aneurysms are those involving arteries of the brain, where they cause strokes, and of the aorta, where a rupture will cause fatal bleeding in a matter of minutes.

Aneurysms are normally caused by chronic high blood pressure, although any trauma or infection, such as syphilis, that weakens vessel walls can lead to an aneurysm. In addition, at least some aortic aneurysms have been linked to inherited disorders, such as *Marfan's syndrome,* which weakens connective tissues in vessel walls.

It is not known whether other genetic factors are involved in the development of other types of aneurysms. Fibrous scar tissue replacing muscle tissue damaged in a heart attack may result in a left ventricular aneurysm. The damaged area doesn't contract and, if large enough, reduces cardiac output. In addition, the area is a site at which a blood clot may form, posing a risk of embolism.

Most aneurysms form gradually as vessel walls become less elastic. When a weak point develops, the arterial pressures distort the wall, creating an aneurysm. Unfortunately, because many aneurysms are painless, they are likely to go undetected until they leak or rupture.

When aneurysms are detected by ultrasound or other scanning procedures, the risk of rupture can sometimes be estimated on the basis of their size. For example, an aortic aneurysm larger than 6 cm has a 50 percent chance of rupturing within 10 years. Depending on size and location, aneurysm treatment includes reducing hypertension if present, watchful waiting to see if it enlarges, and when the risk of rupture exceeds the risk of repair, surgery. A large aneurysm in an accessible area, such as the abdomen, may be surgically removed and the vessel repaired. Figure 44● shows a large aortic aneurysm and the steps involved in its surgical repair with a synthetic patch.

■ PROBLEMS WITH VENOUS VALVE FUNCTION   *HA p. 571*

One of the consequences of aging is an increase in the fragility of connective tissues throughout the body. Blood vessels are no exception: With age, the walls of veins begin to sag. This change generally affects the superficial veins of the legs first, because at these locations gravity opposes blood flow. The situation is aggravated by a lack of exercise and by an occupation requiring long hours of standing or sitting. When there is little muscular activity in the leg to help keep the blood moving, venous blood pools on the proximal (heart) side of each valve. As the venous walls are distorted, the valves become

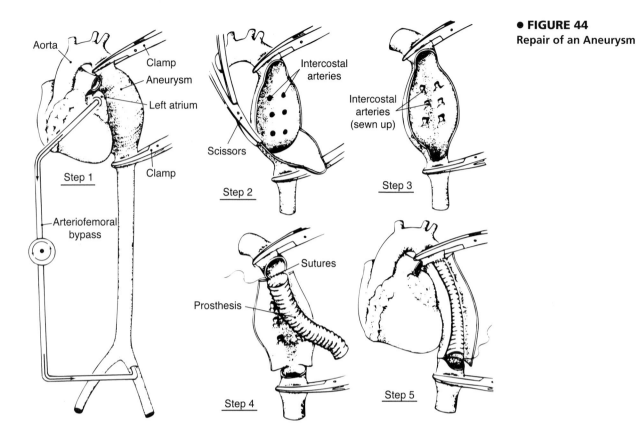

less effective, and gravity can then pull blood back toward the capillaries. This pulling further impedes normal blood flow, and the veins become distended. The sagging, swollen vessels are called **varicose** (VAR-i-kos) **veins**. Though superficial varicose veins are relatively harmless, they may be painful and are unsightly. Surgical procedures are sometimes used to remove or constrict them.

Varicose veins are not limited to the extremities. Another common site involves a network of veins in the walls of the anus. Pressures within the abdominopelvic cavity rise dramatically when the abdominal muscles are tensed. Straining to force defecation can force blood into these veins, and repeated incidents leave them permanently distended. These distended veins, known as **hemorrhoids** (HEM-ō-roydz), can be uncomfortable and, in severe cases, extremely painful. Hemorrhoids and varicose veins are often associated with pregnancy, as a result of changes in blood flow and abdominal pressures on the inferior vena cava. Minor hemorrhoids can be treated by the topical application of drugs that promote the contraction of smooth muscles within the venous walls. More severe cases may benefit from the surgical removal or destruction of the distended veins.

■ CHECKING THE PULSE AND BLOOD PRESSURE *HA p. 575*

You can feel your pulse at any of the large or medium-sized arteries. The usual procedure involves using your fingertips to flatten an artery against a relatively solid mass, preferably a bone. When the vessel is compressed, you feel your pulse as an

episodic pressure against your fingertips. The inside of the wrist is commonly used, because the *radial artery* can easily be pressed against the distal portion of the radius (Figure 45a●). Other accessible arteries include the *external carotid, brachial, temporal, facial, femoral, popliteal, posterior tibial,* and *dorsalis pedis arteries.* Firm pressure exerted at these locations, called **pressure points**, can reduce arterial bleeding distal to the site.

Blood pressure not only forces blood through the circulatory system, but also pushes outward against the walls of the containing vessels, just as air pushes against the walls of an inflated balloon. As a result, we can measure blood pressure indirectly by determining how forcefully the blood presses against the vessel walls.

The instrument used to measure blood pressure is called a **sphygmomanometer** (sfig-mō-ma-NOM-e-ter; *sphygmos,* pulse + manometer, device for measuring pressure). An inflatable cuff is placed around the arm in such a position that its inflation compresses the brachial artery (Figure 45b●). A stethoscope is placed over the artery distal to the cuff, and the cuff is then inflated. A tube connects the cuff to a pressure gauge that reports the cuff pressure (in mm Hg). Inflation continues until the cuff pressure is roughly 30 mm Hg above the pressure sufficient to collapse the brachial artery completely, stop the flow of blood, and eliminate the sound of the pulse.

The air is then slowly let out of the cuff. When the pressure in the cuff falls below systolic pressure, blood can again enter the artery. At first, blood enters only at peak systolic pressures, and the stethoscope picks up the sound of blood pulsing through the artery. As the pressure falls further, the sound changes, because the artery is remaining open for

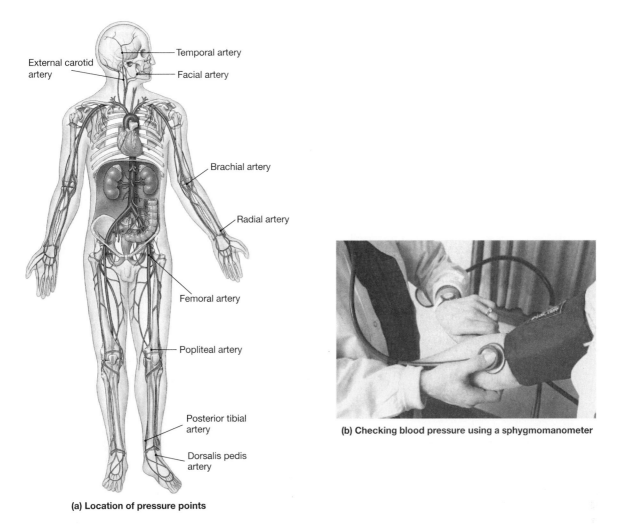

External carotid artery

Temporal artery

Facial artery

Brachial artery

Radial artery

Femoral artery

Popliteal artery

Posterior tibial artery

Dorsalis pedis artery

(a) Location of pressure points

(b) Checking blood pressure using a sphygmomanometer

● **FIGURE 45**
Pressure Points and Blood Pressure Measurement. (a) Pressure points used to monitor the pulse or control peripheral bleeding. (b) Using a sphygmomanometer to measure blood pressure.

longer and longer periods. When the cuff pressure falls below diastolic pressure, blood flow becomes continuous and the sound of the pulse becomes muffled or disappears. Thus, the pressure at which the pulse appears corresponds to the peak systolic pressure; when the pulse fades, the pressure has reached diastolic levels. The distinctive sounds heard during this test, called *sounds of Korotkoff* (sometimes spelled *Korotkov* or *Korotkow*), are produced by turbulence as blood flows past the constricted portion of the artery.

■ THE CAUSES AND TREATMENT OF CEREBROVASCULAR DISEASE

HA p. 579

Most symptoms of cerebrovascular disease appear when atherosclerosis reduces the circulatory supply to the brain. If the blood flow to a portion of the brain is completely shut off, a *cerebrovascular accident (CVA),* or *stroke,* occurs. The most common causes of strokes include **cerebral**

thrombosis (clot formation at a plaque), **cerebral embolism** (drifting blood clots, fatty masses, or air bubbles), and **cerebral hemorrhage** (rupture of a blood vessel, often at the site of an aneurysm). The observed symptoms and their severity vary with the vessel involved and the location of the blockage.

If the circulatory blockage disappears in a matter of minutes, the effects are temporary and the condition is called a **transient ischemic attack (TIA).** A TIA typically indicates that cerebrovascular disease exists, so preventive measures can be taken to forestall more serious incidents. For example, taking aspirin each day slows blood clot formation in patients who experience TIAs and thereby reduces the risks of cerebral thrombosis and cerebral embolism.

If the blockage persists for a longer period, neurons die and the area degenerates. Stroke symptoms are initially exaggerated by swelling and distortion of the injured neural tissues; if the individual survives, in many cases brain function gradually improves. The management and treatment of strokes is difficult. The surgical removal of the offending clot or blood mass caused by a hemorrhagic stroke may be attempted, but the results vary. Recent progress in the emergency treatment of cerebral thromboses and cerebral embolisms has involved the administration of clot-dissolving enzymes such as t-PA. The best results are obtained if the enzymes are administered within an hour after the stroke, although they may still be of use up to 24 hours after. Early recognition of symptoms, with rapid CT scans to distinguish between an ischemic stroke and a hemorrhagic stroke (where anticoagulation would be harmful) is vital. Subsequent treatment of thrombotic or embolic strokes involves clot-dissolving enzymes followed by anticoagulant therapy, typically with heparin (for one to two weeks)

followed by coumadin (for up to one year), to prevent further clot formation. Neither of these treatments is as successful as preventive surgery, in which plaques that are partially blocking one or both of the carotid arteries in the neck are removed before a stroke can occur.

The very best solution is to prevent or restrict plaque formation by controlling the risk factors involved. As public-health knowledge and education about risk factors for strokes have spread, and as treatments for hypertension and high cholesterol have improved, death rates from strokes have dropped 70 percent, from 88 per 100,000 population in 1950 to roughly 60 per 100,000 in 2004.

22

THE LYMPHATIC SYSTEM AND IMMUNITY

The lymphatic system consists of the fluid *lymph,* a network of *lymphatic vessels,* specialized cells called *lymphocytes,* and an array of *lymphoid tissues* and *lymphoid organs* throughout the body. This system has three major functions: (1) to protect the body through the *immune response,* (2) to transport fluid from the interstitial spaces to the bloodstream, and (3) to help distribute hormones, nutrients, and wastes.

The immune response produced by activated lymphocytes is responsible for the detection and destruction of pathogenic microorganisms, wayward cells or toxic substances that can cause illness. For example, viruses, bacteria, and tumor cells are usually recognized and eliminated by cells of the lymphatic system. Immunity is the specific resistance to disease, and all the cells and tissues involved with the production of immunity are considered part of an *immune system.* Note that whereas the lymphatic system is an anatomically distinct system, the immune system is a physiological system that includes the lymphatic system as well as components of the integumentary, cardiovascular, respiratory, digestive, and other systems.

HISTORY AND PHYSICAL EXAMINATION AND THE LYMPHATIC SYSTEM   *HA p. 604*

Individuals with lymphatic system disorders experience a variety of signs and symptoms. The pattern observed depends on whether the problem affects the immune functions or the circulatory functions of the system (Figure 46●). Important symptoms and signs include the following (some will be discussed in more detail in a later section):

- **Recurrent infections** occur for a variety of reasons including multiple exposures, anatomic susceptibility, and inadequate immune response. Tonsillitis and *adenoiditis* are common recurrent infections in children. Serious infections throughout the body are common among people with immunodeficiency disorders such as AIDS (acquired immune deficiency disease) and *severe combined immunodeficiency disease* (SCID). When the immune response is inadequate, the individual may not be able to overcome even a minor infection. Infections of the respiratory system are very common; recurring, chronic gastrointestinal infections can produce chronic diarrhea. The pathogens involved often do not affect people with a normal immune response. Infections are also a problem for individuals who take medications that suppress the immune response. Examples of immunosuppressive drugs include anti-inflammatories such as the *corticosteroids* (for example, *Prednisone*) as well as more-specialized drugs such as *methotrexate* and *cyclosporine.*

- Infections are typically characterized by *enlarged lymph nodes.* Lymph nodes also enlarge in cancers of the lymphatic system,

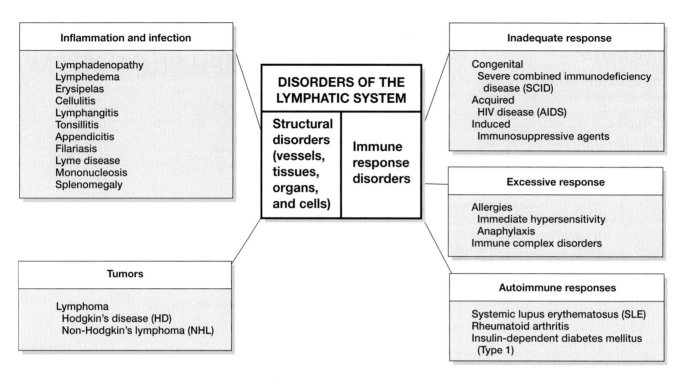

● FIGURE 46
Disorders of the Lymphatic System

such as *lymphoma*, or when primary tumors in other tissues have metastasized to regional lymph nodes. The status of regional lymph nodes is therefore important in the diagnosis and treatment of many cancers. The onset and duration of swelling; the size, texture, and mobility of the nodes; the number of affected nodes; and the degree of tenderness all assist in diagnosis. For example, nodes containing cancer cells tend to be large, hard, fixed in place, and nontender. On palpation,

these nodes feel like dense masses rather than individual lymph nodes. In contrast, infected lymph nodes tend to be large, freely mobile, and very tender.

• **Lymphangitis** consists of *erythematous* (red) streaks on the skin that may develop with an inflammation of superficial lymph vessels. Lymphangitis commonly occurs in the limbs, with reddened streaks that start at an infection site and

TABLE 24   Representative Blood Tests for Disorders of the Lymphatic System and Immunity

Laboratory Test	Significance of Normal Values	Abnormal Values
Complete Blood Count	See laboratory tests for blood disorders, Table 21b,c	
WBC count	Adults: 5000–10,000/mm³ (High-risk values: >30,000/mm³ or <2,500/mm³)	Increased in chronic and acute bacterial infections, tissue death (burns), leukemia, parasitic diseases, and stress; decreased in some viral diseases, aplastic and pernicious anemias, systemic lupus erythematosus (SLE), and use of some medications
Differential WBC count	Neutrophils: 50–70%	Increased in acute bacterial infection, myelocytic leukemia, and stress; decreased in aplastic and pernicious anemias, viral infections, radiation treatment, and use of some medications
	Lymphocytes: 20–30%	Increased in chronic infections, lymphocytic leukemia, infectious mononucleosis, and viral infections; decreased in radiation treatment, AIDS, and corticosteroid therapy
	Monocytes: 2–8%	Increased in chronic inflammation, viral infections, and tuberculosis; decreased in aplastic anemia and corticosteroid therapy
	Eosinophils: 2–4%	Increased in allergies, parasitic infections, and some autoimmune disorders; decreased with steroid therapy
	Basophils: 0.5–1%	Increased in inflammatory processes and during healing; decreased in hypersensitivity reactions and corticosteroid therapy

23

progress toward the trunk. Before the linkage to the lymphatic system was known, this sign was called "blood poisoning."

- **Splenomegaly** is an enlargement of the spleen that can result from acute infections such as *endocarditis* and *mononucleosis,* chronic infections such as *malaria,* and cancers such as *leukemia.* The spleen can be examined through palpation or percussion to detect splenic enlargement. In percussion, an enlarged spleen produces a distinctive dull sound in the left upper quadrant of the abdomen. The patient's history may also reveal important clues. For example, an individual with an enlarged spleen may report a feeling of fullness after eating a small meal, probably because the enlarged spleen limits gastric expansion.

- Prolonged *weakness* and *fatigue* typically accompany immunodeficiency disorders, *Hodgkin's disease* and other *lymphomas,* and *mononucleosis.*

- **Skin lesions,** such as hives (urticaria), or contact dermatitis can develop during allergic reactions. Immune responses to a variety of allergens, including animal dander, pollen, dust, medications, and some foods, may cause such lesions.

- **Respiratory problems,** including rhinitis and wheezing (caused by bronchospasm), may accompany the allergic response to allergens such as pollen, hay, dust from many sources, and mildew. *Bronchospasm* occurs when the bronchial smooth muscles contract and constrict the airways, making breathing difficult. Bronchospasm, which often accompanies severe allergic or asthmatic attacks, may be a response to the appearance of antigens or irritants within the respiratory passageways, or may accompany more widespread exposure to antigens.

When lymphatic circulatory functions are impaired, the most common sign is *lymphedema,* a tissue swelling caused by the buildup of interstitial fluid. Lymphedema can result from trauma and scarring of lymphatic vessels or from a lymphatic blockage due to a tumor or infections. Lymphedema can also be due to congenital malformations.

Diagnostic procedures and laboratory tests used to detect disorders of the lymphatic system are detailed in Tables 24 and 25.

TABLE 25 Representative Diagnostic Tests for Disorders of the Lymphatic System and Immunity

Diagnostic Procedure	Method and Results	Representative Uses
Skin Tests		
Hypersensitivity response	Antigens are extracted in a sterile, diluted form; examples include animal dander, pollen, certain foods, medications (especially penicillin) that cause hypersensitivities, and insect venom	Identifies specific antigens that may cause sensitivity reaction (allergy)
Prick test	A small amount of antigen is applied as a prick to the skin. Positive test: Erythema, hardening, and swelling appear around puncture area; usually, the area affected is measured and must be of minimal size to qualify for a positive test	As above; also a screening test for tuberculosis exposure, but not as accurate as intradermal test
Intradermal test	Antigen is injected into the skin to form a 1–2 mm bleb. Positive test: Within 15 minutes to two days, a reddened wheal is produced that is larger than 5 mm in diameter and larger than those seen in the control group	As above
Patch test	A patch is impregnated with the antigen and applied to the surface of the skin. Positive test: The patch provokes an allergic response, usually over several hours to several days.	As above (nickel allergy may be inadvertently tested for by wearing jewelry or watches)
Tuberculin skin test	Injection of tuberculin protein into the dermis. Positive test: Red, hardened area >710 mm wide appears around injection site 48–72 hours later.	Indicates presence of antibodies to the organism that causes tuberculosis (infection may be active or dormant). May be repeated within a week to increase accuracy.
Nuclear, CT, or ultrasound	Scan images created by radiation from radioisotopes, x-rays, or ultrasound reveal position, shape, and size of spleen	Detects abscesses, tumors of the liver or spleen, and infarcts of liver or splenic tissue
Biopsy of lymphoid tissue	Surgical excision of suspicious lymphoid tissue for pathological examination	Determines potential malignancy and staging of cancers in progress. Usually lymph nodes adjacent to tumors.
Lymphangiography	Dye is injected into a lymphatic vessel in the distal portion of a limb and travels to nodes; x-rays are taken as the dye accumulates in the lymph nodes. Technically difficult.	Identifies and determines the stages of Hodgkin's disease and other lymphomas and detects the cause of lymphedema

23

DISORDERS OF THE LYMPHATIC SYSTEM

Disorders of the lymphatic system that affect the immune response can be sorted into three general categories, as diagrammed in Figure 46●:

1. **Disorders resulting from an insufficient immune response.** This category includes immunodeficiency disorders, such as AIDS and SCID; the immature immune system of infants; and the decreased immune response in the elderly. Individuals with depressed immune defenses can develop life-threatening diseases caused by microorganisms that are harmless to other individuals.

2. **Disorders resulting from an excessive immune response.** Conditions such as allergies and *immune complex disorders* can result from an immune response that is out of proportion to the size of the stimulus.

3. **Disorders resulting from an inappropriate immune response.** Autoimmune disorders result when normal tissues are mistakenly attacked by T cells or the antibodies produced by activated B cells. For instance, in *thrombocytopenic purpura*, the body forms antibodies against its own platelets. We will consider representative disorders from each of these categories.

■ LYMPHEDEMA   *HA p. 607*

Lymphedema is an accumulation of fluid within a tissue due to inadequate lymphatic drainage. Temporary lymphedema can result from tight clothing. Chronic lymphedema can result from the formation of scar tissue owing to repeated bacterial infections or from surgery that cuts or removes lymphatic vessels. Axillary or inguinal lymph node removal (sometimes required during surgery for breast cancer or malignant melanoma) or radiation treatment can lead to lymphedema of the adjacent limb (Figure 47●). Lymphedema can also result from parasitic infections. For example, in **filariasis** (fil-a-RĪ-a-sis), immature stages of a parasitic roundworm (generally *Wuchereria bancrofti*) are transmitted by mosquitoes or black flies. The adult

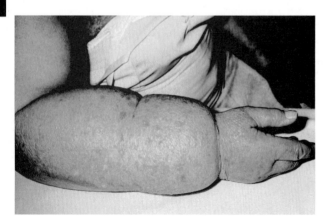

● **FIGURE 47**
Lymphedema

worms accumulate within lymphatic vessels and lymph nodes. Repeated scarring of the passageways eventually blocks lymphatic drainage, producing extreme lymphedema with permanent distension of tissues. The limbs or external genitalia typically become grossly distended; this parasite-induced lymphedema is called **elephantiasis** (el-e-fan-TĪ-a-sis).

Therapy for chronic lymphedema consists of treating infections with antibiotics and (when possible) reducing the swelling. One treatment involves the application of elastic wrappings that squeeze the tissue. This external compression elevates the hydrostatic pressure of the interstitial fluids and opposes the entry of additional fluid from the capillaries.

■ SCID   *HA p. 610*

In **severe combined immunodeficiency disease (SCID)**, the individual fails to develop either cellular or humoral immunity. Lymphocyte populations are reduced, and normal B and T cells are not present. Patients with SCID are unable to provide an immune defense, and even a mild infection can prove fatal.

Total isolation offers protection at great cost and with severe restrictions on the individual's lifestyle. Bone marrow transplants from compatible donors, usually a close relative, have been used to colonize lymphatic tissues with functional lymphocytes.

The most famous SCID patient was the "bubble boy," David. David was kept in physical isolation until age 12, when he received a bone marrow transplant. Before the donor marrow cells established a functional immune system, David died of cancer. Although technology had protected him from external pathogens, without immune surveillance he had no defense against threats from within.

For these patients, whose SCID reflects the lack of a specific enzyme (adenosine deaminase), treatment with the enzyme itself has helped. In 1990, genetic engineering techniques were used to insert normal genes for this enzyme into the lymphocytes of two children suffering from this form of SCID. The experiment was successful, and the children have regained and (thus far) retained normal immune function with periodic infusions of the gene.

■ AIDS   *HA p. 610*

Acquired immune deficiency syndrome (AIDS), or *late-stage HIV disease*, develops after infection by an RNA virus known as the *human immunodeficiency virus (HIV)*. There are at least three types of HIV, designated *HIV-1*, *HIV-2*, and *HIV-3*. Most people with HIV in the United States are infected with HIV-1; HIV-2 infections are most common in West Africa. Because most of those infected with HIV-1 eventually develop AIDS, but not all individuals infected with HIV-2 do so, HIV-2 may be a less dangerous virus. The distribution and significance of HIV-3 infection remain to be determined.

The virus responsible for HIV/AIDS selectively infects helper T cells. Infection of T cells by itself impairs the immune response, because these cells play a central role in coordinating cellular and humoral responses to antigens. To make matters worse, suppressor T cells are relatively unaffected by the virus, and over time the excess of suppressing factors "turns off" the normal immune response. Circulating antibody levels decline, cellular immunity is reduced, and the body is left without defenses against a wide variety of microbial invaders.

Because the immune function is so reduced, ordinarily harmless pathogens can initiate lethal infections, known as *opportunistic infections.* People with AIDS are especially prone to lung infections and pneumonia, often caused by *Pneumocystis carinii* or other fungi, and to a wide variety of bacterial, viral, and protozoan diseases. Because immune surveillance is also depressed, the risk of cancer increases. One of the most common cancers seen in people with AIDS, though very rare in normal individuals, is **Kaposi's sarcoma**, characterized by rapid cell division in endothelial cells of cutaneous blood vessels. This cancer has been linked to repeated infection with a herpes virus, *Herpes type 8.*

Infection with HIV occurs through intimate contact with the body fluids of infected individuals. Although all body fluids carry the virus, the major routes of transmission involve contact with blood, semen, or vaginal secretions. Most people with AIDS become infected through sexual contact with an HIV-infected person (who may *not necessarily* exhibit the clinical symptoms of AIDS). The next largest group consists of intravenous drug users who share blood-contaminated needles. A relatively small number of individuals have become infected with the virus after receiving a transfusion of contaminated blood or blood products. Screening of donated blood has now reduced the risk of HIV infection from donated blood to perhaps one in 40,000 transfusions. Finally, infants are born with the disease, having acquired it from infected mothers prior to or at delivery. Treatment of pregnant HIV-infected women with antiviral drugs can reduce the risk of maternal-fetal transmission of the virus through breast-milk.

The best defense against AIDS consists of avoiding sexual contact with infected individuals. All forms of sexual intercourse carry the potential risk of viral transmission. The use of synthetic condoms greatly reduces the chance of infection (although it does not completely eliminate it). Condoms that are not made of synthetic materials are effective in preventing pregnancy but do not block the passage of viruses.

Clinical symptoms of AIDS may not appear for 5–10 years or more after infection, and when they do appear they are often mild, consisting of *lymphadenopathy* and chronic but nonfatal infections. After a variable period of time, full-blown AIDS develops. AIDS is almost invariably fatal, but the recent use of multiple antiviral drugs in combination has dramatically improved survival in up to 50 percent of patients. Drug resistance is a concern, but hope exists that HIV may be treatable as a chronic infection, if not cured outright. Deaths in the United States have already climbed above 350,000, and estimates of the number of infected individuals range up to 1 million. The numbers worldwide are even more frightening: The World Health Organization estimated in 2003 that more than 40 million people may be infected, with 5–6 million new infections and 3 million deaths each year.

Despite intensive efforts, a vaccine has yet to be developed that will provide immunity from HIV infection. However, the survival rate for people with AIDS has been steadily increasing because new drugs are available that slow the progress of the disease, and improved antibiotic therapies help combat secondary infections. This combination is extending their life span while the search for more-effective treatment continues.

■ INFECTED LYMPHOID NODULES
HA p. 611

Lymphoid nodules can be overwhelmed by a pathogenic invasion. The result is a localized infection accompanied by regional swelling and discomfort. The tonsils are a first line of defense against infection of the pharyngeal walls.

An individual with **tonsillitis** has infected tonsils. Symptoms include a sore throat, high fever, and often leukocytosis (an abnormally high white blood cell count). The affected tonsils (normally, the pharyngeal tonsils) become swollen and inflamed, sometimes enlarging enough to partially block the entrance to the trachea. Breathing then becomes difficult or, in severe cases, impossible. If the infection proceeds, abscesses may develop within the tonsillar or peritonsillar tissues. Bacteria may enter the bloodstream by passing through the lymphatic capillaries and vessels to the venous system.

In the early stages, antibiotics may control the infection, but once abscesses have formed, the best treatment involves surgical drainage of the abscesses and **tonsillectomy**, the removal of the tonsil. Tonsillectomy was once highly recommended prior to the development of antibiotics, in order to prevent recurring tonsillar infections. The procedure does reduce the incidence and severity of subsequent infections, but questions have arisen about the overall cost to the individual, especially now that antibiotics are available to treat severe infections.

Appendicitis generally follows an erosion of the epithelial lining of the vermiform appendix. Several factors may be responsible for the initial ulceration—notably, bacterial or viral pathogens. Bacteria that normally inhabit the lumen of the large intestine then cross the epithelium and enter the underlying tissues. Inflammation occurs, and the opening between the vermiform appendix and the rest of the intestinal tract may become constricted. Mucus secretion and pus formation accelerate, and the organ becomes increasingly distended. Eventually, the swollen and inflamed appendix may rupture, or *perforate.* If it does, bacteria will be released into the warm, dark, moist confines of the peritoneal space, where they can cause a life-threatening peritonitis. The most effective treatment for appendicitis is the surgical removal of the organ, a procedure known as an **appendectomy**.

23

■ LYMPHOMAS   *HA p. 614*

Lymphomas are malignant tumors consisting of cancerous lymphocytes or lymphocytic stem cells. About 61,000 cases of lymphoma are diagnosed in the United States each year. There are many types of lymphoma. One form, called **Hodgkin's disease (HD)**, accounts for roughly 13 percent of all lymphoma cases. Hodgkin's disease most commonly strikes individuals at age

15–35 years or over age 50. The reason for this pattern of incidence is unknown. Although the cause of the disease is uncertain, an infectious agent (probably a virus) may be involved.

Other types of lymphoma are usually grouped together under the heading of **non-Hodgkin's lymphoma (NHL).** These lymphomas are extremely diverse. More than 85 percent of NHL cases are associated with chromosomal abnormalities of the tumor cells, typically involving *translocations,* in which sections of chromosomes have been swapped from one chromosome to another. The shifting of genes from one chromosome to another interferes with the normal regulatory mechanisms, and the cells grow uncontrollably (become cancerous). The nature of the cancer depends on which of the many types of lymphocyte are affected. A combination of inherited and environmental factors may be responsible for specific translocations. For example, one form, called **Burkitt's lymphoma**, develops only after genes from chromosome 8 have been translocated to chromosome 14. (There are at least three variations.) Burkitt's lymphoma normally affects male children in Africa and New Guinea who have been infected with the *Epstein–Barr virus (EBV)*. This highly variable virus is also responsible for infectious mononucleosis (described later).

EBV infects B cells, but under normal circumstances most infected cells are destroyed by the immune system. After recovery, perhaps one in a million memory B cells remain as infected reservoirs of the virus. This virus affects many people, and childhood exposure generally produces lasting immunity. Children who develop Burkitt's lymphoma may have a genetic susceptibility to EBV infection. In addition, the presence of another illness, such as malaria, can weaken the immune system enough that sporadically occurring lymphoma cells are not recognized, and therefore not destroyed.

The first symptom usually associated with any lymphoma is a painless enlargement of lymph nodes. The involved nodes have a firm, rubbery texture. Because the nodes are pain free, the condition is typically overlooked until it has progressed far enough for secondary symptoms to appear. For example, patients seeking help for recurrent fevers, night sweats, gastrointestinal or respiratory problems, or weight loss may be unaware of any underlying lymph node changes. In the late stages of the disease, symptoms can include liver or spleen enlargement, central nervous system dysfunction, pneumonia, a variety of skin conditions, and anemia.

In planning treatment, clinicians consider the histological structure of the nodes and the stage of the disease. In a biopsy, the node is described as *nodular* or *diffuse.* A nodular node retains a semblance of normal structure, with follicles and germinal centers. In a diffuse node, the interior of the node has changed and the follicle structure has broken down. In general, nodular lymphomas progress more slowly than do the diffuse forms, which tend to be more aggressive. Conversely, nodular lymphomas are more resistant to treatment and are more likely to recur even after remission has been achieved.

The most important factor influencing which treatment is selected is the stage of the disease. Table 26 presents a simplified classification of lymphomas in terms of the stage. When the condition is diagnosed early (stage I or stage II), localized therapies may be effective. For example, the cancerous node(s) may be surgically removed and the region(s) irradiated to kill residual cancer cells. Success rates are very high when a lymphoma is detected in an early stage. For Hodgkin's disease, localized radiation can produce remission that lasts 10 years or more in more than 90 percent of patients. The treatment of localized NHL is somewhat less effective. The five-year remission rates average 60–80 percent for all types; success rates are higher for nodular forms than for diffuse forms.

Although these results are encouraging, few lymphomas are diagnosed in the early stages. For example, only 10–15 percent of NHL patients are diagnosed at stage I or stage II. For lymphomas at stages III and IV, most treatments involve chemotherapy. Combination chemotherapy, in which two or more drugs are administered simultaneously, is the most effective treatment.

23

TABLE 26	Cancer Stages in Lymphoma
Stage I	Involvement of a single node or region (or of a single extranodal site). *Typical treatment:* surgical removal or localized irradiation (or both); in slowly progressing forms of non-Hodgkin's lymphoma, treatment may be postponed indefinitely.
Stage II	Involvement of nodes in two or more regions (or of an extranodal site and nodes in one or more regions) on the same side of the diaphragm. *Typical treatment:* surgical removal and localized irradiation that includes an extended area around the cancer site (the *extended field*).
Stage III	Involvement of lymph node regions on both sides of the diaphragm. This is a large category that is subdivided on the basis of the organs or regions involved. For example, in stage III, the spleen contains cancer cells. *Typical treatment:* combination chemotherapy, with or without radiation; radiation treatment may involve irradiating all of the thoracic and abdominal nodes, plus the spleen (*total axial nodal irradiation*, or *TANI*).
Stage IV	Widespread involvement of extranodal tissues above and below the diaphragm; involvement of bone marrow. *Treatment:* highly variable, depending on the circumstances. Combination chemotherapy is always used and may be combined with whole-body irradiation. The "last resort" treatment involves massive chemotherapy followed by a bone marrow transplant.

A **bone marrow** or **hematologic stem cell transplant** is a treatment option for acute, late-stage lymphoma. When suitable donor marrow stem cells are available, the patient receives whole-body irradiation, chemotherapy, or some combination of the two sufficient to kill tumor cells throughout the body. Unfortunately, this treatment also destroys normal bone marrow stem cells. Donor stem cells are then infused. Within two weeks, the donor cells colonize the bone marrow and begin producing red blood cells, granulocytes, monocytes, megakaryocytes, and lymphocytes.

Potential complications of stem cell transplantation include the risk of infection and bleeding while the donor marrow is becoming established. The immune cells of the donor may also attack the tissues of the recipient, a response called **graft-versus-host disease (GVHD)**. For a person with stage I or stage II lymphoma without bone marrow involvement, bone marrow can be removed and stored (frozen) for more than 10 years. If other treatment options fail or if the person comes out of remission at a later date, an autologous marrow transplant can be performed. This procedure eliminates both the need for finding a matched donor and the risk of GVH disease.

■ BREAST CANCER   *HA p. 614*

Breast cancer is the primary cause of death for women between ages 35 and 45, but the disease actually becomes more common after age 50. There will be an estimated 39,800 deaths in the United States from breast cancer in 2004, and roughly 212,500 new cases reported. An estimated 12 percent of women in the United States will develop breast cancer at some point in their lifetimes. The incidence is highest among Caucasians, somewhat lower in African Americans, and the lowest in Asians and Native Americans. Notable risk factors include (1) a family history of breast cancer, (2) a pregnancy after age 30, and (3) early menarche (first menstrual period) or late menopause (last menstrual period). Despite repeated studies (and rumors), there are no proven links between oral contraceptive use, estrogen therapy, breast feeding, fat consumption, or alcohol use and breast cancer. It appears likely that multiple factors are involved; most women never develop breast cancer, even women in families with a history of this disease.

Early detection of breast cancer is the key to reducing the death rate. *Most breast cancers are found through self-examination,* but the use of clinical screening techniques has increased in recent years. **Mammography** involves the use of x-rays to examine breast tissues; the radiation dosage can be restricted because only soft tissues must be penetrated. This procedure gives the clearest picture of conditions within the breast tissues. Ultrasound can provide some information, but the images lack the detail of standard mammograms.

For treatment to be successful the cancer must be identified while it is still relatively small and localized. Once it has grown larger than 2 cm (0.78 in.) the chances for long-term survival worsen. A poor prognosis also follows if the cancer cells have spread through the lymphatic system to the axillary lymph nodes. If the nodes are not yet involved, the chances of five-year survival are about 82 percent, but if four or more nodes are involved the survival rate drops to 21 percent.

Treatment of breast cancer begins with the removal of the tumors and sampling of the axillary lymph nodes. Because the cancer cells usually begin spreading before the condition is diagnosed, surgical treatment involves the removal of part or all of the affected breast.

- In a **segmental mastectomy**, or "lumpectomy," only a portion of the breast is removed. This, combined with irradiation, is the preferred treatment for small, localized tumors.

- In a **total mastectomy**, the entire breast is removed, but other tissues are left intact.

- In a **modified radical mastectomy**, the most common operation, the breast and nodes are removed but the muscular tissue remains intact.

A combination of chemotherapy, radiation treatments, and hormone treatments may be used to supplement the surgical procedures.

■ DISORDERS OF THE SPLEEN
HA p. 619

The spleen responds like a lymph node to infection, inflammation, or invasion by cancer cells. The enlargement that follows is called **splenomegaly** (splen-ō-MEG-a-lē; *megas,* large); the spleen can rupture under these conditions. One relatively common condition that causes temporary splenomegaly is **mononucleosis**. This condition, also known as the "kissing disease," results from acute infection by the Epstein–Barr virus. In addition to enlargement of the spleen, symptoms of mononucleosis may include fever, sore throat with tonsillitis, widespread swelling of lymph nodes, increased numbers of atypical lymphocytes in the blood, and, after recovery, the presence of circulating antibodies to the virus. The condition typically affects young adults (age 15–25) in the spring or fall. Treatment is symptomatic, as no drugs are effective against this virus. The most dangerous aspect of the disease is the risk of rupturing the enlarged spleen, which becomes fragile. Patients are therefore cautioned against heavy exercise and other activities that increase abdominal pressures. If the spleen does rupture, severe hemorrhaging can occur. Death may follow unless a transfusion and an immediate splenectomy are performed.

An individual whose spleen is missing or nonfunctional has **hyposplenism** (hī-pō-SPLEN-izm) Hyposplenism usually does not pose a serious problem. Hyposplenic individuals, however, are more prone to some bacterial infections, including infection by *Streptococcus pneumoniae,* than are individuals with normal spleens, so immunization against *S. pneumoniae* is recommended. In **hypersplenism,** for unknown reasons the spleen becomes overactive; the increased phagocytic activities lead to anemia (low RBC count), leukopenia (low WBC count), and thrombocytopenia (low platelet count). Splenectomy is the only known cure for hypersplenism.

23

■ IMMUNE DISORDERS   *HA p. 619*

Because the immune response is so complex, there are many opportunities for things to go wrong. A great variety of clinical conditions may result from disorders of the immune functions associated with the lymphatic system. In an **immunodeficiency disease** either the immune system fails to develop normally or the immune response is blocked in some way. Examples of immunodeficiency diseases include ***severe combined immunodeficiency* disease (SCID)** and *acquired immune deficiency syndrome* (AIDS). SCID and AIDS are discussed earlier in this section.

Autoimmune disorders develop when the lymphatic system mistakenly targets normal body cells and tissues. Lymphocytes usually recognize and ignore the antigens normally found in the body. The recognition system can malfunction, however, and when it does the activated B cells begin to manufacture antibodies against other cells and tissues. The trigger may be a reduction in suppressor T cell activity, excessive stimulation of helper T cells, tissue damage that releases large quantities of antigens, viral or bacterial toxins, or some combination of factors.

The symptoms produced depend on the identity of the antigen attacked by these misguided antibodies, called **autoantibodies**. Several conditions described in earlier chapters are autoimmune disorders. For example, *rheumatoid arthritis* occurs when autoantibodies form immune complexes within connective tissues, especially around the joints. Many other autoimmune disorders appear to be cases of mistaken identity. For example, proteins associated with the measles, Epstein–Barr, influenza, and other viruses contain amino acid sequences that are similar to those of myelin proteins. As a result, antibodies that target these viruses may also attack myelin sheaths. This mechanism accounts for the neurologic complications that sometimes follow a vaccination or viral infection. Table 27 lists representative autoimmune disorders and the targets of the antibodies responsible for these conditions.

Allergies are inappropriate or excessive immune responses to antigens. The sudden increase in cellular activity or antibody titers can have a number of unpleasant side effects. For example, neutrophils or cytotoxic T cells may destroy normal cells while attacking the antigen, or the antibody-antigen complex may trigger a massive inflammatory response. Antigens that trigger allergic reactions are often called **allergens**.

■ LYME DISEASE   *HA p. 619*

In November 1975, the town of Lyme, Connecticut, experienced an epidemic of adult arthritis and juvenile arthritis. Between June and September, 59 cases were reported, 100 times the statistical average for a town of its size. Symptoms were unusually severe: Affected people experienced chronic fever and a prominent rash that began as a red bull's-eye centered on what appeared to be an insect bite (Figure 48●). Joint degeneration and nervous system problems often occurred somewhat later. It took almost two years to track down the cause of this condition, now known as **Lyme disease**.

TABLE 27   Autoimmune Disorders

Disorder	Antibody Target
Psoriasis	Epidermis of skin
Vitiligo	Melanocytes of skin
Rheumatoid arthritis	Connective tissues at joints
Myasthenia gravis	Synaptic ACh receptors
Multiple sclerosis	Myelin sheaths of axons
Addison's disease	Adrenal cortex
Graves' disease	Thyroid follicles
Hypoparathyroidism	Chief cells of parathyroid
Thyroiditis	Thyroid-binding globulin
Type I diabetes mellitus	Beta cells of pancreatic islets
Rheumatic fever	Myocardium and heart valves
Systemic lupus erythematosus	DNA, cytoskeletal proteins, other tissue components
Thrombocytopenic purpura	Platelets
Pernicious anemia	Parietal cells of stomach
Chronic hepatitis	Hepatocytes of liver

Lyme disease is caused by the bacterium *Borrelia burgdorferi*, which normally lives in white-footed mice. The disease is transmitted to humans and other mammals by the bite of a tick that harbors that bacterium. Deer, which can carry infected adult ticks without becoming ill, have helped spread the ticks through populated areas. The high rate of infection among children reflects the fact that they play outdoors during the summer in fields where deer may also be found. Children are thus more likely to encounter—and be bitten by—infected ticks. After 1975, the Lyme disease problem be-

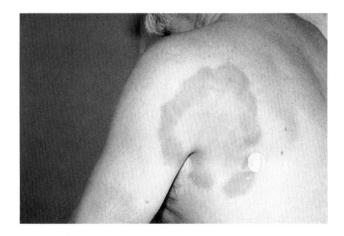

● **FIGURE 48**
Lyme Disease

came regional and then national in scope. The incidence in the United States is now about 12,500 cases per year, and Lyme disease has also been reported in Europe. A vaccine has been developed but is not widely available.

■ SYSTEMIC LUPUS ERYTHEMATOSUS

HA p. 619

Systemic lupus erythematosus (LOO-pus ē-rith-ē-ma-TŌ-sis), or SLE, appears to result from a generalized breakdown in the antigen recognition mechanism. An individual with severe SLE manufactures *autoantibodies* against the body's own nucleic acids (antinuclear antibodies, or ANA), ribosomes, clotting factors, blood cells, platelets, and lymphocytes. The immune complexes form deposits in peripheral tissues, producing anemia, kidney damage, arthritis, and vascular inflammation. CNS function deteriorates if the blood flow through damaged cranial or spinal vessels slows or stops.

The most visible sign of this condition is the presence of a butterfly-shaped rash centered over the bridge of the nose (Figure 49●). SLE affects women nine times as often as it affects men, and the incidence in the United States averages two to three cases per 100,000 population. The disorder has no known cure, but almost 80 percent of people with SLE survive five years or more after diagnosis. Treatment consists of controlling the symptoms and depressing the immune response through the administration of specialized drugs or corticosteroids.

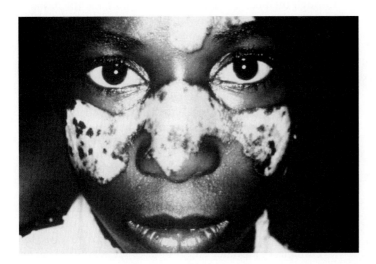

● **FIGURE 49**
Butterfly Rash of Systemic Lupus Erythematosus

23

CHAPTER 24
THE RESPIRATORY SYSTEM

AN INTRODUCTION TO THE RESPIRATORY SYSTEM AND ITS DISORDERS

The anatomical components of the respiratory system can be divided into two parts: an *upper respiratory system,* which includes the nose, nasal cavity, paranasal sinuses, and pharynx; and a *lower respiratory system,* composed of the larynx, trachea, bronchi, and lungs. The *respiratory tract* consists of the airways that carry air to and from the exchange surfaces of the lungs. The respiratory tract can be divided into a *conducting portion* and a *respiratory portion.* The conducting portion begins at the entrance to the nasal cavity and extends through the pharynx and larynx and along the trachea and bronchi to the terminal bronchioles. The respiratory portion of the tract includes the respiratory bronchioles and the alveoli, which are part of the respiratory membrane, where gas exchange occurs.

The respiratory system provides a route for the movement of air into and out of the lungs and supplies a large, warm, moist surface area for the exchange of oxygen and carbon dioxide between the air and circulating blood. Disorders affecting the respiratory system may therefore involve any or all of the following three mechanisms:

1. **Interfering with the movement of air along the respiratory passageways.** Internal or external factors may be involved. Within the respiratory tract, the constriction of small airways, or bronchospasm, as in *asthma,* can reduce airflow to the lungs. The blockage of major airways (for example, by a swollen epiglottis or as a result of choking on a toy or a piece of food) can completely shut off the air supply. External factors that interfere with air movement include (1) the introduction of air *(pneumothorax)* or blood *(hemothorax)* into the pleural cavity, with subsequent lung collapse; (2) the buildup of fluid within the pleural cavities (a *pleural effusion*), which compresses and collapses the lungs; and (3) arthritis, muscular paralysis, or other conditions that prevent the normal skeletal or muscular activities responsible for moving air into and out of the respiratory tract.

2. **Damaging or otherwise impeding the diffusion of gases at the respiratory membrane.** The walls of the alveoli are part of the respiratory membrane, where gas exchange occurs. Any disease that affects the alveolar walls will reduce the efficiency of gas exchange. In *emphysema,* alveoli are destroyed. With the inflammation and obstruction of *lung cancer,* or infection of the lungs, as in the various types of pneumonia, respiratory exchange is disrupted by the buildup of fluid or mucus within the alveoli and bronchi.

3. **Blocking or reducing the normal circulation of blood through the alveolar capillaries.** Blood flow to portions of the lungs may be prevented by the circulatory blockage of a *pulmonary embolism.* Not only does a pulmonary embolism prevent normal gas exchange in the affected regions of a lung, but also it results in tissue damage and, if the blockage persists for several hours, permanent alveolar collapse. Pulmonary blood pressure may then rise (a condition called

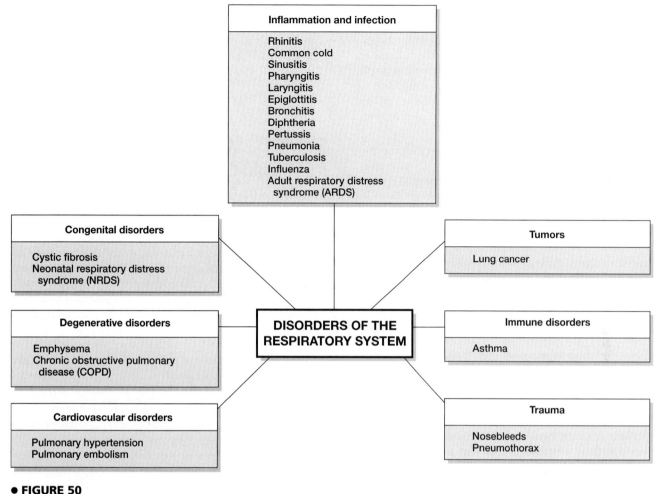

● FIGURE 50
Disorders of the Respiratory System

pulmonary hypertension), leading to pulmonary edema and a reduction in alveolar function in other portions of the lungs.

These types of problems can result from trauma, congenital or degenerative problems, the formation of tumors, inflammation, or infection of the lungs (Figure 50●). Illnesses caused by infections of the upper respiratory tract include some of the most common diseases. Many respiratory infections are transmitted by droplets in the air, typically emitted in a sneeze or cough. Infections of the lower respiratory tract include two of the most deadly diseases in human history: pneumonia and tuberculosis. Table 29 summarizes information about some of the most important infectious diseases of the respiratory system, along with their causative organisms.

Respiratory system disorders also occur secondarily, as a consequence of dysfunctions of other body systems. For instance, asthma may result from problems with immune function, and most pulmonary emboli originate as intravascular clots that embolize, reach the pulmonary arteries, and impair lung perfusion.

■ THE PHYSICAL EXAMINATION AND THE RESPIRATORY SYSTEM   *HA p. 624*

Several components of the physical examination will detect signs of respiratory disorders:

1. **Inspection** can reveal abnormal dimensions, such as the "barrel chest" that develops in emphysema or other chronic obstructive disorders, or *clubbing* of the fingers (p. 21). Clubbing is typically a late sign of disorders such as chronic lung infections or congestive heart failure. *Cyanosis,* a blue color of the skin and mucous membranes, generally indicates hypoxia (low tissue oxygen content). Laboratory testing of arterial blood gases will assist in determining the cause and extent of the hypoxia. Postural changes, such as sitting up, leaning forward, and exhaling through pursed lips or gasping also indicate respiratory difficulties.

2. **Palpation** of the bones, muscles, and surface of the thoracic cage can detect structural problems or asymmetry. For example, a one-sided pleural effusion blunts the vibration of speech passed through the bronchi to the chest wall on that side.

24

3. **Percussion** on the surface of the thoracic cage over the lungs should yield sharp, resonant sounds. Dull or flat sounds can indicate structural changes that decrease air in the lungs, such as those accompanying pneumonia, or the collapse of part of a lung *(atelectasis)*. Increased resonance can result from obstructive disorders, such as emphysema, due to hyperinflation of the lungs as the individual attempts to improve alveolar ventilation, or from air in the pleural space collapsing the lung (pneumothorax).

4. **Auscultation** of the lungs with a stethoscope yields the distinctive sounds of inhalation and exhalation. These sounds vary in intensity, pitch, and duration. Abnormal breath sounds accompany several pulmonary disorders:

   • **Rales** (rahlz) are hissing, whistling, scraping, or rattling sounds associated with increased airway resistance. The sounds are created by turbulent airflow past accumulated pus or mucus or through airways narrowed by inflammation. Descriptions and interpretations of these sounds are highly subjective, but in general, *moist rales* are gurgling sounds produced as air flows over fluids within the respiratory tract. They are heard in conditions such as *bronchitis, tuberculosis,* and *pneumonia*. *Dry rales* are produced as air flows over thick masses of mucus, through inflamed airways, or into fluid-filled alveoli. Dry rales are characteristic of *asthma,* congestive heart failure, and *pulmonary edema*. *Rhonchi* are loud dry rales produced by mucus buildup in the air passages, which often clear after coughing.

   • **Stridor** is a very loud, high-pitched sound that can be heard without a stethoscope. Stridor generally indicates acute airway obstruction, such as the partial blockage of the glottis by a foreign object.

   • **Wheezing** is a whistling, musical sound that can occur with inhalation or exhalation. It generally indicates airway obstruction due to mucus buildup or bronchospasms.

   • **Coughing** is a familiar sign of several respiratory disorders. Although primarily a reflex mechanism that clears the airway, coughing may also indicate irritation of the lining of the respiratory passageways. The duration, pitch, causative factors, and productivity associated with coughing may be important clues in the diagnosis of a respiratory disorder. (A *productive cough* ejects *sputum,* a mixture of mucus, cell debris, and pus; a *nonproductive cough,* or *dry cough,* does not.) If the cough is productive, the sputum can be collected and examined microscopically for cells, microorganisms, and debris. If epithelial cells are rare the sputum can be cultured to identify specific microorganisms. (Epithelial cells are a marker for saliva, which contains oral microorganisms that make sputum analysis unreliable.)

   • A **friction rub** is a distinctive crackling sound produced by abrasion (which is frequently painful) between abnormal serous membranes. A *pleural rub* accompanies respiratory movements and indicates problems with the pleural membranes, such as *pleurisy*. A *pericardial rub* accompanies the heartbeat and indicates inflammation of the pericardium, as in *pericarditis*.

5. During the assessment of a person's vital signs, the respiratory rate (number of breaths per minute) is recorded, along with notations about the general rhythm and depth of respiration. *Tachypnea* is a respiratory rate faster than 20 breaths per minute in an adult; *bradypnea* is an adult respiratory rate below 12 breaths per minute.

Table 28 introduces important procedures and laboratory tests that are useful in diagnosing respiratory disorders.

SYMPTOMS OF LOWER RESPIRATORY DISORDERS

Lower respiratory disorders generally cause one or two major symptoms—in particular, *chest pain* and *dyspnea*:

1. Chest pain associated with a respiratory disorder frequently worsens when the person takes a deep breath or coughs. This pain with breathing is somewhat distinct from the chest pain experienced by individuals with the heart problems of angina (pain appears during exertion) or myocardial infarction (pain is continuous, even at rest). Disorders affecting the pleural membranes may cause localized, inspiratory (pleuritic) chest pain in the specific region of the thorax where they occur. A person with such pleuritic pain will usually press against the sensitive area and avoid coughing or deep breathing in an attempt to reduce the local movement that brings on pain.

2. *Dyspnea,* or difficulty in breathing, can be a symptom of pulmonary disorders, cardiovascular disorders, metabolic disorders, or environmental factors such as hypoxia at high altitudes. Dyspnea may be a chronic problem, or it may develop only during exertion or when the person is lying down.

Dyspnea due to respiratory problems generally indicates one of the following classes of disorders:

• **Obstructive disorders** result from increased resistance to airflow along the respiratory passageways. The individual usually struggles to breathe, even at rest, and exhalation is more difficult than inhalation. Examples of obstructive disorders include *asthma* and *emphysema*.

• **Restrictive disorders** include (1) arthritis; (2) paralysis or weakness of respiratory muscles as a result of trauma, muscular dystrophy, myasthenia gravis, multiple sclerosis, polio, or other factors; (3) physical trauma or congenital structural disorders, such as scoliosis, that limit lung expansion; and (4) pulmonary fibrosis, in which abnormal fibrous tissue in the alveolar walls slows oxygen diffusion into the bloodstream. Individuals with restrictive disorders initially experience dyspnea only during exertion, because pulmonary ventilation cannot increase enough to meet the respiratory demand.

Cardiovascular disorders that produce dyspnea include *coronary artery disease, congestive heart failure,* and *pulmonary embolism*. In *paroxysmal nocturnal dyspnea,* a person awakens at night, gasping for air. In most cases, the underlying cause is a reduced cardiac output due to advanced heart disease or heart failure. Periodic, or, *Cheyne–Stokes respiration,* consists of alternating cycles of rapid, deep breathing and periods of respiratory arrest. This breathing pattern is most commonly seen in people with CNS disorders or congestive heart failure.

24

TABLE 28   Representative Diagnostic and Laboratory Tests for Respiratory Disorders

Diagnostic Procedure	Method and Result	Representative Uses
Pulmonary function studies	A spirometer is used to determine lung volumes and capacities, including V_T, IC, ERV, IRV, FRC, vital capacity, and total lung capacity on exertion. Forced vital capacity (FVC) is the amount of air forcibly exhaled after maximal inhalation.	Differentiate obstructive from restrictive lung diseases; determine extent of pulmonary disease. Increased functional residual capacity (FRC) occurs in obstructive diseases, such as emphysema and chronic bronchitis FRC is normal or decreased in restrictive diseases, such as pulmonary fibrosis
Peak expiratory flow (PEF)	Measured by forceful exhalation into a PEF meter	Useful in evaluating asthma; permits self-monitoring guides appropriate medication of airway obstruction
Bronchoscopy	Fiber-optic tubing is inserted through the mouth into the trachea, larynx, and bronchus for inspection.	Detects abnormalities such as inflammation and tumors. Used to remove aspirated foreign objects or mucus from bronchi and to obtain samples of secretions or tissue for examination.
Mediastinoscopy	An instrument is inserted into an incision made at the jugular notch superior to the manubrium of the sternum.	Detects abnormalities of mediastinal lymph nodes; permits removal of biopsy specimen; detects pulmonary disorders and metastasis of lung cancer
Lung biopsy	Lung tissue is removed for pathological analysis via bronchoscopy, needle aspiration, or during exploratory surgery of the thoracic area.	Differentiates pulmonary pathologies and determines the presence of malignancy
Chest x-ray study	Standard x-ray produces film sheet with radiodense tissues shown in white on a negative image.	Detects abnormalities of lungs, such as tumors, inflammation of the lungs, rib or sternal fractures, or pneumothorax; detects fluid accumulation (pulmonary edema), pneumonia, and atelectasis; determines heart size
Thoracentesis	A needle is inserted into the intrapleural space for removal of fluid for analysis or for relieving pressure due to fluid accumulation.	See Pleural fluid analysis
Pulmonary angiography	A catheter is inserted in the femoral vein and threaded through the right ventricle and into the pulmonary arteries. Contrast dye is intermittently injected as x-ray films are taken.	Detects pulmonary embolism and other pulmonary vascular abnormalities; measures right atrial, ventricular, and pulmonary artery pressures
Lung scan	Radionuclides are inhaled and injected intravenously; radiation that is emitted is captured to create an image of the lungs.	Determines areas with decreased blood flow or ventilation due to pulmonary embolism or other pulmonary disease.
Computerized tomography (CT) scan	Standard CT; intravenous contrast media are often used.	Detects tumors, cysts, or other structural abnormalities

Laboratory Test	Normal Values in Blood Serum or Plasma	Significance of Abnormal Values
Arterial blood gases and pH		
pH	7.35–7.45	Lower than 7.35 indicates acidosis; higher than indicates alkalosis.
P_{CO_2}	35–45 mm Hg	higher than 45 mm Hg with pH lower than 7.35 indicates respiratory acidosis present in disorders such as chronic bronchitis, in chronic obstructive pulmonary disease, and in CNS depression leading to irregular breathing; lower than 35 mm Hg with a pH higher than 7.45 indicates respiratory alkalosis that occurs during prolonged hyperventilation; also seen in pulmonary embolism.
P_{O_2}	75–100 mm Hg	lower than 75 mm Hg may occur in pneumonia, asthma, and COPD.

24

TABLE 28   Representative Diagnostic and Laboratory Tests for Respiratory Disorders (*Continued*)

Diagnostic Procedure	Method and Result	Representative Uses
HCO_3^-	22–28 mEq/L	Lower than 28 mEq/L (with elevated P_{CO_2} and decreased pH) indicates renal compensation for respiratory acidosis; lower than 22 mEq/L (with decreased P_{CO_2} and elevated pH) is characteristic of renal compensation for respiratory alkalosis.

Laboratory Test	Normal Results	Significance of Abnormal Values
Sputum studies		
Cytology	No malignant cells are present.	Sloughed malignant cells indicate cancerous process in lungs
Culture and sensitivity (C&S)	Sputum sample is placed on growth medium.	Identifies pathogenic organism and the organism's susceptibility to antibiotics
Acid-fast stain for bacilli	Staining technique reveals no acid-fast bacilli.	Presence of stained rod-shaped microbes may indicate active infection with tuberculosis bacteria
Tuberculin skin test	Skin wheal, produced by intradermal application of tuberculin, at 48–72 hours should be <10 mm (area that is hardened, as opposed to the reddened area).	A skin papule at injection site that measures >10 mm after 48–72 hours is a positive test for immunity for tuberculosis. Does not prove active infection.
Pleural fluid analysis		
Fluid color and clarity	No pus; fluid is clear, with WBC <1000 mm³	WBC >1000 per mm³ indicates probable infectious or inflammatory process; detects lung or pleural malignancy.
Culture	No bacterial growth (fluid sample is placed on growth medium)	Presence of bacteria indicates infection; fungi can be present in immunocompromised person

Dyspnea may also be related to metabolic problems, such as the acute acidosis associated with *diabetes mellitus* and *uremia*. The fall in blood pH can trigger *Kussmaul breathing*, which consists of rapid, deep breaths.

■ OVERLOADING THE RESPIRATORY DEFENSES   *HA p. 625*

Large quantities of airborne particles can overload the respiratory defenses and produce a variety of illnesses. Chemical or physical irritants that reach the lamina propria or underlying tissues promote the formation of scar tissue (*fibrosis*), reducing the elasticity of the lung, and can restrict airflow along the passageways. Irritants or foreign particles may also enter the lymphatic vessels of the lung, producing inflammation of the regional lymph nodes. Chronic irritation and the stimulation of the epithelium and its defenses cause changes in the epithelium that increase the likelihood of lung cancer.

Severe symptoms of such disorders develop slowly; they may take 20 years or more to appear. **Silicosis** (sil-i-KŌ-sis; produced by the inhalation of silica dust), **asbestosis** (as-bes-TŌ-sis; from the inhalation of asbestos fibers), and **anthracosis** (an-thra-KŌ-sis; the "black lung disease" of coal miners, caused by the inhalation of coal dust) are conditions caused by the overloading of the respiratory defenses (Figure 51●). Rescue workers at the World Trade Center towers site were exposed to enormous amounts of dust and smoke of undetermined composition. Although the primary component was undoubtedly concrete dust, samples have also been found to contain asbestos, heavy metals, and other products of combustion from the fires that burned for weeks. Many of

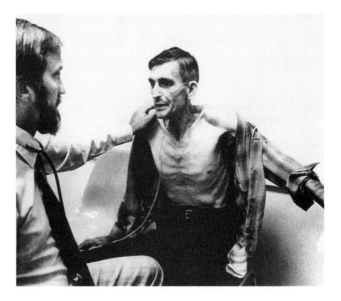

● **FIGURE 51**
A Person with Anthracosis (Black Lung Disease)

TABLE 29	Examples of Infectious Diseases of the Respiratory System	
Disease	**Organism(s)**	**Description**
Bacterial diseases		
Sinusitis	*Streptococcus pneumoniae* or *Haemophilus influenzae*	Inflammation of the paranasal cavities; headaches, pain, purulent nasal discharge, pressure in facial bones
Pharyngitis	*Streptococcus pyogenes*	Inflammation of the throat; strep throat; sore throat, fever, no cough or sputum
Laryngitis	*Streptococcus pneumoniae* or *Haemophilus influenzae*	Inflammation of the larynx; dry, sore throat; hoarse voice or loss of voice
Epiglottitis	*Haemophilus influenzae*	Inflammation of the epiglottis; most common in children; fever, noisy breathing, drooling; swelling can block trachea and lead to suffocation
Bronchitis	*Streptococcus pneumoniae* or *Mycoplasma pneumoniae*	Inflammation of the bronchial passageways; productive cough with sputum; may enter alveoli and cause pneumonia
Diphtheria	*Corynebacterium diphtheriae*	Inflammation of the pharynx; pseudomembrane in pharynx; bacterial toxin affects heart and other tissues
Pertussis (whooping cough)	*Bordetella pertussis*	Highly contagious disease of children; nasal mucus production; severe coughing ends in a "whoop" sound during inhalation
Pneumonia	*Streptococcus pneumoniae, Staphylococcus aureus,* or *Klebsiella pneumoniae*	Inflammation of the lungs; alveoli fill with fluids; chills, high fever; cough with sputum of mucus, pus, and blood
Tuberculosis	*Mycobacterium tuberculosis*	Highly contagious infection of the lungs; can spread to other tissues; abscesses, or tubercles, form in lungs; bacteria multiply in WBCs and alveolar macrophages and break down alveoli
Viral diseases		
Common cold	Rhinoviruses or coronaviruses	Mucosal edema, nasal obstruction, nasal discharge (coryza), sneezing, and headache
Influenza	Orthomyxoviruses (influenza virus A, B, C)	Frequent variations in Type A and B viruses can produce new epidemics; fever, headache, sore throat, nasal discharge, muscle pain (*myalgia*), fatigue, chest pain, and cough.

the rescue workers have developed a persistent cough and hyperreactive airways that are prone to bronchospasm.

■ NOSEBLEEDS *HA p. 626*

The extensive vascularization of the nasal cavity and the relatively vulnerable position of the nose make a nosebleed, or **epistaxis** (ep-i-STAK-sis), relatively common. Bleeding usually involves vessels of the mucosa covering the anterior cartilaginous portion of the septum. Packing the vestibule with gauze or pinching the external nares together to squeeze the vessels against the septum will often control the bleeding until clotting occurs. More severe bleeding originating elsewhere in the nasal cavity may require packing the posterior portion of the cavity via the internal nares.

Epistaxis can result from any factor affecting the integrity of the epithelium or the underlying vessels. Examples would include trauma, such as a punch in the nose; drying; infec-

tions; allergies; or clotting disorders. Hypertension may also provoke a nosebleed by rupturing small vessels of the lamina propria.

24

■ DISORDERS OF THE LARYNX
HA p. 630

Infection or inflammation of the larynx is known as **laryngitis** (lar-in-JĪ-tis). This condition often affects the vibrational qualities of the vocal cords; hoarseness is the most familiar symptom. Mild cases are temporary and seldom serious, but bacterial or viral infection of the epiglottis or upper trachea in children can be very dangerous because the swelling may close the glottis or trachea and cause suffocation. **Acute epiglottitis** (ep-i-glo-TĪ-tis) can develop relatively rapidly following a bacterial infection of the throat. Although most common in children, it does occur in adults, especially those with

Hodgkin's disease or *leukemia,* two cancers described earlier. Serious inflammation and edema of the trachea can also occur in small children following infection with one of the parain-fluenza viruses. The condition, **laryngotracheobronchitis,** is more commonly called **croup** (kroop).

■ EXAMINING THE LIVING LUNG

HA p. 634

A chest x-ray remains the standard diagnostic screening test for chest conditions. This procedure can detect abnormalities in lung structure including scar tissue formation, fluid accu-mulation, or distortion of the conducting passageways. CT scans show much greater definition of internal structures and can clarify the nature of abnormalities initially spotted on a chest x-ray. CT scans are particularly helpful in diagnosing and tracking the progression of lung cancers. Lung scans are made when radioactive tracers are injected or radioactive gases are inhaled. These procedures can detect abnormalities in air-flow or pulmonary blood circulation.

One method of investigating the status of the respira-tory passageways is the use of a **bronchoscope,** a fiber-optic bundle small enough to be inserted through the mouth and larynx into the trachea and steered along the conducting passageways to the level of the smaller bronchioles. This pro-cedure is called **bronchoscopy** (brong-KOS-kō-pē). In ad-dition to permitting direct visualization of bronchial structures, the bronchoscope can collect tissue or mucus samples from the respiratory tract. In **bronchography** (brong-KOG-ra-fē), a bronchoscope or catheter introduces a radiopaque material into the bronchi. This technique can permit detailed x-ray analysis of bronchial masses, such as tumors or other obstructions, but has been largely replaced by lung CT scans.

■ COPD: ASTHMA, BRONCHITIS, AND EMPHYSEMA   *HA p. 637*

Chronic obstructive pulmonary disease (COPD) is a pro-gressive disorder of the airways that restricts airflow and re-duces alveolar ventilation. Three patterns of symptoms may appear in individuals with COPD. *Asthma,* or *asthmatic bron-chitis,* is the term that is commonly applied when symptoms are acute and intermittent. The terms *bronchitis* and *emphysema* are usually applied when symptoms are chronic and only slowly progressive toward an acute stage. However, the symptoms and causes of these conditions overlap to such an extent that the term *COPD* is being used in clinical prac-tice to refer to all three disorders.

Asthma (AZ-ma) affects an estimated 3–6 percent of the U.S. population. There are several forms of asthma, each char-acterized by unusually sensitive and irritable air-conducting passageways that respond to irritants by constricting, an event described as a *bronchospasm.* In many cases, the trigger appears to be an immediate hypersensitivity reaction to an allergen in the inhaled air. Drug reactions, air pollution,

chronic respiratory infections, exercise, and emotional stress can also induce an asthma attack in sensitive individuals.

The most obvious and potentially dangerous symptoms include (1) the constriction of smooth muscles all along the bronchial tree, (2) edema and swelling of the mucosa of the respiratory passageways, and (3) the accelerated production of mucus. The combination makes breathing very difficult. Exhalation is affected more than inhalation; the narrowed passageways often collapse before exhalation is completed. Although mucus production increases, mucus transport slows, so fluids accumulate along the passageways. Coughing and wheezing then develop. The bronchoconstriction and mucus production occur in a few minutes, in response to the release of histamine and prostaglandins by mast cells. The activated mast cells also release interleukins, leukotrienes, and platelet-activating factors. As a result, over a period of hours, neutrophils and eosinophils migrate into the area, which then becomes inflamed, further reducing airflow and damaging respiratory tissues. Because the inflammation compounds the problem, antihistamines, bronchodilators, and anti-inflammatory steroids may be needed to control a severe asth-ma attack.

When a severe attack occurs, it reduces the functional ca-pabilities of the respiratory system. Peripheral tissues become starved for oxygen, a condition that can prove fatal. Asthma fatalities have been increasing in recent years. The annual death rate from asthma in the United States is approximate-ly four deaths per million population among those age 5–34. Mortality among asthmatic African Americans is twice that of Caucasian Americans.

The treatment of asthma involves the dilation of the respi-ratory passageways by administering **bronchodilators** (brong-kō-DĪ-la-torz) (drugs that relax bronchial smooth muscle) and by reducing inflammation and swelling of the respiratory mu-cosa through antinflammatory medication. Important bron-chodilators include *theophylline, epinephrine, albuterol,* and other beta-adrenergic drugs. Although the strongest beta-adrenergic drugs are quite useful in a crisis, they are effective only for relatively brief periods, and overuse of them can lead to reduced efficiency. Asthmatic individuals must be closely monitored, due to the drugs' potential effects on cardiovascu-lar function. Anti-inflammatory medications, such as inhaled or ingested steroids, are becoming increasingly important.

Bronchitis (brong-KĪ-tis) is an inflammation and swelling of the bronchial lining, leading to overproduction of mucous secretions. The most characteristic symptom is frequent coughing with copious sputum production. An estimated 20 percent of adult males have *chronic bronchitis,* a condition that is most commonly related to cigarette smoking, but that also results from other environmental irritants, such as chem-ical vapors. Over time, the increased mucus production can block smaller airways and reduce air exchange and respiratory efficiency. Chronic bacterial infections leading to more lung damage are common. Treatment involves stopping smoking and the administration of bronchodilators, antibiotics, and supplemental oxygen as necessary.

Emphysema (em-fi-SĒ-muh) is a chronic, progressive con-dition characterized by shortness of breath and an inability

24

to tolerate physical exertion. The underlying problem is the destruction of respiratory exchange surfaces. In essence, respiratory bronchioles and alveoli are functionally eliminated. The alveoli gradually expand and capillaries deteriorate, leaving large nonfunctional cavities in the lungs where gas exchange is severely decreased or eliminated.

Emphysema has been linked to the inhalation of air that contains fine particulate matter or toxic vapors, such as those in cigarette smoke. Early in the disease, local regulation shunts blood away from the damaged areas, and the individual may not notice problems, even with strenuous activity. As the condition progresses, the reduction in exchange surface limits the ability to provide adequate oxygen. However, obvious clinical symptoms typically fail to appear until the damage is extensive.

Alpha₁-antitrypsin, an enzyme that is normally present in the lungs, helps prevent degenerative changes in lung tissue. Most people requiring treatment for emphysema are adult smokers, a group that includes both individuals with alpha₁-antitrypsin deficiency and those with normal tissue enzymes. In the United States, one person in 1000 carries two copies of a gene that codes for an abnormal, inactive form of this enzyme. A single change in the amino acid sequence of the enzyme appears responsible for this defect. At least 80 percent of nonsmokers with abnormal alpha₁-antitrypsin will develop emphysema, generally at age 45–50. *All* smokers will develop at least some emphysema, typically by age 35–40.

Unfortunately, the loss of alveoli and bronchioles in emphysema is permanent and irreversible. Further progression can be limited by the cessation of smoking. Other aspects of pulmonary structure and function may be relatively normal. The only effective treatment for severe cases is the administration of supplemental oxygen, but lung transplants have helped some patients, as has the surgical removal of nonfunctional lung tissue. For people with alpha₁-antitrypsin deficiency who are diagnosed early, attempts are under way to provide enzyme supplements by daily infusion or periodic injection.

On physical exam, individuals with COPD commonly expand their chests permanently in an effort to enlarge their lung capacities and make the best use of the remaining functional alveoli. This adaptation may give them a distinctive "barrel-chested" appearance. The respiratory rate with advanced disease increases dramatically. The lungs are fully inflated at each breath, and expirations are forced. Two clinical patterns of advanced disease occur.

People with emphysema tend to maintain near-normal arterial P_{O_2}. Their respiratory muscles work hard, and they use a lot of energy just breathing. As a result, they tend to be thin. Because their blood oxygenation is near normal, skin color in Caucasians in this group is pink. The combination of heavy breathing and pink coloration has led to the descriptive term *pink puffers* for individuals with this condition.

People with chronic bronchitis may have symptoms of heart failure, including widespread edema. Their blood oxygenation is low, and their skin may have a bluish color. The combination of widespread edema and bluish coloration has led to the descriptive term *blue bloaters* for individuals with this condition.

RESPIRATORY DISTRESS SYNDROME (RDS) *HA p. 637*

Septal cells among the lining cells of the alveoli begin producing surfactant at the end of the sixth fetal month. By the eighth month, surfactant production has risen to the level required for normal respiratory function. **Neonatal respiratory distress syndrome** (NRDS), also known as *hyaline membrane disease (HMD),* develops when surfactant production fails to reach normal levels. Although some forms of HMD are inherited, the condition most commonly accompanies premature delivery.

In the absence of surfactant, the alveoli tend to collapse during exhalation. Although the conducting passageways remain open, the newborn infant must then inhale with extra force to reopen the alveoli on the next breath. In effect, every breath must approach the power of the first, so the infant rapidly becomes exhausted. Respiratory movements become progressively weaker; eventually, the alveoli fail to expand and gas exchange ceases.

One method of treatment involves assisting the infant by administering air under pressure so that the alveoli are held open. This procedure, known as **positive end-expiratory pressure** (PEEP), can keep the newborn alive until surfactant production increases to normal levels. Surfactant from other sources administered via the trachea on the first day of life has reduced the death rate from NRDS from nearly 100 percent to less than 10 percent. Stable surfactants can be (1) extracted from cows' lungs (*Survanta*), (2) obtained from the liquid (amniotic fluid) that surrounds full-term infants, or (3) synthesized by gene-splicing techniques (*Exosurf*). These preparations are usually administered in the form of a fine mist of surfactant droplets. Clinical trials are under way to deliver oxygen to the respiratory membrane by means of a perfluorocarbon solution (*Liquivent*). The lungs are filled with the solution, which readily absorbs oxygen. Because the alveoli are filled with fluid rather than with air, they remain open despite the lack of surfactant. This procedure avoids the necessity of administering air under pressure, which can have undesirable side effects.

Surfactant abnormalities can also develop in adults as the result of severe respiratory infections or other sources of pulmonary injury. Alveolar collapse follows, producing a condition known as **adult respiratory distress syndrome** (ARDS). The PEEP procedure is typically used to maintain life until the underlying problem can be corrected, but at least 50–60 percent of ARDS cases result in fatalities.

24

TUBERCULOSIS *HA p. 637*

Tuberculosis (TB) is a major health problem throughout the world. With roughly 3 million deaths each year, it is the leading cause of death from infectious diseases. An estimated 2 *billion* people are infected, with 8 million cases of active, potentially contagious disease each year. Unlike other deadly diseases, such as AIDS, TB can be transmitted through casual contact, and has been spread from one airplane passenger to other passengers. Anyone who breathes is at risk of contracting this disease; all it takes is exposure to the causative bacterium. Coughing, sneezing, or speaking by an actively infected individual spreads the pathogen through the air in the form of tiny droplets that can be inhaled by other people. Crowded living conditions and prolonged contact (usually days to weeks) increase the risk of infection.

Untreated, the disease progresses in stages. The primary site of infection is usually in the lungs. At the site of infection, macrophages and fibroblasts wall off the area, forming an abscess. For most people in otherwise good health this containment leaves them with latent, noncontagious TB diagnosed by a few scarred nodules visible on chest x-ray, a positive TB skin test (see Table 28) and no lab or clinical signs of active disease. In 10 percent of latent TB cases, at some point after primary infection the scar tissue barricade fails, the bacteria move into the surrounding tissues, and the process repeats itself. The resulting masses of fibrous tissue distort the conducting passageways, increasing resistance and decreasing airflow. In the alveoli, the surfaces that have been attacked are destroyed. The combination severely reduces the area available for gas exchange. Other organs may be infected, and without antibiotic treatment the mortality rate is approximately 33 percent.

Treatment for TB is complex, because the bacteria is slow-growing and it can spread through the bloodstream to many different tissues. Prolonged treatment is required and resistance can develop to standard antibiotics, particularly if infected people fail to take their antibiotics at appropriate intervals over the entire recommended treatment period. To combat antibiotic resistance, several drugs are used in combination over a period of six to nine months, with "directly observed therapy" (DOT) wherein health personnel watch the patient swallow the medications. The most effective drugs now available include *isoniazid,* which interferes with bacterial replication, and *rifampin,* which blocks bacterial protein synthesis.

Tuberculosis was extremely common in the United States early in the 20th century. An estimated 80 percent of Americans born around 1900 became infected with tuberculosis during their lives. Although most were able to recover from the primary bacterial infection, it was the leading cause of death in 1906. These statistics have been drastically changed with the advent of effective antibiotics in the 1940s and techniques for early detection of infection. Between 1906 and 1999 the death rate fell from 200 deaths per 100,000 population to 0.3 deaths per 100,000 population. Strong public health measures, including TB surveillance by TB skin testing and chest x-rays, lab identification and drug susceptibility testing, directly observed therapy, isolation of infectious patients, and investigation and treatment of close contacts has significantly reduced the number of new cases of TB to 15,991 in 2001. The TB problem is much more severe in developing nations. As of 2000 an estimated 10–15 million people in the United States are infected with the bacterium, 50 percent of them immigrants who had latent infection when they arrived. These individuals with TB are not infectious unless the disease reactivates, causing symptoms of coughing, fever, and weight loss. Reactivation occurs when resistance is lowered by conditions that weaken the immune system, including aging, malnutrition, and infections such as HIV/AIDS.

■ PNEUMONIA   *HA p. 637*

Pneumonia (noo-MŌ-nē-a) is an infection of the lobules of the lung. As inflammation occurs within the lobules, respiratory function deteriorates as a result of fluid leakage into the alveoli and/or swelling and constriction of the respiratory bronchioles. When bacteria are involved, they are frequently species normally found in the mouth and pharynx that have somehow managed to evade the respiratory defenses. As a result, pneumonia becomes more likely when the respiratory defenses have been compromised by other factors, such as epithelial damage from smoking or the breakdown of the immune system in AIDS. The most common pneumonia that develops in people with AIDS results from infection by the fungus *Pneumocystis carinii.* These organisms are normally found in the alveoli, but in healthy individuals the respiratory defenses are able to prevent infection and tissue damage.

■ PULMONARY EMBOLISM
HA p. 637

The lungs are the only organs that receive the entire cardiac output. Blood pressure in the pulmonary circuit is usually relatively low. Pulmonary vessels can easily become blocked by blood clots, fat masses, or air bubbles in the pulmonary arteries. Blockage of a vessel stops blood flow to all of the alveoli serviced by the obstructed vessel. This condition is called a **pulmonary embolism**. A condition described earlier, *venous thrombosis,* can promote development of a pulmonary embolism. If the blockage remains in place for several hours, the alveoli will permanently collapse. If the blockage occurs in a major pulmonary vessel, rather than a minor tributary, pulmonary resistance increases. This resistance places extra strain on the right ventricle, which may be unable to maintain cardiac output. Congestive heart failure may be the result. Thrombolytic therapy with heparin or other drugs can be lifesaving.

■ PNEUMOTHORAX   *HA p. 637*

Any injury to the thorax that penetrates the parietal pleura or damages the alveoli and the visceral pleura can allow air into the pleural cavity. This condition, called **pneumothorax** (noo-mō-THŌR-aks), breaks the fluid bond between the pleurae and allows the elastic fibers to contract. The result is a partially collapsed lung, a condition termed **atelectasis** (at-e-LEK-ta-sis; *ateles,* imperfect + *ektasis,* expansion).

A pneumothorax may develop because of imperfections and air leakage in the walls of superficial alveoli. These problems often have a congenital basis. A sudden lung collapse, called a **spontaneous pneumothorax**, may be triggered by heavy exercise, coughing, or other activities that increase the pressures inside the alveoli of the lungs.

■ THORACENTESIS   *HA p. 637*

The delicate pleural membranes can become inflamed because of chronic irritation or infection. Inflammation produces symptoms of pleuritis, or pleurisy. Acute and chronic inflammation often causes a change in the permeability of the pleura, leading to a **pleural effusion** (an abnormal accumulation of fluid within the pleural cavities). When a pleural effusion is detected on an x-ray, samples of pleural fluid may be obtained, using a long needle inserted between the ribs. This sampling procedure is called **thoracentesis** (or *thoracocentesis*). The fluid collected is usually checked for the presence of blood, white blood cells, bacteria, proteins, and glucose.

■ AIR OVEREXPANSION SYNDROME

HA p. 637

Swimmers descending 1m or more beneath the surface experience significant increases in pressure, due to the weight of the overlying water. Air spaces throughout the body decrease in volume. The increase in pressure normally produces mild discomfort in the middle ear, but some people experience acute pain and disorientation. The water pressure first collapses the auditory tube. As the volume of air in the middle ear decreases, the tympanic membrane is forced inward. This uncomfortable situation can be remedied by closing the mouth, pinching the nose, and exhaling gently; elevating the pressure in the nasopharynx forces air through the auditory tube and into the middle ear space. As the volume of air in each middle ear increases, the tympanic membrane returns to its normal position. When the swimmer returns to the surface, the pressure drops and the air in the middle ear expands. This expansion usually goes unnoticed, because the air simply forces its way out along the auditory tube and into the nasopharynx.

Scuba divers breathe air under pressure, and that air is delivered at the same pressure as that of their surroundings. A descent to a depth of 10 m doubles the pressure on a diver, due to the weight of the overlying water. This pressure reduces the volume of air in the lungs to half of what it was at the surface. What happens if that diver then takes a full breath of pressurized air from a scuba tank and heads for the surface? As the water pressure decreases, the air expands (this is an example of a principle known as *Boyle's law*). Thus, at the surface, the volume of air in the lungs will be twice what it was at 10 m. Such a drastic increase in lung volume cannot be tolerated; the usual result is a tear in the wall of the lung. The symptoms and severity of this *air overexpansion syndrome* depend on where the air ends up. If the air flows into the pleural cavity, the lung may collapse; if it enters the mediastinum, it may compress the pericardium and produce symptoms of cardiac tamponade. Worst of all, the air may rupture blood

vessels and enter the bloodstream. The air bubbles then form emboli that can block blood vessels in the heart or brain, producing a heart attack or stroke. These are all serious conditions, so divers are trained to avoid holding their breath and to exhale when swimming toward the surface.

■ LUNG CANCER   *HA p. 645*

Lung cancers now account for 13 percent of new cancer cases and 28 percent of all cancer deaths, making this condition the primary cause of cancer death in the U.S. population. It kills more people than colon, breast, and prostate cancer combined. Despite advances in the treatment of other forms of cancer, the survival statistics for lung cancer have not changed significantly. Even with early detection, the five-year survival rates are only 30 percent for men and 50 percent for women, and more than 50 percent of people with lung cancer die within a year of diagnosis.

Detailed statistical and experimental evidence has shown that *85–90 percent of all lung cancers are the direct result of cigarette smoking.* Claims to the contrary are simply unjustified and insupportable. The data are far too extensive to detail here, but the incidence of lung cancer for nonsmokers is 3.4 per 100,000 population, whereas the incidence for smokers ranges from 59.3 per 100,000 for those who smoke between a half-pack and a pack per day to 217.3 per 100,000 for those who smoke one to two packs per day. Before about 1970, this disease affected primarily middle-aged men, but as the number of female smokers has increased (a trend that started in the 1940s), so has the number of women who die from lung cancer.

Smoking changes the quality of the inspired air, making it drier and contaminated with several carcinogenic compounds and particulate matter. The combination overloads the respiratory defenses and damages the epithelial cells throughout the respiratory system. The risk of developing lung cancer appears to be related to the total cumulative exposure to the carcinogens. The more cigarettes smoked, the greater the risk, whether those cigarettes are smoked over a period of weeks or years. Up to the point at which tumors form, the histological changes induced by smoking are reversible; a normal epithelium will return if the carcinogens are removed. At the same time, the statistical risks decline to significantly lower levels. Ten years after quitting, a former smoker stands only a 10 percent greater chance of developing lung cancer than does a nonsmoker.

The fact that cigarette smoking typically causes cancer is not surprising in view of the toxic chemicals contained in the smoke. What is surprising is that *more* smokers do not develop lung cancer. Evidence suggests that some smokers have a genetic predisposition to developing one form of lung cancer. Dietary factors may also play a role in preventing lung cancer, although the details are controversial. In terms of their influence on the risk of lung cancer, there is a general agreement that (1) vitamin A has no effect; (2) vegetables containing beta-carotene reduce the risk, but pills of beta-carotene may increase the risk; and (3) a high-cholesterol, high-fat diet increases the risk.

24

CHAPTER 25
THE DIGESTIVE SYSTEM

AN INTRODUCTION TO THE DIGESTIVE SYSTEM AND ITS DISORDERS

■ SYMPTOMS AND SIGNS OF DIGESTIVE SYSTEM DISORDERS
HA p. 650

The digestive system consists of the *digestive tract* and *accessory digestive organs.* The digestive tract is divided into the oral cavity, pharynx, esophagus, stomach, small intestine, and large intestine. Most of the absorptive functions of the digestive system occur in the small intestine, with lesser amounts in the stomach and large intestine. The accessory digestive organs provide acids, enzymes, and buffers that assist in the chemical breakdown of food. The accessory organs include the salivary glands, the liver, the gallbladder, and the pancreas. The salivary glands produce saliva, a lubricant that contains enzymes that aid in the digestion of carbohydrates. The liver produces bile, which is concentrated in the gallbladder and released into the small intestine for fat emulsification. The pancreas secretes enzymes and buffers that are important to the digestion of proteins, carbohydrates, and lipids.

The activities of the digestive system are controlled through a combination of local reflexes, autonomic innervation, and the release of gastrointestinal hormones such as *gastrin, secretin,* and *cholecystokinin.*

The functions of the digestive organs are varied, and the symptoms and signs of digestive system disorders are equally diverse. Common symptoms of digestive disorders include the following:

- **Pain** may occur in a number of locations. Widespread pain in the oral cavity can result from (1) trauma; (2) infection of the oral mucosa by bacteria, fungi, or viruses; or (3) a deficiency in vitamin C (*scurvy*) or in one or more of the B vitamins. Focal pain in the oral cavity accompanies (1) the infection or blockage of salivary gland ducts; (2) tooth disorders, such as fractures, *dental caries, pulpitis,* abscesses, and *gingivitis;* and (3) oral lesions, such as those produced by the *herpes simplex* virus.

Abdominal pain is characteristic of a variety of digestive disorders. The pain may be distressing, but tolerable; but if the pain is acute and severe, a surgical emergency may exist. Abdominal pain can cause rigidity in the abdominal muscles in the painful area. The rigidity is easily felt on palpation. The muscle contractions (*guarding*) may be voluntary, in an attempt to protect a painful area, or an involuntary spasm resulting from irritation of the peritoneal lining, as in *peritonitis.* People with peritoneal inflammation also experience *rebound tenderness,* in which pain appears when fingertip pressure on the abdominal wall is suddenly removed.

Abdominal pain can result from disorders of the digestive, circulatory, urinary, or reproductive system. Digestive tract disorders producing abdominal pain include *appendicitis, peptic ulcers, pancreatitis, cholecystitis, hepatitis, intestinal diverticulosis, diverticulitis, peritonitis,* and certain cancers.

- **Dyspepsia**, or indigestion, is pain or discomfort in the epigastric region. Digestive tract disorders associated with dyspepsia include *esophagitis, gastritis, peptic ulcers, gastroesophageal reflux,* and *cholecystitis.*

- **Nausea** is a sensation that usually precedes or accompanies vomiting. Nausea results from digestive disorders or from disturbances of central nervous system function.

- **Anorexia** is a decrease in appetite that, if prolonged, is accompanied by weight loss. Digestive disorders that cause anorexia include *stomach cancer,* pancreatitis, hepatitis, and several forms of *diarrhea.* Anorexia may also accompany disorders that involve other systems.

- **Dysphagia** is difficulty in swallowing. The difficulty may result from trauma, infection, inflammation, or a blockage of the posterior oral cavity, pharynx, or esophagus. For example, the infections of *tonsillitis, pharyngitis,* and *laryngitis* may cause dysphagia.

■ THE PHYSICAL EXAMINATION AND THE DIGESTIVE SYSTEM   *HA p. 650*

Physical assessment can provide information that is useful in the diagnosis of digestive system disorders. The abdominal region is particularly important, because most of the digestive system is located within the abdominopelvic cavity. Four methods of physical assessment of the digestive system are inspection, palpation, percussion, and auscultation:

1. **Inspection** can provide a variety of diagnostic clues:

   - Bleeding of the gums, as in gingivitis, and characteristic oral lesions can be seen on inspection of the oral cavity. Examples of distinctive lesions include those of oral herpes simplex infections and *thrush*—lesions produced by infection of the mouth by *Candida albicans.* This fungus (also called a yeast) is a normal resident of the digestive tract. However, the fungus may cause widespread oral and esophageal infections in immunodeficient people, such as individuals who have AIDS or who are undergoing immunosuppressive therapies. (Healthy infants can also get a mild oral *thrush* infection.)

   - Peristalsis in the stomach and intestines may be seen as waves passing across the abdominal wall in people who do not have a thick layer of abdominal fat. The waves become very prominent during the initial stages of intestinal obstruction.

   - A general yellow discoloration of the skin and sclera, a sign called *jaundice,* results from excessive levels of bilirubin in body fluids. Jaundice is commonly seen in individuals with cholecystitis or liver diseases such as hepatitis and *cirrhosis.*

   - Abdominal distention is caused by (1) accumulation of fluid in the peritoneal cavity, as in *ascites;* (2) air or gas (flatus) within the digestive tract; (3) obesity; (4) abdominal masses, such as tumors, or enlargement of visceral organs; (5) pregnancy; (6) the presence of an abdominal *hernia;* or (7) fecal impaction, as in severe and prolonged constipation.

   - **Striae** are multiple scars, 1–6 cm in length, that are visible through the epidermis. Striae develop in damaged dermal tissues after stretching; they are typically seen in the abdominal region after a pregnancy or some other rapid weight gain. Abnormal striae may develop after ascites or in cases of subcutaneous edema. Purple striae are a sign of *Cushing's disease.*

2. **Palpation** of the abdomen may reveal specific details about the status of the digestive system, including the following:

   - The presence of abnormal masses, such as tumors, within the abdominal cavity

   - Abdominal distention from (1) excess fluid within the digestive tract or peritoneal cavity or (2) gas within the digestive tract

   - Herniation of digestive organs through the inguinal canal or weak spots in the abdominal wall.

   - Changes in the size, shape, or texture of visceral organs. For example, in several liver diseases, the liver becomes enlarged and firm, and these changes can be detected on palpation of the right upper quadrant.

   - Voluntary or involuntary abdominal muscle contractions (called guarding)

   - Rebound tenderness.

   - Specific areas of tenderness and pain. For example, someone with acute hepatitis, a liver disease, generally experiences pain on palpation of the right upper quadrant. In contrast, a person with appendicitis generally experiences pain when the right *lower* quadrant is palpated.

3. **Percussion** of the abdomen is less instructive than percussion of the chest, because the visceral organs do not contain extensive air spaces that reflect the sounds conducted through surrounding tissues. However, the stomach usually contains a small air bubble, and percussion over this area produces a sharp, resonant sound. The sound becomes dull or disappears when the stomach fills, the spleen enlarges, or the peritoneal cavity contains abnormal quantities of peritoneal fluid, as in ascites.

4. Auscultation can detect gurgling abdominal sounds or bowel sounds, produced by peristaltic activity along the digestive tract. Increased bowel sounds occur in people with acute diarrhea, and bowel sounds may disappear in people with (1) advanced intestinal obstruction; (2) peritonitis, an infection of the peritoneum; or (3) spinal-cord injuries that prevent normal innervation.

Diagnostic procedures, such as endoscopy, are commonly used to provide additional information. Information on representative diagnostic procedures is given in Table 30.

One common method of categorizing digestive disorders uses a combination of four anatomical and functional characteristics (Figure 52●). The largest category includes inflammation and infection of the digestive organs. Table 31 summarizes information about infectious diseases of the digestive system. Digestive system cancers are also relatively common and diverse. Malabsorption disorders are characterized by problems with the absorption of one or more nutrients. Congenital disorders result from developmental problems affecting the structure of the digestive tract or accessory organs.

25

TABLE 30 Examples of Tests Used in the Diagnosis of Gastrointestinal Disorders

Diagnostic Procedure	Method and Result	Representative Uses
Oral Cavity Sialography	X-ray film is taken after contrast medium is injected into salivary ducts while patient's salivary glands are stimulated	Identifies calculi (stones), inflammation, and tumors in salivary duct or gland
Periapical (PA) x-rays	Periapical film is taken of crown and root area of several teeth	Detects tooth decay, tooth impactions, fractures, progression of bone loss with periodontal disease, inflammation of periodontal ligament, and periapical abscesses (at end of root)
Bitewing x-rays of teeth	Bitewing film is taken. X-ray cone is pointed perpendicularly toward the spaces between the crowns of adjacent posterior teeth	Reveal tooth decay between teeth and early bone loss in periodontal disease
Upper GI (esophagus, stomach, and duodenum)		
Esophagogastroduodenoscopy, esophagoscopy, gastroscopy	Fiber-optic endoscope is inserted into oral cavity and further into esophagus, stomach, and duodenum to permit visualization of lining and lumen	Detects tumors, ulcerations, polyps, esophageal varices, inflammation, and obstructions; provides biopsy of tissues in upper GI tract
Barium swallow	Series of x-rays of esophagus is taken after barium sulfate is ingested to increase contrast of structures; normally done with upper GI series	Determines abnormalities of pharynx and esophagus, especially to determine cause of dysphagia; identifies tumors, hiatal hernia, and diverticuli
Upper GI series	Series of x-rays of stomach and duodenum is taken, after barium sulfate is ingested to increase contrast	Investigates epigastric pain; detects ulcers, polyps, gastritis, tumors, and inflammation in upper GI tract
Esophageal function studies Acid reflux and Acid clearing	Slender tube inserted from oral cavity into stomach through which dilute hydrochloric acid solution is passed. A pH probe is placed in the esophagus. The pH at the level of the gastroesophageal sphincter is measured periodically to detect acid reflux. Hydrochloric acid solution is then given again, and patient swallows until the acid is cleared from the esophagus.	Detects gastroesophageal reflux Acid clearing normally occurs in less than 10 swallows. If more are required, patient may have severe esophagitis from chronic gastric acid reflux (gastroesophageal reflux disease, or GERD).
Gastric analysis Basal gastric secretion	When patient is fasting, tube is inserted into nose and routed along esophagus to sample stomach contents. Acid levels are periodically monitored through this *nasogastric (NG) tube* to determine basal levels of gastric secretion.	Identifies increased acid levels, a cause of ulceration of gastric lining. Ulceration in the absence of increased acidity may indicate a malignancy.
Gastric acid stimulation	Chemical stimulus is administered and acid levels are monitored to determine results of gastric acid stimulation test	As above
Arteriography (celiac and mesenteric)	Catheter is inserted into femoral artery and threaded to celiac trunk or a mesenteric artery, where contrast medium is injected; x-ray films are then taken	Determines site of bleeding in gastrointestinal tract or blockage of blood supply
Lower GI (colon)		
Barium enema	Barium sulfate enema is administered to provide contrast in intestinal lumen, and x-ray film series is taken	Detects causes of abdominal pain and bloody stools; detects tumors and obstructions of bowel; identifies polyps and diverticula
Sigmoidoscopy Colonoscopy	Fiber-optic tube is inserted into rectum for viewing of anus, rectum, sigmoid colon; colonoscopy allows viewing of the entire large intestine to cecum; biopsy can be performed	Detects tumors, polyps, and ulcerations of intestinal lining. Polyps and biopsies may be removed with scope. Recommended screening test for colon cancer in persons over 50

25

TABLE 30	Examples of Tests Used in the Diagnosis of Gastrointestinal Disorders (*Continued*)	
Diagnostic Procedure	**Method and Result**	**Representative Uses**
Liver, Pancreas, Gallbladder		
Cholangiography Intravenous or percutaneous	Intravenously administered dye is concentrated in liver and released into bile duct; x-ray films are taken Percutaneous method involves insertion of needle with catheter into bile duct, using ultrasonography for guidance; dye is then administered through catheter	Detects biliary calculi and other obstructions of biliary tract
Liver biopsy	Needle is inserted through small skin incision and into liver; small plug of liver tissue is removed	Liver tissue is examined for evidence of cirrhosis, hepatitis, tumors, and granuloma. Often guided by CT scan during procedure.
Endoscopic retrograde cholangiopancreatography (ERCP)	Duodenoscope (fiber-optic tube) is inserted into oral cavity and threaded through stomach to the duodenum; through scope, catheter is inserted into duodenal ampulla for dye injection; x rays are then taken to visualize the bile duct system	Detects calculi, tumors, or cysts of pancreatic or bile ducts; determines presence of pancreatic tumor
Abdomen		
X-ray of abdomen	Radiodense tissues appear in white on negative film image	Detects abdominal masses, obstructions, and presence of free air in peritoneal cavity (usually from perforated ulcer)
Computerized tomography	Cross-sectional radiation is applied to provide three-dimensional images of liver, gallbladder, pancreas, and bowel	Identifies tumors, pancreatitis (acute, chronic), hepatic cysts and abscesses, and biliary calculi (stones); may diagnose appendicitis
Ultrasonography	Liver, pancreas, gallbladder, and bowel are examined by means of sound waves emitted by transducer placed on abdomen. Sound waves deflect off dense structures and produce echoes, which are amplified and graphically recorded.	Detects polyps, tumors, abscesses, hepatic cysts, and gallstones; may diagnose appendicitis

DIGESTIVE SYSTEM DISORDERS

■ DENTAL PROBLEMS AND SOLUTIONS

HA p. 660

The mass of a plaque deposit protects the bacteria that normally reside in the mouth from salivary secretions. As the pathogenic bacteria digest nutrients, they generate acids that erode the enamel and dentin of the teeth. The result is **dental caries**, or "cavities." Vaccines are now being developed to prevent dental caries by promoting specific resistance to *Streptococcus mutans*, the most abundant bacterium at these sites.

If *S. mutans* (or another bacterium) reaches the pulp and infects it, *pulpitis* (pul-PĪ-tis) results. Treatment generally involves the complete removal of the pulp tissue, especially the sensory innervation and all areas of decay; the pulp cavity is then packed with appropriate materials. This procedure is called a *root canal.*

Brushing the exposed surfaces of your teeth after you eat helps prevent the settling of bacteria and the entrapment of food particles, but bacteria between your teeth (in the region known as the *interproximal space*) and within the gingival sulcus may elude the brush. Dentists therefore recommend the daily use of dental floss to clean these spaces and stimulate the gingival epithelium. If bacteria and plaque remain within the gingival sulcus for extended periods, the acids generated begin eroding the connections between the neck of the tooth and the gingiva. The gums appear to recede from the teeth, and **periodontal disease** develops. As this disease progresses, the bacteria attack the cementum, eventually destroying the periodontal ligament and eroding the bone of the alveolus (Figure 53●). This deterioration loosens the tooth, and it falls out or must be pulled. Periodontal disease is the most common cause of tooth loss.

Lost or broken teeth have commonly been replaced by "false teeth" attached to a plate or frame inserted into the mouth. An alternative that uses *dental implants* was developed in the 1980s. A ridged titanium cylinder is inserted into

25

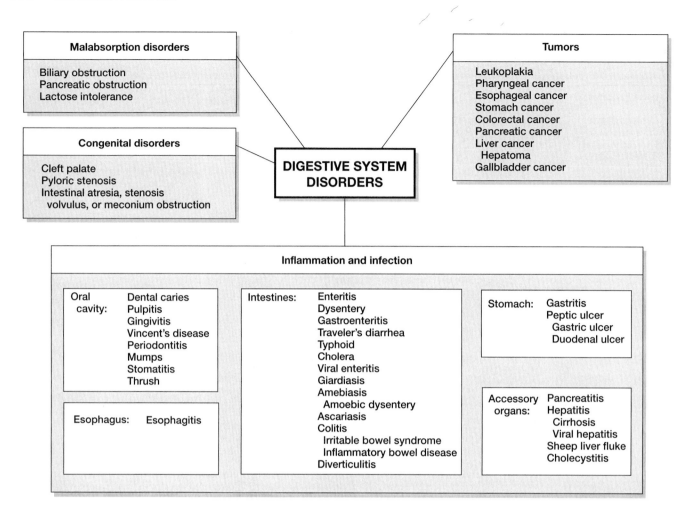

Malabsorption disorders

Biliary obstruction
Pancreatic obstruction
Lactose intolerance

Congenital disorders

Cleft palate
Pyloric stenosis
Intestinal atresia, stenosis
 volvulus, or meconium obstruction

DIGESTIVE SYSTEM DISORDERS

Tumors

Leukoplakia
Pharyngeal cancer
Esophageal cancer
Stomach cancer
Colorectal cancer
Pancreatic cancer
Liver cancer
 Hepatoma
Gallbladder cancer

Inflammation and infection

Oral cavity:	Dental caries Pulpitis Gingivitis Vincent's disease Periodontitis Mumps Stomatitis Thrush
Esophagus:	Esophagitis

Intestines:	Enteritis Dysentery Gastroenteritis Traveler's diarrhea Typhoid Cholera Viral enteritis Giardiasis Amebiasis Amoebic dysentery Ascariasis Colitis Irritable bowel syndrome Inflammatory bowel disease Diverticulitis

Stomach:	Gastritis Peptic ulcer Gastric ulcer Duodenal ulcer
Accessory organs:	Pancreatitis Hepatitis Cirrhosis Viral hepatitis Sheep liver fluke Cholecystitis

● **FIGURE 52**
Disorders of the Digestive System

the alveolus, and osteoblasts lock the ridges into the surrounding bone. After four to six months, an artificial tooth is screwed into the cylinder.

Dental implants are not suitable for everyone. Enough alveolar bone must be present to provide a firm attachment, for example, and complications may occur during or after surgery. Nevertheless, as the technique evolves, dental implants will become increasingly important. Roughly 42 percent of individuals over age 65 have lost all their teeth; the rest have lost an average of 10 teeth.

25

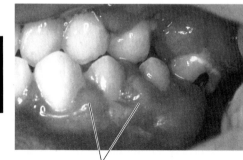

(a) Inflamed and
swollen gingivae

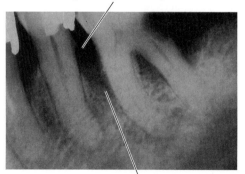

Enlarged interproximal spaces

(b) Eroding alveolar bone

● **FIGURE 53**
Periodontal Disease. (a) Notice the inflammation and swelling of the gingivae. This is an indication of serious periodontal disease. (b) An x-ray of three teeth in a person with severe periodontal disease. The gums have receded from the necks of the teeth, enlarging the interproximal spaces, and the alveolar bone is being eroded.

TABLE 31   Examples of Infectious Diseases of the Digestive System

Disease	Organism(s)	Description
Bacterial Diseases		
Dental caries	*Streptococcus mutans* and other oral bacteria	Tooth decay; bacteria in dental plaque on teeth produce acids that dissolve tooth enamel, leading to cavities
Pulpitis	As above	Infection of the pulp of the tooth
Gingivitis	As above	Infection of the gums
Vincent's disease	As above	Acute necrotizing ulcerative gingivitis, or trench mouth; bacterial infection and ulcer formation
Periodontitis	As above	Infection of gums and bone; results in loosening and loss of teeth
Peptic ulcers	*Helicobacter pylori*	Ulcers in gastric lining
Traveler's diarrhea	Enterotoxic form of *Escherichia coli*	Mild to severe watery diarrhea, nausea, vomiting, abdominal pain and general lack of energy
Typhoid	*Salmonella typhi*	Infection of the intestines and gallbladder; abdominal pain, abdominal distention, low WBC count, and enlarged spleen
Cholera	*Vibrio cholerae*	Intestinal infection; symptoms include nausea, vomiting, abdominal pain, and diarrhea; causes severe dehydration
Viral Diseases		
Mumps	Mumps virus (paramyxovirus)	Infection of the salivary glands; can spread to the pancreas meninges, and gonads
Viral enteritis	Rotaviruses	Intestinal infection; causes watery diarrhea, especially in young children
Viral hepatitis	Hepatitis A virus (HAV)	Infectious hepatitis; transmitted by fecal-contaminated water, milk, shellfish, or other food, no chronic disease
	Hepatitis B virus (HBV)	Serum hepatitis; transmitted by exchange of body fluid through transfusions, wounds, shared needles (intravenous drug use), or sexual contact; pregnant carrier of the virus can pass it on to her baby. May become chronic
	Hepatitis C virus (HCV)	Formerly non-A, non-B hepatitis; transmitted through blood, injecting drug use, and sexual contact; may be chronic
	Hepatitis D virus	Occurs only in persons infected with HBV; transmission as for HBV
	Hepatitis E virus	Transmission as for HAV, no chronic disease
Fungal Diseases		
Candidiasis (thrush)	*Candida albicans*	Yeast infection of the oral mucosa and/or esophagus; forms white, milky patches in the mouth
Parasitic Diseases		
Protozoa		
Giardiasis	*Giardia lamblia*	Intestinal infection; symptoms include diarrhea, bloating, and weight loss; more common in children than adults
Amebiasis, or amoebic dysentery	*Entamoeba histolytica*	Infection of the large intestine; can produce ulcers and peritonitis; diarrhea contains blood
Helminths		
Ascariasis	*Ascaris lumbricoides*	Roundworm infestation of the intestines; larval movement to pharynx causes cough and damage to intestinal wall; adult worms eat contents of the intestine
Enterobiasis, or pinworms	*Enterobius vermicularis*	Small (1 cm) roundworm, common in small children, lives in colon, deposit eggs on perianal skin, cause itching. Eggs survive in the environment for up to two weeks. Ingested eggs continue the life cycle.

25

■ ACHALASIA, ESOPHAGITIS, AND GERD *HA p. 661*

In the condition known as **achalasia** (ak-a-LĀ-zē-uh), a swallowed bolus descends along the esophagus relatively slowly, as a result of abnormally weak peristaltic waves, and its arrival does not trigger the opening of the lower esophageal sphincter. Materials then accumulate at the base of the esophagus like cars at a stoplight. Secondary peristaltic waves may occur repeatedly, causing the individual discomfort. The most successful treatment involves weakening the lower esophageal sphincter muscle by either cutting the circular muscle layer at the base of the esophagus or expanding a balloon in the lower esophagus until the muscle layer tears.

Brief, limited reflux of stomach contents into the lower part of the esophagus often occurs after meals, and the incidence is increased with abdominal obesity, pregnancy, and reclining after large meals. A weakened or permanently relaxed sphincter can cause frequent, prolonged reflux, leading to **esophagitis** (ē-sof-a-JĪ-tis), or inflammation of the esophagus from contact with powerful stomach acids. The esophageal epithelium has few defenses against attack by acids and enzymes; inflammation, epithelial erosion, and intense discomfort result. Frequent episodes of backflow result in **gastroesophageal reflux disease (GERD),** which causes the symptoms commonly known as heartburn. This condition supports a multimillion-dollar industry devoted to producing and promoting antacids and medications that suppress acid production. Simply elevating the head of one's bed reduces GERD symptoms as much as medication in many people. Some people with GERD may cough and suffer throat and lung problems, presumably from flow of stomach fluids up the esophagus and aspiration into the trachea. However, coughing may promote esophageal reflux, so the cause-and-effect link is uncertain.

■ ESOPHAGEAL VARICES *HA p. 661*

The veins draining the inferior portion of the esophagus empty into tributaries of the hepatic portal vein. If the venous pressure in the hepatic portal vein becomes abnormally high due to liver damage or the constriction of the vessels, blood will pool in the submucosal veins of the esophagus. The veins become grossly distended and create bulges in the esophageal wall. These distorted **esophageal varices** (VAR-i-sēz; *varices,* dilated veins) may rupture, causing massive bleeding into the submucosal tissues or into the lumen of the esophagus and stomach. Esophageal varices commonly develop in individuals with advanced *cirrhosis,* a chronic liver disorder that restricts hepatic blood flow (p. 146).

■ STOMACH CANCER *HA p. 665*

Stomach, or *gastric,* **cancer** is one of the most common lethal cancers, responsible for roughly 12,000 deaths in the United States in 2003. The incidence is higher in Japan, Korea, and other countries where the typical diet includes large quantities of pickled, fermented, or smoked foods. Because the signs and symptoms can resemble those of gastric ulcers, the condition may not be recognized in its early stages. Diagnosis generally involves examining x-rays of the stomach at various degrees of distension. The mucosa can also be visually inspected by using a flexible instrument called a *gastroscope.* Attachments permit the collection of tissue samples for analysis. Treatment of stomach cancer starts with *gastrectomy* (gas-TREK-tō-mē), the surgical removal of part or all of the stomach. People can survive even a total gastrectomy, because the loss of such functions as food storage and acid production is not life-threatening.

Protein breakdown can still be performed by the small intestine, although at reduced efficiency, and the diet can be supplemented with Vitamin B_{12}, which usually requires intrinsic factor from the stomach in order to be absorbed.

■ DRASTIC WEIGHT-LOSS TECHNIQUES *HA p. 667*

At any moment, an estimated 20 percent of the U.S. population is dieting to promote weight loss. In addition to the appearance of "fat farms" and exercise clubs, the use of surgery to promote weight loss has been on the rise.

■ SURGICAL REMODELING OF THE GASTROINTESTINAL TRACT

Gastric stapling attempts to correct an overeating problem by reducing the size of the stomach. A large portion of the gastric lumen is stapled shut, leaving only a small pouch in contact with the esophagus and duodenum. After this surgery, the individual can eat only a small amount before the stretch receptors in the gastric wall become stimulated and a feeling of fullness results. Gastric stapling is a major surgical procedure with many potential complications. In addition, the smooth muscle in the wall of the functional portion of the stomach gradually becomes increasingly tolerant of distension, and the operation may have to be repeated.

A *gastric bypass* is a surgical procedure that connects the proximal small intestine to a small pouch formed by a superior portion of the stomach, with the same goal of reducing gastric capacity. This procedure seems more effective than gastric stapling, but involves more complicated surgery.

An earlier, more drastic approach involves the surgical bypass of a large portion of the jejunum. This procedure reduces the effective absorptive area, producing a marked weight loss. After the operation, the individual must take dietary supplements to ensure that all the essential nutrients and vitamins can be absorbed before the chyme enters the large intestine. Because chronic diarrhea and serious liver disease are potential complications, ileal-jejunal bypass surgery has largely been replaced by gastric bypass or stapling. Abdominal surgery and anesthesia in obese people has higher risks of complications and death than in people with normal body weights, but obesity has its own increased risks as well.

■ LIPOSUCTION

One much-publicized method of battling obesity is **liposuction,** a surgical procedure for the removal of unwanted adipose tissue. Adipose tissue is flexible, but not as elastic as areolar tissue, and it tears relatively easily. In liposuction, a small incision is made through the skin, and a tube is inserted into the underlying adipose tissue. Suction is then applied. Because adipose tissue tears easily, chunks of tissue containing adipocytes, other cells, fibers, and ground substance can be vacuumed away. Liposuction is among the most common cosmetic surgeries performed today, with approximately 375,000 procedures performed in 2002.

Liposuction has received a lot of news coverage, and many advertisements praise the technique as easy, safe, and effective. In fact, it is not always easy, and it can be dangerous and have limited effectiveness. The density of adipose tissue varies from place to place in the body and from individual to individual, and it is not always easy to suck through a tube. An anesthetic must be used to control pain, and anesthesia always poses risks; also, blood vessels are stretched and torn, and extensive bleeding can occur. Heart attacks, pulmonary embolism, and fluid balance problems can develop, with fatal results. The death rate for this procedure is one in 5000. Finally, adipose tissue can repair itself, and adipocyte populations recover over time. The only way to ensure that fat lost through liposuction will not return is to adopt a lifestyle that includes a proper diet and adequate exercise. Over time, such a lifestyle can produce the same weight loss, *without liposuction,* eliminating the surgical expense and risk.

■ VOMITING   *HA p. 669*

The responses of the digestive tract to chemical or mechanical irritation are predictable: Fluid secretion accelerates all along the digestive tract, and the intestinal contents are eliminated as quickly as possible. The *vomiting reflex* occurs in response to irritation of the fauces, pharynx, esophagus, stomach, or proximal portions of the small intestine. These sensations are relayed to the *vomiting center* of the medulla oblongata, which coordinates the motor responses. In preparation for vomiting, the pylorus relaxes and the contents of the duodenum and proximal jejunum are discharged into the stomach by strong peristaltic waves that travel toward the stomach rather than toward the ileum.

Vomiting, or *emesis* (EM-e-sis), then occurs as the stomach regurgitates its contents through the esophagus and pharynx. During regurgitation, the uvula and soft palate block the entrance to the nasopharynx. Increased salivary secretion assists in buffering the stomach acids, thereby slowing erosion of the teeth. In conditions marked by repeated vomiting, severe tooth damage can occur; this is one sign of the eating disorder *bulimia.* Most of the force of vomiting comes from expiratory movements that elevate intra-abdominal pressures and force the stomach against the tensed diaphragm.

■ COLON INSPECTION AND CANCER
   HA p. 672

Colon cancers are relatively common. Approximately 105,500 cases were diagnosed in the United States in 2003 (in addition to 2,000 cases of rectal cancers). An estimated 57,000

deaths occurred from colon and rectal cancers combined. The mortality rates for these cancers remains high, and the best defense appears to be early detection and prompt treatment. It is believed that most colorectal cancers begin as small, localized mucosal tumors, or **polyps** (POL-ips), that grow from the intestinal wall. The prognosis improves dramatically if cancerous polyps are removed before metastasis has occurred. An early sign of polyp formation may be the appearance of blood in the feces. Unfortunately, many people ignore small amounts of blood in fecal materials, because they attribute the bleeding to "harmless" hemorrhoids. This offhand diagnosis should always be professionally verified.

One screening test involves checking the feces for blood, a simple procedure that can easily be performed on a stool (fecal) sample as part of a routine physical. For those at increased risk because of family history, associated disease, or age (over 50), visual inspection of the colon lumen by fiberoptic colonoscopy is recommended. A flexible **colonoscope** (kō-LON-ō-skōp) permits direct visual inspection of the lining of the large intestine. The colonoscope can take a biopsy of the mucosal lining and remove small polyps.

Primary risk factors for colorectal cancer include (1) a diet rich in animal fats and low in fiber and vegetables, (2) *inflammatory bowel disease,* and (3) a number of inherited disorders that promote epithelial tumor formation along the intestines.

■ DIVERTICULOSIS AND COLITIS
   HA p. 673

In **diverticulosis** (dī-ver-tik-ū-LŌ-sis), pockets (*diverticula*) form in the mucosa, generally in the sigmoid colon. The pockets get forced outward, probably by the pressures generated during defecation. If they push through weak points in the muscularis externa, the pockets form semi-isolated chambers that are subject to recurrent infection and inflammation. The infections cause pain and occasional bleeding, a condition known as *diverticulitis* (dī-ver-tik-ū-LĪ-tis).

Irritable bowel syndrome is characterized by diarrhea, constipation, or an alternation between the two. When constipation is the primary problem, the condition is sometimes called a *spastic colon,* or *spastic colitis.* Irritable bowel syndrome may have a partly neurological basis; the mucosa is usually normal in appearance, but the motility can be abnormal. Hypersensitivity and reactivity of the "enteric nervous system" of the bowel has been implicated. Bulking agents such as fiber or psyllium (*Metamucil*) and drugs that affect the enteric nervous system can help control symptoms.

Inflammatory bowel disease, or **colitis** (kō-LĪ-tis), is much more serious than irritable bowel syndrome. It involves chronic inflammation of the digestive tract. In Crohn's disease, all layers of the intestinal wall may be involved, particularly in the ileum, but the disease is focal, with areas of healthy bowel between areas of disease. In *ulcerative colitis* the colonic mucosa becomes diffusely inflamed, and ulcerated, with areas of tissue death and deterioration of colonic function. Acute bloody diarrhea, cramps, fever, and weight loss may occur with either disease.

25

The treatment of inflammatory bowel disease involves anti-inflammatory drugs and corticosteroids that reduce inflammation. In cases that do not respond to other therapies, immunosuppressive drugs, such as *cyclosporine,* may be effective. People with inflammatory bowel disease also have an increased risk of intestinal cancer.

The treatment of prolonged severe inflammatory bowel disease may involve a **colectomy** (kō-LEK-tō-mē)—the removal of all or a portion of the colon. If a large part or even all of the colon must be removed, normal connection with the anus cannot be maintained. Instead, the end of the intact digestive tube is sutured to the abdominal wall, and wastes then accumulate in a plastic pouch or sac attached to the opening, or a surgically created internal pouch. If the attachment involves the colon, the procedure is a **colostomy** (kō-LOS-tō-mē); if the ileum is involved, it is an **ileostomy** (il-ē-OS-tō-mē). Surgeons can sometimes construct an internal pouch of healthy intestine attached to the rectum, preserving bowel function at the cost of diarrhea.

■ DIARRHEA   *HA p. 673*

In **diarrhea** (dī-a-RĒ-uh), an individual has frequent, watery bowel movements. The condition results when the mucosa of the colon becomes unable to maintain normal levels of absorption, or when the rate of fluid entry into the colon exceeds the organ's maximum reabsorptive capacity. Bacterial, viral, or protozoan infection of the colon or small intestine can cause acute bouts of diarrhea lasting several days. Severe diarrhea is life-threatening due to cumulative fluid and ion losses.

Many conditions result in diarrhea, and we will consider only a representative sampling here. Most infectious diarrhea involves organisms that are shed in the stool and then spread from person to person (or from person or animal to food to person). Proper food preparation and storage, hand washing after defecation, clean drinking water, and good sewage disposal are the best preventive measures.

■ GASTROENTERITIS

An irritation of the small intestine can lead to a series of powerful peristaltic contractions that eject the contents of the small intestine into the large intestine. An extremely powerful irritating stimulus produces a "clean sweep" of the absorptive areas of the digestive tract. Vomiting clears the stomach, duodenum, and proximal jejunum, and peristaltic contractions evacuate the distal jejunum, ileum, and colon. Bacterial toxins, viral infections, and various poisons may produce these extensive gastrointestinal responses. Conditions affecting primarily the small intestine are usually referred to as a form of **enteritis** (en-ter-Ī-tis). If both vomiting and diarrhea are present, the term **gastroenteritis** (gas-trō-en-ter-Ī-tis) may be used instead.

■ TRAVELER'S DIARRHEA

Traveler's diarrhea, a form of infectious diarrhea generally caused by a bacterial or viral infection, develops because the irritated or damaged mucosal cells are unable to maintain normal absorption levels. The irritation stimulates the production of mucus, and the damaged cells and mucous secretions add to the volume of feces produced. Despite the inconvenience, this type of diarrhea is usually temporary, and limited diarrhea is probably a reasonably effective method of rapidly removing an intestinal irritant. Drugs, such as *Lomotil,* that prevent peristaltic contractions in the colon slow the diarrhea, but leave the irritant intact, and the symptoms may return with a vengeance when the effects of the drug fade. A three- to five-day course of antibiotics may be effective in controlling diarrhea due to bacterial infection.

■ GIARDIASIS

Giardiasis is an infection caused by the protozoans *Giardia intestinalis* and *G. lamblia* (Figure 54●). These pathogens can colonize the duodenum and jejunum and interfere with the normal absorption of lipids and carbohydrates. Many people do not develop acute symptoms but act as carriers who can spread the disease. Acute symptoms usually appear within three weeks of initial exposure. Diarrhea, abdominal cramps, nausea, and vomiting are the primary complaints. These symptoms persist for five to seven days or longer, and some patients are subject to relapses, with chronic bloating, diarrhea, and weight loss. Treatment typically consists of the oral administration of drugs, such as *metronidazole,* that can kill the protozoan.

The transmission of giardiasis requires that food or water be contaminated with feces that contain *cysts*—resting stages of the protozoan that are produced during its passage through the large intestine. Rates of infection are highest (1) in developing countries with poor sanitation, (2) among campers drinking surface water, (3) among individuals with impaired immune systems (as in AIDS), and (4) among toddlers and young children. The cysts can survive in cold, fresh water for months, and they are not killed by the chlorine treatment used to kill bacteria in drinking water. Travelers are advised to boil or ultrafilter water and to heat food properly before eating it, as these preventive measures will destroy the cysts.

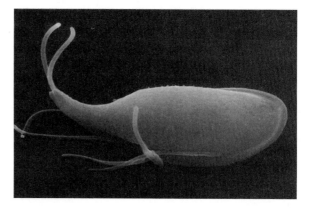

● **FIGURE 54**
Giardia. *Giardia* is a flagellated protozoan that may infect the small intestine.

25

■ CHOLERA EPIDEMICS

Cholera (KOL-e-ruh) epidemics are most common in areas where sanitation is poor, drinking water is contaminated by fecal wastes, and eating raw fish or shellfish is popular. After an incubation period of one to two days, the symptoms of nausea, vomiting, and diarrhea persist for two to seven days. The cholera bacteria bind to the intestinal lining and release toxins that stimulate a massive secretion of fluids across the intestinal epithelium. Fluid loss during the worst stage of the disease can approach 1 liter per hour. This dramatic loss causes a rapid drop in blood volume, leading to acute hypovolemic shock and damage to the kidneys and other organs. Without treatment, a person with cholera can die of acute dehydration in a matter of hours. Treatment consists of oral or intravenous fluid replacement while the disease runs its course. Antibiotic therapy may also prove beneficial. A vaccine is available, but its low success rate (40–60 percent) and short-lived effectiveness (four to six months of protection) make it relatively ineffective in preventing or controlling cholera outbreaks. Many hundreds of thousands of cases of cholera were reported during an epidemic that began in Peru in 1991 and has since spread through Central and South America. The death rate was 0.5 percent—a remarkably low rate, compared with death rates in other outbreaks in the 20th century, which were as high as 60 percent.

■ LACTOSE INTOLERANCE

Lactose intolerance is a malabsorption syndrome, revealed by eating dairy foods, that results from the lack of the enzyme *lactase* at the brush border of the intestinal epithelium. The condition poses more of a problem than would be expected if the only outcome were the inability to use the disaccharide lactose. Undigested lactose provides a particularly stimulating energy source for the bacterial inhabitants of the colon. The result is increased intestinal gas, cramps, and diarrhea, problems that can develop if the individual drinks more than 8 oz of milk or eats similar amounts of other dairy products.

Lactose intolerance appears to have a genetic basis. Infants produce lactase to digest milk, but older children and adults may stop producing this enzyme. In some populations, lactase production continues throughout adulthood. Only about 15 percent of Caucasians develop lactose intolerance, whereas estimates ranging from 80 to 90 percent have been suggested for the adult African and Asian populations. These differences reflect and affect dietary preferences in those groups, and food relief efforts must take these facts into account. For example, shipping powdered milk to famine-stricken areas in Africa can make matters worse if supplies are distributed to adults rather than to children.

■ CONSTIPATION   *HA p. 673*

Constipation is infrequent defecation of small, dry, hard feces, usually less than three times a week. It is not the frequency of defecation that indicates constipation, and there is not a "right" number of bowel movements a week. Constipation occurs when fecal materials move through the colon so slow-

ly that excessive reabsorption of water occurs. The feces become extremely compact, difficult to move, and highly abrasive. Inadequate dietary fiber and water, coupled with a lack of exercise, is a common cause of constipation.

Constipation can generally be treated by more fiber and water in the diet and the oral administration of stool softeners such as Colace. This promotes movement of water into the feces, increasing fecal mass and softness. Indigestible fiber adds bulk to the feces, retaining moisture and stimulating stretch receptors that promote peristalsis. Thus, the promotion of peristalsis is a benefit of high-fiber cereals. Active movement during exercise also assists in the movement of fecal materials through the colon. Laxatives, or **cathartics** (ka-THAR-tiks), promote defecation by irritating the lining of the colon to stimulate peristalsis. These compounds should be used sparingly to avoid nerve damage in the colon that may occur with prolonged use and ultimately interferes with normal peristalsis.

■ LIVER DISEASE   *HA p. 676*

■ ASSESSING LIVER STRUCTURE AND FUNCTION

The liver is the largest and most important visceral organ, and liver disorders affect almost every other vital system in the body. A variety of clinical tests are used to check the functional and physical state of the liver:

- **Liver scans** involve the injection of radioisotope-labeled compounds into the bloodstream. Compounds are chosen that will be selectively absorbed by Kupffer cells, liver cells, or abnormal liver tissues.
- **CT scans** of the abdominal region are commonly used to provide information about cysts, abscesses, tumors, or hemorrhages in the liver.
- A **liver biopsy** can be taken by a long needle, commonly guided by CT scans to avoid large blood vessels. The needle is inserted into and through the abdominal wall. Laparoscopic examination can also reveal gross structural changes in the liver or gallbladder, and collect a biopsy as well.
- Liver function tests can assess specific functional capabilities, and serum and plasma protein assays can detect changes in the liver's rate of plasma protein synthesis. Serum enzyme tests can reveal liver damage by detecting intracellular enzymes in the circulating blood.

■ HEPATITIS

Hepatitis (hep-a-TĪ-tis) is inflammation of the liver. Viruses that target the liver are responsible for most cases of hepatitis, although some environmental toxins can cause similar symptoms. Five forms of viral hepatitis have been identified:

1. **Hepatitis A**, or *infectious hepatitis,* typically results from the ingestion of water, milk, shellfish, or other food contaminated by infected fecal wastes. The disease has a relatively short incubation period of two to six weeks and generally runs its course in a matter of months. Fatalities are rare among individuals under age 40. There is no ongoing chronic infection.

25

2. **Hepatitis B**, or *serum hepatitis,* is transmitted by the exchange of body fluids during intimate contact. For example, infection can occur through the transfusion of unscreened blood products, through a break in the skin or mucosa, or by sexual contact. (Blood products have been screened for hepatitis B since the 1970s.) The incubation period for hepatitis B ranges from one to six months. If a pregnant woman is infected, the newborn baby may become infected at birth. Presumably because of the newborn's weaker immune system, up to 90 percent of infants infected at birth become chronically infected or chronic carriers, whereas only 1 percent of newly infected adults become chronic carriers. Hepatitis B vaccinations are recommended for all newborns to break the cycle of maternal–infant transmission. Chronic carriers are infectious and may experience cumulative liver damage, including increased risk of liver cancer. Treatment with interferon and antiviral medication may slow the progression of this and other liver diseases.

3. **Hepatitis C**, originally designated *non-A, non-B hepatitis,* was most commonly transferred from individual to individual through the collection and transfusion of unscreened contaminated blood. Since 1990, blood-screening procedures have been used to lower the incidence of transfusion-related hepatitis C to one case per 100,000 units transfused. The disease can also be transmitted among injecting drug users by the sharing of needles, and evidence suggests that it is rarely sexually transmitted. The chronic infectious carrier state of hepatitis C infection produces significant liver damage in roughly half the individuals infected with the virus. Interferon treatment, alone or (preferably) combined with antiviral medication early in the course of the disease, can reduce the viral load to undetectable levels and slow the progression of the disease in many patients.

4. **Hepatitis D** is caused by a virus that produces symptoms only in people already infected with hepatitis B. The transmission of hepatitis D resembles that of hepatitis B. In the United States, the disease is most common among intravenous drug users. The combination of hepatitis B and hepatitis D causes progressive and severe liver disease.

5. **Hepatitis E** resembles hepatitis A in that it is transmitted by the ingestion of contaminated food or water and is not a chronic disease. Hepatitis E is the most common form of hepatitis worldwide, but cases seldom occur in the United States. Hepatitis E infections are most acute and are potentially lethal for pregnant women.

The hepatitis viruses disrupt liver function by attacking and destroying liver cells. An infected individual may develop a high fever, and the liver may become inflamed and tender. As the disease progresses, several hematological parameters change markedly. For example, enzymes normally confined to the cytoplasm of functional liver cells appear in the circulating blood. Normal metabolic regulatory activities become less effective, and blood glucose levels decline. Plasma protein synthesis slows, and the clotting time becomes unusually long. The injured hepatocytes stop removing bilirubin from the circulating blood, and symptoms of jaundice appear.

Hepatitis is either acute or chronic. *Acute hepatitis* is characteristic of hepatitis A and E. Almost everyone who contracts hepatitis A or hepatitis E (except in pregnancy) eventually recovers, although full recovery can take several months. Once recovered, infected individuals cannot transmit the disease. Symptoms of acute hepatitis include severe fatigue and jaundice. *Chronic hepatitis* is a progressive disorder that can lead to severe medical problems as liver function deteriorates and cirrhosis develops. Common complications include the following:

- The formation of *esophageal varices,* due to portal hypertension
- Ascites, caused by increased peritoneal fluid production
- Bacterial peritonitis, which may recur for unknown reasons
- Hepatic encephalopathy, which is characterized by disorientation and confusion, probably caused by high blood levels of ammonia and the presence of abnormal concentrations of fatty acids, amino acids, and waste products

Hepatitis B, C, and D infections can produce both acute and chronic hepatitis. Individuals with chronic forms are potentially infectious and may eventually experience fatal liver failure or liver cancer. Roughly 10 percent of hepatitis B patients develop potentially dangerous complications; the percentage is higher for hepatitis C.

Passive immunization with pooled immunoglobulins is available for people exposed to the hepatitis A and B viruses. (Active immunization for hepatitis A and B is also available and is preferred.) No vaccines are available for hepatitis C, D, or E.

■ CIRRHOSIS

The underlying problem in **cirrhosis** (sir-Ō-sis) appears to be the widespread destruction of hepatocytes and scarring of the liver by exposure to drugs (especially alcohol), viral infection, ischemia, or a blockage of the hepatic ducts. Initially, the damage to hepatocytes leads to the formation of extensive areas of scar tissue that branch throughout the liver. The surviving hepatocytes then undergo repeated cell divisions, but the fibrous tissue prevents the new hepatocytes from achieving a normal arrangement of lobules. As a result, the liver gradually converts from an organized assemblage of lobules to a fibrous aggregation of poorly functioning cell clusters. Jaundice, ascites, and other symptoms appear as the condition progresses.

25

AN INTRODUCTION TO THE URINARY SYSTEM AND ITS DISORDERS

The urinary system consists of the kidneys, where urine production occurs, and the conducting system, which transports and stores urine prior to its elimination from the body. The conducting system comprises the ureters, the urinary bladder, and the urethra. Although the kidneys perform all the vital functions of the urinary system, problems with the conducting system can have direct and immediate effects on renal function.

■ THE HISTORY AND PHYSICAL EXAMINATION OF THE URINARY SYSTEM   *HA p. 688*

The primary symptoms of urinary system disorders are pain and changes in the frequency of urination.

1. **Pain:** The nature and location of pain can provide clues to the source of the problem. For example:

   - Pain in the superior pubic region may be associated with urinary bladder disorders.

   - Pain in the superior lumbar region or in the flank that radiates to the right upper quadrant or left upper quadrant can be caused by kidney infections such as *pyelonephritis,* or by kidney stones.

   - **Dysuria** (painful or difficult urination) can occur with *cystitis, urethritis,* or *urinary obstructions.* In males, enlargement of the prostate gland can lead to compression of the urethra and dysuria.

2. **Urgency and/or frequency:** Individuals with urinary system disorders may urinate more or less frequently than usual and may produce normal or abnormal amounts of urine. An irritation of the lining of the ureters or urinary bladder can lead to the desire to urinate with increased frequency, although the total amount of urine produced each day remains normal. Unsuppressable detrusor muscle contractions may also lead to urinary urgency and frequency. When these problems exist, the individual feels the urge to urinate when the urinary bladder volume is very small. The irritation may result from trauma, urinary bladder infection (*cystitis*) or tumors, increased acidity of the urine, or hormonal and aging changes causing detrusor hyperreflexia.

   - **Incontinence**, an inability to control urination voluntarily, may involve periodic involuntary leakage or inability to delay urination—a continual, slow trickle of urine from a bladder that is always full. Incontinence results from urinary bladder or urethral problems, damage or weakening of the muscles of the pelvic floor, or interference with normal sensory or motor innervation in the region. Renal function and daily urinary volume are normal.

   - In **urinary retention**, renal function is normal, at least initially, but urination does not occur. Urinary retention in males commonly results from enlargement of the prostate

CHAPTER 26
THE URINARY SYSTEM

and compression of the prostatic urethra. In both sexes, urinary retention can result from the obstruction of the outlet of the urinary bladder or from central nervous system damage involving control of the detrusor muscle, such as might be caused by a stroke or damage to the spinal cord.

- Changes in the volume of urine produced by a normally hydrated person indicate problems either at the kidneys or with the control of renal function. **Polyuria**, the production of excessive amounts of urine, results from hormonal or metabolic problems, such as those associated with *diabetes* (p. 86), or from damage to the glomeruli, as in *glomerulonephritis*. **Oliguria** (a urine volume of 50–500 ml/day) and **anuria** (0–50 ml/day) are conditions that indicate serious kidney problems and potential renal failure. Renal failure can occur with *heart failure,* renal ischemia, *circulatory shock,* burns, pyelonephritis, hypovolemia, and a variety of other disorders.

Important clinical signs of urinary system disorders include the following:

- **Edema:** Renal disorders often lead to protein loss in the urine (proteinuria) and, if severe, result in generalized edema in peripheral tissues. Facial swelling, especially around the eyes, is common.

- **Fever:** A fever commonly develops when the urinary system is infected by pathogens. Urinary bladder infections (cystitis) typically result in a low-grade fever; kidney infections, such as pyelonephritis, can produce very high fevers.

During the physical assessment, percussion or palpation can be used to check the status of the kidneys and urinary bladder. The kidneys lie in the costovertebral area, the region bounded by the lumbar spine and the 12th rib on either side. To detect tenderness due to kidney inflammation, the examiner gently thumps a fist over each flank posterior to the kidneys. This usually does not cause pain, unless the underlying kidney is inflamed.

The urinary bladder can be palpated just superior to the pubic symphysis. However, on the basis of palpation alone, urinary bladder enlargement due to urine retention can be difficult to distinguish from the presence of an abdominal mass.

LABORATORY AND IMAGING TESTS AND THE URINARY SYSTEM *HA p. 688*

Many procedures and laboratory tests are used in the diagnosis of urinary system disorders. The functional anatomy of the urinary system can be examined with the use of a variety of sophisticated procedures. For example, an x-ray of the kidneys, ureters, and urinary bladder is taken after the administration of a radiopaque compound that will enter the urine. The resulting image is called an **intravenous pyelogram** (PĪ-el-ō-gram), or **IVP**. This procedure, sometimes called an *excretory urogram (EU),* permits the detection of unusual kidney, ureter, or urinary bladder structures and masses. Computerized tomography (CT) scans or ultrasound scans may also provide useful information about localized abnormalities.

Unusual laboratory findings may also provide clues as to the nature of a urinary system disorder:

- **Hematuria**, the presence of red blood cells in urine, indicates bleeding at either the kidneys or the conducting system. Hematuria producing dark red or tea-colored urine typically indicates bleeding in the kidney, and hematuria producing bright red urine indicates bleeding in the inferior portion of the urinary tract (the urinary bladder or the urethra). Hematuria most commonly occurs with trauma to the kidneys, calculi (kidney stones), tumors, or urinary tract infections.

TABLE 32 **Examples of Tests Used in the Diagnosis of Urinary System Disorders**

Diagnostic Procedure	Method and Result	Representative Uses
Cystoscopy	A small tube (cystoscope) is inserted through the urethra into the urinary bladder to view the lining of the urethra, urinary bladder, and ureteral openings within the bladder	Used to obtain a biopsy specimen or to remove stones (calculi) and small tumors; provides direct visualization of urethra, urinary bladder, and ureteral openings
Retrograde pyelography	Radiopaque dye is injected into ureters through a catheter in cystoscope inserted into urinary bladder; x-ray films are then taken	Detects obstructions of ureter caused by tumors, calculi, or strictures; visualizes renal pelvis and ureters without relying on renal filtration (useful if renal function is impaired)
Renal biopsy	Using ultrasound as a guide, biopsy needle is inserted through back and into kidney. Specimen is then removed for analysis.	Determines cause of renal disease; detects rejection of transplanted kidney; used to perform tumor biopsy
Intravenous pyelography (IVP)	Dye injected intravenously is filtered at kidney and excreted into urinary tract; x-rays are then taken to view kidneys, ureters, and urinary bladder	Determines size of kidney, obstructions such as calculi or tumors, or anatomical abnormalities; relies on renal filtration of contrast medium
Cystography	Dye is inserted through catheter placed in urethra and threaded into urinary bladder. X-rays are then taken.	Identifies tumors of urinary bladder and rupture of bladder by trauma; if x-rays are taken as patient voids, detects reflux of urine from urinary bladder to ureters

26

- **Hemoglobinuria** is the presence of hemoglobin in urine. Hemoglobinuria indicates increased hemolysis of red blood cells in the bloodstream due to cardiovascular or metabolic problems. Conditions that result in hemoglobinuria include the *thalassemias, sickle cell anemia, hypersplenism,* and some autoimmune disorders.

- Changes in the color of urine accompany some renal disorders. For example, urine becomes (1) cloudy due to the presence of bacteria, lipids, crystals, or epithelial cells; (2) red or brown from hemoglobin or myoglobin; (3) blue-green from bilirubin; or (4) brown-black from excessive concentration. Not all color changes are abnormal, however; some foods and several prescription drugs can cause changes in urine color. For example, a serving of beets can give urine a red-

dish color, whereas eating rhubarb can give urine an orange tint, and B vitamins turn it a vivid yellow.

Other diagnostic procedures used to examine the functional anatomy of the urinary system are detailed in Table 32. Figure 55● outlines the major classes of disorders of the urinary system.

■ URINALYSIS   *HA p. 697*

A variety of sophisticated laboratory tests can be performed on urine samples to determine levels of various electrolytes and nutrients. Several basic screening tests can also be performed by recording changes in the color of test strips that are dipped in the sample. Urine pH and approximate urinary concentrations

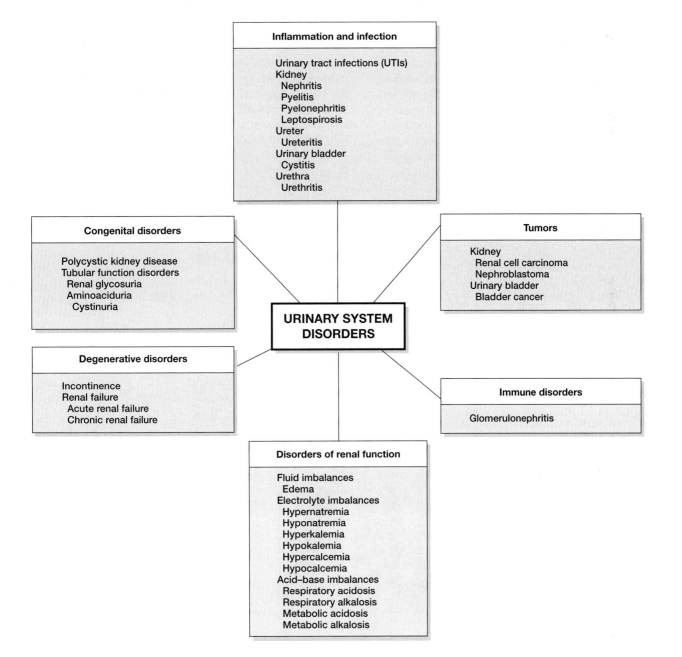

● **FIGURE 55**
Disorders of the Urinary System

of glucose, ketones, bilirubin, urobilinogen, plasma proteins, and hemoglobin, as well as the density, or *specific gravity,* of the urine, can be monitored by this technique. The specific gravity may also be determined by floating a simple device known as a **urinometer** (ū-ri-NOM-e-ter) or **densitometer** (den-si-TOM-e-ter) in a urine sample. A urine sample may also be spun in a centrifuge, and any sediment examined under the microscope. Mineral crystals, bacteria, red or white blood cells, and deposits, known collectively as **casts**, can be detected in this way. Figure 56● provides an overview of the major categories of urinary casts. During a urinary tract infection, bacteria may be cultured to determine their identities. New test strips can detect WBCs and nitrites found in urine infections.

More comprehensive analyses can determine the total osmolarity of the urine and the concentration of individual electrolytes and minor metabolites, metabolic wastes, vitamins, and hormones. A test for one hormone in the urine, *human chorionic gonadotropin* (hCG), provides an early and reliable proof of pregnancy.

The information provided by urinalysis can be especially useful when correlated with the data obtained from blood tests. The term **azotemia** (a-zō-TĒ-mē-uh) refers to the presence of excess metabolic wastes in the blood. This condition may result from the overproduction of urea or other nitrogenous wastes by the liver ("prerenal syndrome"). Conversely, in **uremia** (u-RĒ-mē-uh), all normal kidney functions are adversely affected.

The total volume of urine produced in a 24-hour period may also be of interest. *Polyuria* (pol-ē-Ū-rē-uh) refers to excessive production of urine—well over 2 liters per day. Polyuria most commonly results from endocrine disorders (such as the various forms of diabetes), metabolic disorders, or damage to the filtration apparatus, as in glomerulonephritis. **Oliguria** (o-li-GŪ-rē-uh) refers to inadequate urine production (50–500 ml/day). In **anuria** (a-NŪ-rē-uh), a negligible amount of urine is produced (0–50 ml/day), a potentially fatal problem.

DISORDERS OF THE URINARY SYSTEM

■ ADVANCES IN THE TREATMENT OF RENAL FAILURE *HA p. 697*

One normal kidney is sufficient to filter the blood and maintain homeostasis. As a result, renal failure will not develop unless both kidneys are damaged. The management of chronic renal failure typically involves restricting water and salt intake and minimizing dietary protein intake. This combination reduces strain on the urinary system by minimizing the volume of urine produced and preventing the generation of large quantities of nitrogenous wastes. Acidosis, a common problem in people with renal failure, can be countered by ingesting bicarbonate ions.

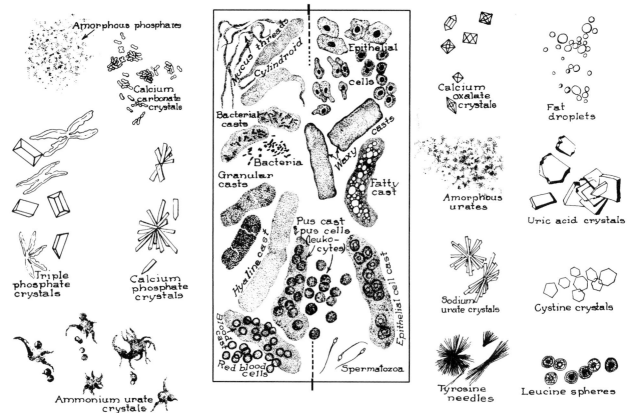

● **FIGURE 56**
Microscopic Examination of Urine Sediment. [Redrawn after Todd and Sanford.]

If drugs and dietary controls cannot stabilize the composition of blood, more drastic measures are taken. In **hemodialysis** (hē-mō-dī-AL-i-sis), a *dialysis machine* containing an artificial membrane is used to regulate the composition of blood (Figure 57a●). The basic principle involved in this process, called **dialysis**, is passive diffusion across a selectively permeable membrane. The patient's blood flows past an artificial *dialysis membrane*, which contains pores large enough to permit the diffusion of small ions, but small enough to prevent the loss of plasma proteins. On the other side of the membrane flows a special **dialysis fluid**.

As diffusion occurs across the membrane, blood composition changes. Potassium ions, phosphate ions, sulfate ions, urea, creatinine, and uric acid diffuse across the membrane into the dialysis fluid. Bicarbonate ions and glucose diffuse into the bloodstream. In effect, diffusion across the dialysis membrane replaces normal glomerular filtration, and the characteristics of the dialysis fluid (which can be modified for each patient) ensure that important metabolites remain in the bloodstream rather than diffusing across the membrane.

For temporary kidney dialysis, a silicone rubber tube called a *shunt* is inserted into a medium-sized artery and vein (Figure 57b●). (The typical location is the forearm, although the lower leg is sometimes used.) The shunt can be used like a tap in a wine barrel, to draw a blood sample or to connect the individual to a dialysis machine.

While connected to the dialysis machine, the individual sits quietly as blood circulates from the arterial shunt, through the machine, and back through the venous shunt. In the machine, the blood flows within a tube composed of dialysis membrane, and diffusion occurs between the blood and the surrounding dialysis fluid.

One alternative to the use of hemodialysis is **peritoneal dialysis**, in which the peritoneal lining is used as a dialysis membrane. Dialysis fluid is introduced into the peritoneum through a catheter in the abdominal wall, and the fluid is removed and replaced at intervals. One procedure, for example, involves cycling 2 liters of fluid in an hour—15 minutes for infusion, 30 minutes for exchange, and 15 minutes for fluid reclamation. This procedure may be performed in a hospital or

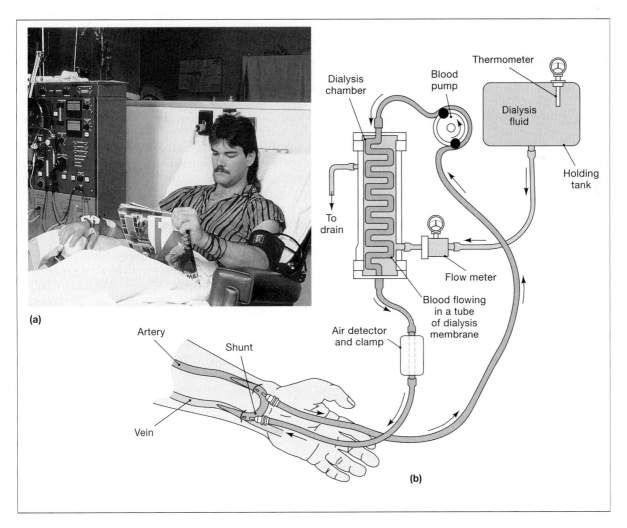

(a)

(b)

● **FIGURE 57**

Hemodialysis. (a) A patient is hooked up to a dialysis machine. **(b)** The path of blood during dialysis. Preparation for hemodialysis typically involves the implantation or surgical creation of a shunt that permits normal blood flow when the patient is not hooked up to the machine.

at home. An interesting variation is **continuous ambulatory peritoneal dialysis (CAPD)**, in which patients self-administer 2 liters of dialysis fluid through the catheter and then continue normal activity until four to six hours later, when the fluid is removed and replaced with fresh dialysis fluid.

Probably the most satisfactory solution, in terms of overall quality of life, is *kidney transplantation.* This procedure involves the implantation of a new kidney obtained from a living donor or a cadaver. More than half of the 15,127 kidneys transplanted in 2003 were obtained from living, related donors. In most cases, the damaged kidney is removed and its blood supply is connected to the transplant. An arterial graft is inserted to carry blood from the iliac artery or the aorta to the transplant, which is placed in the pelvic or lower abdominal cavity.

The success rate for kidney transplantation varies, depending on how aggressively the recipient's immune system attacks the donated organ and whether infection develops. The one-year success rate for kidney transplantation is now 85–95 percent. The use of kidneys taken from close relatives significantly improves the chances that the transplant will succeed for five years or more. Immunosuppressive drugs are administered to reduce tissue rejection, but unfortunately, this treatment also lowers the individual's resistance to infection.

■ BLADDER CANCER   *HA p. 701*

In the United States in 2004, there will be an estimated 60,000 new cases of **bladder cancer**, and 12,700 deaths. The incidence among males is four times that among females, and most patients are age 60–70. Environmental factors, especially exposure to *2-naphthylamine* or related compounds, are responsible for most bladder cancers. For this reason, the bladder cancer rate is highest among cigarette smokers (with twice the risk of nonsmokers) and employees of chemical and rubber companies. The mechanism responsible appears to involve damage to tumor suppressor genes (such as *p53*) that regulate cell division. The prognosis is reasonably good for localized superficial cancers (five-year survival: 94 percent), but poor for people with severe metastatic bladder cancer (five-year survival: 6 percent). Treatment of metastasized bladder cancer is very difficult, because the cancer spreads rapidly through adjacent lymphatic vessels and through the bone marrow of the pelvis.

■ PROBLEMS WITH THE CONDUCTING SYSTEM   *HA p. 701*

Local blockages of the collecting tubules, collecting ducts, or ureters may result from the formation of small blood clots, pithelial cells, lipids, or other materials, collectively called **casts.** Casts are often excreted in the urine and visible in microscopic analysis of urine samples. Calculi (KAL-kū-lē), or "kidney stones," form from calcium deposits, magnesium salts, or crystals of uric acid. This condition is called **nephrolithiasis** (nef-rŌ-li-THō-a-sis). The blockage of the urinary passage by a stone or other factors, such as external compression, results in **urinary obstruction.** Urinary obstruction is a painful and serious problem because it will reduce or eliminate filtration in the affected kidney. Kidney stones are usually visible on an x-ray, and if peristalsis and fluid pressures are insufficient to dislodge them, they must be surgically removed or destroyed. One interesting nonsurgical procedure involves breaking kidney stones apart with a lithotripter, similar in principle to the one used to destroy gallstones.

■ PROBLEMS WITH THE MICTURITION REFLEX   *HA p. 704*

Infants lack voluntary control over micturition because the necessary corticospinal connections have yet to be established. Accordingly, "toilet training" before age 2 often involves training the parent to anticipate the timing of the reflex rather than training the child to exert conscious control. **Incontinence** (in-KON-ti-nens) is the inability to control urination voluntarily. Trauma to the internal or external urethral sphincter can contribute to incontinence in otherwise healthy adults. For example, some mothers develop stress incontinence if childbirth overstretches and damages the sphincter muscles. In this condition, elevated intraabdominal pressures—caused, for example, by a cough or sneeze—can overwhelm the sphincter muscles, causing urine to leak out. Incontinence can also develop in older individuals due to a general loss of muscle tone. Damage to the central nervous system, the spinal cord, or the nerve supply to the urinary bladder or external urethral sphincter can also produce incontinence. For example, incontinence commonly accompanies Alzheimer's disease or spinal cord damage. In most cases, the affected individual develops an **automatic bladder.** The micturition reflex remains intact, but voluntary control of the external urethral sphincter is lost, so the person cannot prevent the reflexive emptying of the urinary bladder. Damage to the pelvic nerves can abolish the micturition reflex entirely, because those nerves carry both afferent and efferent fibers of this reflex arc. The urinary bladder then becomes greatly distended with urine and remains filled to capacity while the excess urine flows into the urethra in an uncontrolled stream. The insertion of a catheter is often needed to facilitate the discharge of urine.

26

THE PHYSICAL EXAMINATION AND THE REPRODUCTIVE SYSTEM *HA p. 709*

The male reproductive system consists of the gonads (testes), a series of specialized ducts (the epididymis, ductus deferens, ejaculatory duct, and urethra), accessory glands (the seminal vesicles, prostate gland, and bulbourethral glands), and the external genitalia (penis and scrotum). The female reproductive system consists of the gonads (ovaries), derivatives of an embryonic system of ducts (the uterine tubes, uterus, and vagina), accessory glands (the greater and lesser vestibular glands), the external genitalia (the clitoris, labia majora, and labia minora), and secondary sexual organs (the mammary glands of the breasts).

■ ASSESSMENT OF THE MALE REPRODUCTIVE SYSTEM

An assessment of the male reproductive system begins with a physical examination. Common signs and symptoms of male reproductive disorders include the following:

- **Testicular pain** may result from *testicular torsion* (twisting of the spermatic cord, with resulting ischemia and intense pain). Significant pain may result from various infections, including *gonorrhea* or other sexually transmitted diseases and mumps, or the presence of an inguinal hernia. Testicular cancer and cryptorchidism are less likely to cause pain initially. Pain may be perceived as coming from the testicle while actually originating elsewhere in the reproductive tract, such as along the ductus deferens or within the prostate gland, or in other systems, as in *appendicitis* or a urinary obstruction (for example, a ureteral kidney stone).

- **Urethral discharge** and associated pain (*dysuria*) are commonly associated with STDs. These symptoms may also accompany disorders, such as *epididymitis* or *prostatitis,* that may or may not be infectious.

- **Impotence**, or *erectile dysfunction (ED),* is the failure to achieve a sexually satisfactory erection. It can occur as a result of psychological factors, such as fear, anxiety, medications, or alcohol abuse. It can also develop secondarily to cardiovascular, hematological, hormonal, or nervous system problems that affect blood pressure or blood flow to the penile arteries.

- **Infertility**, which may be related to hormonal or other testicular disorders affecting sperm production, or a variety of anatomical problems along the reproductive tract.

Inspection of the male reproductive system normally involves the examination of the external genitalia and palpation of the prostate gland. Inspection of the external genitalia entails the following observational steps:

1. Inspection of the penis and scrotum for skin lesions, such as vesicles, chancres, warts, and condylomas (wartlike growths). For example, painful small vesicles in clusters appear with the herpes simplex virus infections. A chancre is a painless ulceration associated with early-stage syphilis. In the

CHAPTER 27
THE REPRODUCTIVE SYSTEM

examination of uncircumcised males, the foreskin is retracted to observe the lining of the prepuce. **Phimosis**, an inability to retract the foreskin in an adult uncircumcised male, generally indicates an inflammation of the prepuce and adjacent tissues.

2. Palpation of each testis, epididymis, and ductus deferens to detect the presence of abnormal masses, swelling, or tumors. Possible abnormal findings include the following:

   • **Scrotal swelling** due to distortion of the scrotal cavity by blood (a *hematocele*), lymph (a *chylocele*), or serous fluid (a *hydrocele*).

   • **Testicular swelling** due to an enlargement of the testis or the presence of a nodular mass. **Orchitis** is a general term for inflammation of the testis. Inflammation can result from an infection, such as syphilis, mumps, or tuberculosis. Testicular swelling may also accompany testicular cancer or testicular torsion.

   • **Epididymal swelling** due to cyst formation *(spermatocele)*, tumor formation, or infection. **Epididymitis** is an acute inflammation of the epididymis that may indicate an infection of the reproductive or urinary tract. The condition may also develop from irritation caused by the backflow, or *reflux*, of urine into the ductus deferens.

   • **Swelling of the spermatic cord** may indicate (1) an inflammation of the ductus deferens *(deferentitis)*, (2) an accumulation of serous fluid in a pocket of serous membrane (a *hydrocele*), (3) bleeding within the spermatic cord, (4) testicular torsion, or (5) the formation of *varicose veins* (p. 114) within the pampiniform plexus, a condition known as a *varicocele*.

3. A digital rectal examination (DRE) can screen for prostate enlargement, prostatitis, tumors (including prostate cancer), or an inflammation of the seminal vesicles. In this procedure, a gloved, lubricated finger is inserted into the rectum and pressed against the anterior rectal wall to palpate the posterior walls of the prostate gland and seminal vesicles.

If urethral discharge is present or if discharge occurs in the course of any of these procedures, the fluid can be cultured to check for the presence of pathogens. Table 33 summarizes information about pathogens that are responsible for infections of the reproductive system. Other potentially useful diagnostic procedures and laboratory tests (for both males and females) are given in Table 34.

■ ASSESSMENT OF THE FEMALE REPRODUCTIVE SYSTEM

Important signs and symptoms of female reproductive disorders include the following:

• **Acute pelvic pain**, a symptom that may accompany disorders such as pelvic inflammatory disease (PID), ruptured tubal pregnancy, a ruptured ovarian cyst, or inflammation of the uterine tubes *(salpingitis)*.

• **Bleeding outside normal menses**, which can result from oral contraceptive use, tumors, hormonal fluctuation, PID, or endometriosis.

• **Amenorrhea**, which may occur in women with anorexia nervosa, women who overexercise and are underweight, extremely obese women, postmenopausal women, and pregnant women.

• **Abnormal vaginal discharge**, which may be the result of a bacterial, fungal, or protozoan infection, including some STDs.

• Dysuria, which may accompany an infection of the reproductive system due to the migration of the pathogen to the urethral entrance, even though the female reproductive and urinary tracts are distinct.

• **Infertility**, which may be related to hormonal disturbances, a variety of ovarian disorders, or anatomical problems along the reproductive tract.

A physical examination generally includes the following steps:

1. Inspection of the external genitalia for skin lesions, trauma, or related abnormalities. Swelling of the labia majora results from (a) regional cellulitis with lymphedema, (b) a *labioinguinal hernia* (rare), (c) bleeding within the labia as the result of local trauma, or (d) *bartholinitis*, an abscess that develops after infection of one of the greater vestibular glands *(Bartholin's glands)*.

2. Inspection or palpation of the perineum, vaginal opening, labia, clitoris, urethral meatus, and vestibule to detect lesions, abnormal masses, or discharge from the vagina or urethra. Samples of any discharge present can be tested to detect and identify any pathogens involved.

3. Inspection of the vagina and cervix by using a speculum—an instrument that retracts the vaginal walls to permit direct visual inspection. Changes in the color of the vaginal walls may be important diagnostic clues. For example:

   • Cyanosis of the vaginal and cervical mucosa normally occurs during pregnancy, but it may also occur when a pelvic tumor exists or in women with congestive heart failure.

   • Reddening of the vaginal walls occurs in vaginitis (p. 155), bacterial infections such as gonorrhea, protozoan infection by *Trichomonas vaginalis*, and yeast infections. It can also appear postmenopausally in some women (a condition known as *atrophic vaginitis*).

The cervix is inspected to detect lacerations, ulceration, polyps, or cervical discharge. A spatula and brush are then used to collect cells from the cervical os and to transfer them to a glass slide or preservative solution. After the sample is fixed with a chemical spray or dip, or is centrifuged, cytological examination is performed. This technique is the best-known example of a *Papanicolaou (Pap) test* (Table 34, p. 156), and the sampling process is commonly called a *Pap smear*. The test screens for the presence of cervical cancer. Ratings of Pap smear results are given in Table 34.

4. A bimanual examination *for the palpation of the uterus, uterine tubes, and ovaries.* The physician inserts two fingers vagi-

27

TABLE 33 Examples of Infectious Diseases of the Reproductive System

Disease	Organism(s)	Description
Bacterial diseases		
Bacterial vaginitis	Varied, but often *Gardnerella vaginalis*	Vaginitis not associated with STDs, caused by resident bacteria; symptoms include watery discharge
Chancroid	*Haemophilus ducreyi*	A relatively rare STD; symptoms include soft chancres, which become ulcerated lesions, and enlarged lymph nodes in the groin
Chlamydia	*Chlamydia trachomatis*	Chlamydial infections cause PID, nongonococcal urethritis, and LGV.
Gonorrhea	*Neisseria gonorrhoeae*	Infection of the epithelial cells of the male and female reproductive tracts; majority of females show no symptoms, but others may develop PID; majority of males develop painful urination (dysuria) and produce a viscous urethral discharge
Lymphogranuloma venereum (LGV)	*Chlamydia trachomatis*	More invasive serotypes of chlamydia STD; symptoms include enlarged lymph nodes, which may abscess and form ulcers
Pelvic inflammatory disease (PID)	*Neisseria gonorrhoeae* *Chlamydia trachomatis*	An infection of the uterine tubes (salpingitis); symptoms include fever, abdominal pain, and elevated WBC counts; can cause peritonitis in severe cases; sterility can result from formation of scar tissue in the uterine tubes
Syphilis	*Treponema pallidum*	STD with a long period of chronic illness; symptoms of primary syphilis include chancres and enlarged lymph nodes; symptoms of secondary syphilis involve a reddish skin rash, fever, and headaches; tertiary syphilis affects CNS and cardiovascular system. Infection during pregnancy may cause severe fetal infection and malformations
Viral diseases		
Genital herpes	Herpes simplex viruses (HSV-1 and HSV-2)	Most cases caused by HSV-2; ulcers develop on external genitalia, heal, and recur.
Genital warts	Human papillomavirus	Warts appear on external genitalia, perineum, and anus, and on vagina and cervix of females; associated with cervical cancer.
Fungal diseases		
Candidiasis	*Candida albicans*	Yeast infection that causes vaginitis; symptoms include itching, burning, and lumpy white discharge.
Parasitic diseases		
Trichomoniasis	*Trichomonas vaginalis*	Flagellated protozoan parasite of both male and female urinary and reproductive tracts; infection produces white or greenish-gray discharge in both sexes and intense vaginal itching in females; may be asymptomatic.

nally and places the other hand against the lower abdomen to palpate the uterus and surrounding structures. The contour, shape, size, and location of the uterus can be determined, and any swellings or masses will be apparent. Abnormalities in other reproductive organs, such as ovarian masses, endometrial growths, or tubal masses, can also be detected in this way.

■ NORMAL AND ABNORMAL SIGNS ASSOCIATED WITH PREGNANCY

Pregnancy imposes a number of stresses on maternal body systems. Several clinical signs may be apparent in the course of a physical examination, including the following:

- **Chadwick's sign** is a normal cyanosis of the vaginal wall seen with the increase in pelvic blood flow of pregnancy.
- The size of the uterus changes drastically during pregnancy; at full term, the uterus extends almost to the level of the xiphoid process.
- Significant uterine bleeding, causing vaginal discharge of blood, most commonly occurs in *placenta previa*, in which the placenta forms near the cervix. Subsequent cervical stretching leads to tearing and bleeding of the vascular channels of the placenta. Vaginal bleeding may also occur prior to miscarriage, or if the placenta suddenly separates from the uterine wall (*abruptio placentae*). In all these cases there is significant risk of maternal or fetal death.

27

TABLE 34   Examples of Tests Used in the Diagnosis of Reproductive Disorders

Diagnostic Procedure	Method and Result	Representative Uses
Females		
Mammography	X-ray film is taken of breast.	Detects cysts or tumors of breast; effective in detecting early breast cancer
Thermography	Heat energy emitted from breast is detected by infrared camera and recorded.	Tumors, cysts, fibrocystic disease, and infection cause localized hot areas; less precise than mammography
Laparoscopy	Fiber-optic tubing is inserted through incision in abdominal wall to view pelvic organs, remove tissue for biopsy, or perform surgical procedures.	Detects pelvic organ abnormalities such as cysts or adhesions; determines cause of pelvic pain; enables diagnosis of pelvic inflammatory disease and endometriosis
Papanicolaou (Pap) smear	Cells from cervix are removed for cytological analysis.	Detects cervical cancer; reported results: *Class I:* no abnormal cells; *Class II:* some abnormal cells, but none that suggest a malignancy (normally due to inflammation); *Class III:* some abnormal cells, possible malignancy; *Class IV:* some abnormal cells, probable malignancy; *Class V:* definite malignancy
Colposcopy	Special instrument is used to view cervical tissue microscopically in situ and to guide removal of tissue for biopsy.	Detects areas of dysplasia and malignancies of cervix; follow-up to abnormal PAP smear
Cervical biopsy	Tissue is removed from cervix for examination.	Detects dysplasia and malignancy
Transvaginal sonography	An ultrasonic probe is used suprapubically or inserted into the vagina.	Obtains high-definition echograms of the ovaries, endometrium; detects tumors, cysts
Males		
Transrectal ultrasonography	Ultrasound transducer is inserted rectally and scan is performed.	Detects prostatic tumor and nodules or abnormalities of seminal vesicles and surrounding structures; used to guide biopsy of nodules
Semen analysis		
Volume	2–5.0 ml	Decreased sperm count causes infertility; infertility could also result if >40% of sperm are immotile or >30% of sperm are abnormal.
Sperm count	60–150 million/ml	
Motility	60–80% are motile	
Sperm morphology	70–90% normal structure	
Testosterone (serum)	Adult male: 175–781 ng/dl Adult female: 10–75 ng/dl	Decreased level could indicate testicular disorder, alcoholism, or pituitary hypofunction.

Laboratory Test	Normal Values in Blood Plasma or Serum	Significance of Abnormal Values
Females		
Estrogen (serum)	Early uterine cycle: 60–400 pg/ml Middle: 100–600 pg/ml Late: 150–350 pg/ml Postmenopausal:<30 pg/ml	Detects hypofunctioning ovaries and helps determine timing of ovulation. Note the large range of normal values, due to variations in the levels of several different estrogens.
Estradiol (serum)	Follicular phase: 20–150 pg/ml Ovulation: 100–500 pg/ml Luteal phase: 60–260 pg/ml	Decreased levels occur in ovarian dysfunction and in amenorrhea. Estradiol is the primary estrogen produced by ovaries.
FSH (serum)	Before and after ovulation. 4–20 mIU/ml Midcycle: 10–40 mIU/ml	Helps determine the cause of infertility and menstrual dysfunction; increased levels occur in absence of estrogens (as during menopause); decreased levels occur with anorexia nervosa or hypopituitarism.
LH (serum)	Follicular phase: 3–30 mIU/ml Midcycle: 30–150 mIU/ml	Determines timing of ovulation; LH is increased in ovarian hypofunction and polycystic ovary syndrome.
Progesterone (serum)	Before ovulation: <70 ng/dl Midcycle: 250–2800 ng/dl	Levels are increased after ovulation and in early pregnancy.

TABLE 34	Examples of Tests Used in the Diagnosis of Reproductive Disorders (Continued)	
Diagnostic Procedure	**Method and Result**	**Representative Uses**
Prolactin (serum)	Nonlactating females: 0–23 ng/ml	Values >100 ng/ml in a nonlactating female may indicate pituitary tumor
Females and Males		
Serologic test for syphilis	Negative	Presence of antibodies indicates past or present infection with syphilis.
Gonorrhea culture	Negative	Positive test indicates gonorrheal infection.
Herpes simplex virus culture	Negative	Positive test indicates presence of virus in culture.
Chlamydia culture	Negative	Positive culture indicates presence of *Chlamydia*.

- Nausea and vomiting tend to occur in pregnancy, especially during the first three months.

- Edema of the limbs, especially the legs, is typical, because the increased total blood volume and the weight of the uterus compress the inferior vena cava and its tributaries. As venous pressures rise in the lower limbs and inferior trunk, varicose veins and hemorrhoids may develop.

- Back pain due to increased stress on muscles of the lower back is common. These muscles are strained as the weight of the uterus accentuates the lumbar curvature.

- A weight gain of 10–12.5 kg (22–27.5 lb) is now considered desirable, although 20 years ago weight increases of 20–25 kg (44–55 lb) were considered acceptable, and 45 years ago, before the linkage between prematurity with low maternal weight gain was recognized, weight gains of 7–10 lbs were sometimes considered desirable. Except for women who start pregnancy very overweight (who with careful monitoring may benefit from little or no weight gain), failure to gain adequate weight during pregnancy can indicate or lead to serious neonatal problems.

- The combination of estrogen, progesterone, prolactin, human placental lactogen (HPL), and other hormones that are elevated during pregnancy appears to promote the development of *insulin resistance,* a decrease in target cell sensitivity to insulin. As a result, pregnant diabetic women are at an increased risk of ketoacidosis. Previously diabetic women who become pregnant are at an increased risk of ketoacidosis. During diabetic pregnancies, glucose levels must be monitored and stabilized to prevent maternal problems as well as fetal death and developmental defects that can occur. Gestational diabetes develops in 1–3 percent of pregnancies.

- In some cases, a dangerous combination of hypertension, proteinuria, edema, and seizures occurs. We will consider this condition, called *eclampsia,* in a later section (p. 172).

DISORDERS OF THE REPRODUCTIVE SYSTEM

Representative disorders of the reproductive system are diagrammed in Figure 58●.

■ TESTICULAR TORSION   HA p. 709

Because the testes are only loosely attached to the scrotal walls, they may become twisted within the scrotal cavity. This condition, called **testicular torsion**, usually occurs in children or adolescents. Symptoms include pain in the groin and inguinal region, local inflammation, and swelling of the scrotum on the affected side. Treatment involves prompt external or surgical manipulation of the testis to relieve the twisting and securing the testes to the scrotal wall. When unilateral torsion is found, both testes are at risk and both treated. Because any kinks in the spermatic cord severely restrict the arterial supply to the testis, corrective measures must be taken within four to six hours, before the testicular tissues become permanently damaged. If the tissues are deprived of circulation for longer periods, the damage will be irreversible, and the affected testis may have to be removed. This surgical procedure is called an **orchidectomy** (ōr-ki-DEK-tō-mē); *orchis,* testis); one testis produces sufficient testosterone and sperm for normal male reproductive function. The common term *castration* indicates that both testes have been removed.

■ CRYPTORCHIDISM   HA p. 712

In **cryptorchidism** (kript-OR-ki-dizm; crypto, hidden), one or both of the testes have not descended into the scrotum by the time of birth. Typically the testes are lodged in the abdominal cavity or within the inguinal canal. This condition occurs in about 3 percent of full-term deliveries and in roughly 30 percent of premature births. In most instances normal descent occurs a few weeks later, but the condition can be surgically corrected if it persists. Corrective measures are usually taken before puberty because cryptorchid (abdominal) testes will not produce sperm, as the temperature inside the peritoneal cavity is too high for sperm production (1–2°C higher than the temperature within the scrotal cavity). The individual will therefore be sterile (infertile). If the testes cannot be moved into the scrotum, they will usually be removed, because about 10 percent of those with uncorrected cryptorchid testes eventually develop testicular cancer.

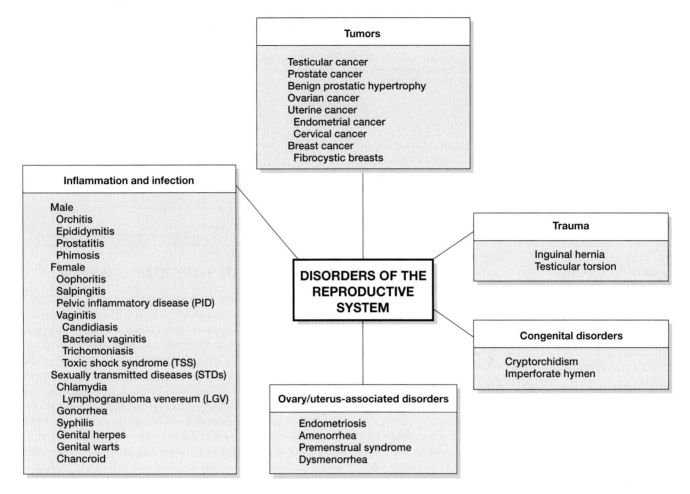

● **FIGURE 58**
Representative Disorders of the Reproductive System

■ TESTICULAR CANCER   *HA p. 713*

Testicular cancer occurs at a relatively low rate: about three cases per 100,000 males per year. Although only about 7,200 new cases are reported each year in the United States, with 400 deaths, testicular cancer is the most common cancer among males age 15–35. The incidence among Caucasian males has more than doubled since the 1930s, but the incidence among African American males has remained unchanged. The reason for this difference is not known.

More than 95 percent of testicular cancers result from abnormal spermatogonia or spermatocytes, rather than abnormal sustentacular cells, interstitial cells, or other testicular cells. Treatment generally consists of a combination of orchiectomy and chemotherapy. The survival rate increased from about 10 percent in 1970 to about 95 percent in 1999, primarily as a result of earlier diagnosis and improved treatment protocols. Cyclist Lance Armstrong has won the grueling Tour de France five consecutive times after successful treatment for advanced testicular cancer.

■ PROSTATITIS AND PROSTATE CANCER
HA p. 717

Prostatic inflammation, or **prostatitis** (pros-ta-TĪ-tis), can occur at any age, but it most often afflicts older men. Prosta-

titis usually results from bacterial infections, but the condition may also develop in the apparent absence of known pathogens. Individuals with prostatitis complain of pain in the lower back, perineum, or rectum, sometimes accompanied by painful urination and the discharge of mucous secretions containing numerous white blood cells (leukocytes) from the external urethral orifice. Antibiotic therapy is usually effective in treating cases resulting from bacterial infection, but in other cases, antibiotics may not provide relief. Prostatic enlargement, or **benign prostatic hypertrophy,** usually occurs spontaneously in men over age 50. The increase in size occurs at the same time that hormonal changes are under way within the testes. Androgen production by the interstitial cells decreases over this period, and at the same time, these endocrine cells begin releasing small quantities of estrogens into the circulation. The combination of lower testosterone levels and the presence of estrogen probably stimulates prostatic growth. Blood pressure drugs that relax smooth muscle can relieve symptoms while vasoconstrictors, such as cold decongestants, can increase obstruction. Drugs and herbs that inhibit formation of testosterone are also effective in reducing symptoms. In severe cases, prostatic swelling can constrict and block the urethra and even the rectum. The urinary obstruction can cause permanent kidney damage if not corrected. Partial surgical removal is the most effective treatment.

In the procedure known as a TURP (transurethral prostatectomy), an instrument pushed along the urethra restores normal function by cutting away the swollen prostatic tissue. Most of the prostate remains in place, and there are no external scars. See pronunciations and boldfacing from first pages of HA5; prostatitis, benign prostatic hypertrophy, TURP are boldfaced

Prostate cancer, a malignancy of the prostate gland, is the second most common cancer and the second most common cause of cancer deaths in males. In 2003, approximately 220,900 new cases of prostate cancer were diagnosed in the United States, and about 28,900 deaths resulted from the ailment. Most patients are elderly. (The average age at diagnosis is 72.) For reasons that are poorly understood, prostate cancer rates for Asian American males are relatively low compared with those of either Caucasian or African Americans. For all age and ethnic groups, the rates of prostate cancer are rising sharply. The reason for the increase is not known. Aggressive diagnosis and treatment of localized prostate cancer in elderly patients is controversial because many of these men have nonmetastatic tumors, and even if untreated are more likely eventually to die of some other disease.

Prostate cancer normally originates in one of the secretory glands. As the cancer progresses, it produces a nodular lump or swelling on the surface of the prostate gland. Palpation of this gland through the rectal wall—a procedure known as a *digital rectal exam* (DRE)—is the easiest diagnostic screening procedure. *Transrectal prostatic ultrasound* (TRUS) can be used to obtain more detailed information about the status of the prostate, but at significantly higher cost to the patient. Blood tests are also used for screening purposes. The most sensitive is a blood test for *prostate-specific antigen (PSA)*. Elevated levels of this antigen, normally present in low concentrations, may indicate the presence of prostate cancer. The *serum enzyme assay,* which checks the level of the isozyme *prostatic acid phosphatase,* detects prostate cancer in later stages of development. Screening with periodic PSA tests is now being recommended for men over age 50.

If cancer is detected before it has spread to other organs, the usual treatment is localized radiation or surgical removal of the prostate gland. This operation, a **prostatectomy** (pros-ta-TEK-tō-mē), can be effective in controlling the condition, but both surgery and radiation can have undesirable side effects, including urinary incontinence and loss of sexual function. Modified surgical procedures can reduce the risks and maintain normal sexual function in perhaps three out of four patients.

The prognosis is much worse for prostate cancer diagnosed after metastasis has occurred, because metastasis rapidly involves the lymphatic system, lungs, bone marrow, liver, or adrenal glands. Survival rates at this stage are relatively low. Treatments for metastasized prostate cancer include widespread irradiation, hormonal manipulation, lymph node removal, and aggressive chemotherapy. Because the cancer cells are stimulated by testosterone, treatment may involve castration or administering hormones that depress GnRH or LH production. Despite these interesting advances in treatment, the average survival time for patients diagnosed with advanced prostate cancer is only 2.5 years.

■ UTERINE TUMORS AND CANCERS
HA p. 731

Uterine tumors are the most common tumors in women. It has been estimated that 40 percent of women over age 50 have benign uterine tumors involving smooth muscle and connective tissue cells. When they are small, these *leiomyomas,* or *fibroids,* generally cause no problems. Stimulated by estrogens, however, they can grow quite large, reaching weights as great as 13.6 kg (30 lb). Occlusion of the uterine tubes, distortion of adjacent organs, and compression of blood vessels may then lead to complications. In symptomatic young women, observation or conservative treatment with drugs or restricted surgery may be utilized to preserve fertility. In older women, a decision may be made to remove the uterus.

Benign epithelial tumors in the uterine lining are called *endometrial polyps.* Roughly 10 percent of women probably have polyps, but because the polyps tend to be small and cause no symptoms, the condition passes unnoticed. If bleeding occurs, if the polyps become excessively enlarged, or if they protrude through the cervical os, they can be removed.

Uterine cancers are less common, affecting approximately 11.9 per 100,000 women. In 2003 in the United States, an estimated 50,800 new cases were reported, and 11,000 women died from the disease. There are two types of uterine cancers: (1) *endometrial* and (2) *cervical.*

Endometrial cancer is an invasive cancer of the endometrium. The condition most commonly affects women age 50–70. Estrogen therapy, used to treat osteoporosis in postmenopausal women, increases the risk of endometrial cancer by 2–10 times. Adding progesterone therapy to the estrogen therapy seems to reduce this risk.

There is no perfect screening test for endometrial cancer, although endometrial biopsy and measuring endometrial thickness by ultrasound may be used. The most common symptom is irregular bleeding, and diagnosis typically involves examination of tissue from a biopsy of the endometrium obtained by suction or scraping. The prognosis varies with the degree of spread. The treatment of early-stage endometrial cancer involves a hysterectomy, perhaps followed by localized radiation therapy. In advanced stages, more aggressive radiation treatment is recommended. Chemotherapy has *not* proved to be very successful in treating endometrial cancers; only 30–40 percent of patients benefit from this approach.

Cervical cancer is the most common reproductive system cancer in women age 15–34. Most women with cervical cancer develop no symptoms until late in the disease. At that stage, vaginal bleeding—especially after intercourse—pelvic pain, and vaginal discharge may appear. Early detection is the key to reducing the mortality rate from cervical cancer. The standard screening test is the Pap smear, named for Dr. George Papanicolaou, a Greek-born, naturalized-American anatomist and cytologist. The cervical epithelium normally sheds its superficial cells, and a sample of cells scraped or brushed from the epithelial surface can be examined for abnormal or cancerous cells. The American Cancer Society recommends yearly Pap tests at ages 20 and 21, followed by smears at one- to three-year intervals until age 65.

27

The primary risk factor for cervical cancer is a history of multiple sexual partners. It appears likely that the condition frequently develops after genital viral infection by one of several *human papillomaviruses* (HPVs), which are transmitted through sexual contact.

Early treatment of abnormal, but not cancerous, lesions detected by mildly abnormal Pap smears may prevent the progression to cancer. The treatment of localized, noninvasive cervical cancer involves the removal of the affected portion of the cervix. The treatment of more advanced cancers typically involves a combination of radiation therapy, hysterectomy, removal of lymph nodes, and chemotherapy.

■ SEXUALLY TRANSMITTED DISEASES   *HA p. 731*

Close physical contact can spread infectious diseases from person to person, and sexual contact is as close as two individuals can get. Some infections are spread almost exclusively by sexual contact. These infections are called **sexually transmitted diseases**, or **STDs**. A variety of bacterial, viral, and fungal infections are included in this category. At least two dozen STDs are currently recognized, and roughly 15 million people become infected each year in the United States. All STDs are unpleasant, and some are deadly. Here, we will discuss six of the most common STDs: *chlamydia, gonorrhea, syphilis, herpes, genital warts,* and *chancroid.* The deadliest, human immune deficiency disease (HIV/AIDS), was discussed on p. 120.

■ CHLAMYDIA

As diagnostic procedures improve, infections by the bacterium *Chlamydia trachomatis* are proving to be the most frequent cause of STDs. Roughly 4 million people become infected each year in the United States, most in the 15–24 age group. Chlamydial infection can have a variety of clinical effects. It is responsible for the majority of cases of pelvic inflammatory disease, and even asymptomatic infections may result in uterine tube blockage infertility. In males it is probably the commonest cause of *nongonococcal urethritis.* A few sereotypes of chlamydia are more invasive, producing **lymphogranuloma venereum (LGV)** and conjunctivitis in the newborn. In LGV, the lymph nodes in the groin become enlarged and inflamed. Abscesses and ulcers then develop over lymph nodes in the region. Chlamydial infections can be treated with antibiotics (tetracycline and sulfa drugs), but all of a person's sexual partners must be treated to prevent reinfection.

■ GONORRHEA

The bacterium *Neisseria gonorrhoeae* is responsible for gonorrhea, one of the most common STDs in the United States. Nearly 2 million cases were reported in the early 1970s; in 2003, 311,922 cases were reported. The bacteria normally invade epithelial cells that line the male or female reproductive tract. In relatively rare cases, they also colonize the pharyngeal or rectal epithelium.

The symptoms of genital infection differ according to the gender of the individual. It has been estimated that up to 80 percent of women infected with gonorrhea experience no symptoms or symptoms so minor that medical treatment is not sought. As a result, these women act as carriers, spreading the infection through their sexual contacts. An estimated 10–15 percent of women infected with gonorrhea experience more acute symptoms, because the bacteria invade the epithelia of the uterine tubes. This infection probably accounts for many of the cases of pelvic inflammatory disease (PID) in the U.S. population. As many as 80,000 women become infertile each year as a result of the formation of scar tissue along the uterine tubes after gonorrheal or chlamydial infection.

Seventy to 80 percent of infected males develop symptoms painful enough to make them seek antibiotic treatment. The asymptomatic 20–30 percent are male carriers who unknowingly spread the infection. The urethral invasion is accompanied by pain on urination (dysuria) and, typically, by a viscous urethral discharge. A sample of the discharge can be cultured to positively identify the organism involved.

■ SYPHILIS

Syphilis (SIF-i-lis) results from infection by the bacterium *Treponema pallidum.* The first reported syphilis epidemics occurred in Europe during the 16th century, possibly introduced by early explorers returning from the New World. The death rate from the "Great Pox," as syphilis was called then, was appalling, far greater than it is today, even after we take into account the absence of antibiotic therapies at that time. It appears likely that the syphilis bacterium has mutated during the interim, to a form that reduces the immediate mortality rate but prolongs the period of chronic illness and increases the likelihood of successful transmission. Syphilis still remains a life-threatening disease. Untreated syphilis can cause serious cardiovascular and neurological illness years after infection, and it can be spread to a fetus during pregnancy, producing congenital malformations.

Primary syphilis begins as the bacteria cross the mucous epithelium and enter the lymphatic vessels and bloodstream. At the site of the invasion, the bacteria multiply. After an incubation period of 1.5 to 6 weeks, their activities produce a painless raised ulcerated lesion, or **chancre** (SHANG-ker) (Figure 59●). The chancre is infectious and persists for several weeks before fading away, even without treatment. In heterosexual men, the chancre tends to appear on the penis; in women, it may develop on the labia, vagina, or cervix. Lymph nodes in the region often enlarge and remain swollen even after the chancre has disappeared.

Symptoms of *secondary syphilis* appear roughly six weeks later. The skin lesions of secondary syphilis are also infectious. It generally involves a diffuse, reddish skin rash. Like the chancre, the rash fades over a period of two to six weeks. These symptoms may be accompanied by fever, headaches, and malaise. The combination is so vague that the disease may easily be overlooked or misdiagnosed. In a few instances, more serious complications, such as meningitis, hepatitis, or arthritis, develop. During primary and secondary syphilis the

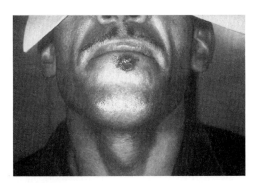

● **FIGURE 59**
A Syphilitic Chancre

bacterium circulates in the blood and can be spread by un-screened transfusion.

The individual then enters the *latent phase,* which is non-infectious. The duration of the latent phase varies widely. Fifty to 70 percent of untreated individuals with latent syphilis fail to develop the symptoms of *tertiary syphilis,* or *late syphilis,* although the bacterial pathogens remain within their tissues. Those who develop tertiary syphilis may do so 10 or more years after primary infection.

The most severe symptoms of tertiary syphilis involve the central nervous system and the cardiovascular system. **Neurosyphilis** may result from a bacterial infection of the meninges or the tissues of the brain or spinal cord. **Tabes dorsalis** (TĀ-bēz dor-SAL-is) results from the invasion and de-myelination of the posterior columns of the spinal cord and the sensory ganglia and nerves. In the cardiovascular system, the disease affects the major vessels, leading to aortic stenosis, aneurysms, or focal calcification.

Equally disturbing are the effects of transmission from mother to fetus across the placenta. In cases where infection does not lead to fetal death, congenital syphilis is marked by infections of the developing bones and cartilages of the skeleton and progressive damage to the spleen, liver, bone marrow, and kidneys. The risk of fetal transmission may be as high as 95 percent, so maternal blood testing is recommended early in pregnancy. Blood donations are screened to prevent transfer through transfusion. The treatment of syphilis involves the administration of *penicillin* or other antibiotics.

■ HERPES

Genital herpes results from infection by herpes viruses. Two different viruses are involved. Eighty to 90 percent of genital herpes cases are caused by the virus known as HSV-2 (herpes simplex virus Type 2), which usually attacks the skin of the external genitalia. The remaining cases are caused by HSV-1, the virus that most commonly causes "cold sores" or "fever blisters" around the mouth. Typically, within a week of the initial herpes infection, the individual develops multiple small, painful, infectious ulcerated lesions on the affected skin or mucous membrane, with associated lymphadenopathy. In women, ulcerations may also appear on the cervix. These

ulcerations gradually heal over the next two to three weeks. Recurring lesions are common, although subsequent incidents are less severe.

During delivery, infection of the newborn with herpes viruses present in the mother's vagina can lead to serious illness, because the infant has few immunological defenses. The antiviral agent *acyclovir* has helped in treating initial infections and in reducing recurrences.

■ GENITAL WARTS

Genital warts, or *condyloma acuminata,* result from infection by one of a number of strains of *human papillomavirus* (HPV). Several of these strains are thought to be responsible for cases of cervical, anal, vaginal, and penile cancer. Roughly 1.2 million cases of genital warts are diagnosed each year in the United States. There is no satisfactory treatment for this problem. The traditional treatments have included cryosurgery, erosion by caustic chemicals, surgical removal, and laser surgery to remove the warts. These treatments expunge the visible signs of infection, but the virus remains within the epidermis. Alpha-interferons have been tried with limited success.

■ CHANCROID

Chancroid is an STD caused by the bacterium *Haemophilus ducreyi.* Chancroid was rarely seen in the United States before 1984, but since then the number of cases has risen dramatically, reaching 4000–5000 per year. Fewer than 200 cases were reported in 2002; however, chancroid is difficult to detect by laboratory test and may be underdiagnosed. The primary sign of this disease is the development of *soft chancres*—soft lesions otherwise resembling those of syphilis. The majority of chancroid patients also develop prominent inguinal lymphadenopathy.

■ ENDOMETRIOSIS   *HA p. 732*

Some conditions interfere with the normal menstrual cycle. In **endometriosis** (en-dō-mē-trē-Ō-sis), an area of endometrial tissue begins to grow outside the uterus. The cause is unknown. Because this condition is most common in the inferior portion of the peritoneum, one possibility is that pieces of endometrium sloughed off during menstruation in some way travel through the uterine tubes into the peritoneal cavity, where they reattach. The severity of the condition depends on the size and location of the abnormal mass. Cyclic or persistent abdominal pain, bleeding, pressure on adjacent structures, and infertility are common signs and symptoms. As the island of endometrial tissue enlarges, the condition becomes more severe.

Endometriosis can generally be diagnosed by inserting a laparoscope through a small opening in the abdominal wall. Using this device, a physician can inspect the outer surfaces of the uterus and uterine tubes, the ovaries, and the lining of the pelvic cavity. Treatment of endometriosis may involve hormonal therapy to suppress uterine cycles, or surgical removal of the endometrial mass. If the condition is widespread, a *hysterectomy* (removal of the uterus) or *oophorectomy* (removal of the ovaries) may be required.

CONTROLLING REPRODUCTIVE FUNCTION

■ BIRTH CONTROL STRATEGIES

HA p. 738

For physiological, logistical, financial, or emotional reasons, most adults practice some form of conception control during their reproductive years. Two out of three U.S. women age 15–44 employ some method of contraception. When the simplest and most obvious method, sexual abstinence, is unsatisfactory for some reason, another method of contraception must be used to avoid unwanted pregnancies. Many methods are available, so the selection process can be quite involved. Because each method has specific strengths and weaknesses, the potential risks and benefits must be carefully analyzed on an individual basis. We will consider only a few of the available contraception methods here. See Table 35 for comparative data on contraceptive methods and effectiveness.

A variety of nonsurgical contraceptive methods are available (Figure 60a●). **Hormonal contraceptives** manipulate the female hormonal cycle so that ovulation does not occur. The contraceptive pills produced in the 1950s used combined estrogen and progestins sufficient to suppress pituitary production of GnRH, so FSH was not released and ovulation did not occur. Most of the oral contraceptives developed subsequently contain much smaller amounts of estrogens, or only progesterone. Current *combination* hormone contraceptives are administered in a cyclic fashion, using medication for three weeks. Then, over the fourth week, no medication is taken. For user convenience, monthly injections, weekly skin patches, and insertable vaginal rings containing combined estrogen/progesterone hormone products are available.

At least 20 brands of combination oral contraceptives are now available, and more than 200 million women are using them worldwide. In the United States, 33 percent of women under age 45 use a combination pill to prevent conception. The failure rate for combination oral contraceptives, when used as prescribed, is 0.24 percent over a two-year period. (*Failure* for a birth control method is defined as a resulting pregnancy.) Birth control pills are not risk free: Combination pills can worsen problems associated with severe hypertension, diabetes mellitus, epilepsy, gallbladder disease, heart trouble, and acne. Women taking oral contraceptives are also at increased risk of venous thrombosis, strokes, pulmonary embolism, and (for women over 35) heart disease. However, pregnancy has similar or higher risks.

Hormonal postcoital contraception, or the emergency "morning after" pill, involves taking either combination estrogen/progesterone birth control pills or progesterone only pills in two large doses 12 hours apart within 72 hours of unprotected sexual intercourse. Particularly useful when barrier methods malfunction or coerced intercourse occurs, it reduces expected pregnancy rates by up to 89 percent. The progesterone-only version is considered safe for nonprescription use and may become available for purchase over the counter in the future.

Progesterone-only forms of birth control are now available: Depo-Provera, the Norplant system, and the proges-

terone-only pill. *Depo-Provera* is injected every three months. Uterine cycles are initially irregular, and eventually cease in roughly 50 percent of women using this product. The most common problems with this contraceptive method are (1) a tendency to gain weight and (2) a slow return to fertility (up to 18 months) after injections are discontinued. The Silastic (silicone rubber) tubes of the *Norplant system* are saturated with progesterone and inserted under the skin. This method provides birth control for approximately five years. Fertility returns immediately after the removal of a Norplant device. Initial high costs and problems with removal have limited its use and its manufacturer has stopped supplying it to the U.S. market. The progesterone-only pill is taken daily and may cause irregular uterine cycles. Skipping just one pill may result in pregnancy.

Barrier contraceptive methods include male **condoms**, also called *prophylactics* or "rubbers," which cover the glans and shaft of the penis during intercourse and keep spermatozoa from reaching the female reproductive tract. Latex condoms also reduce the spread of sexually transmitted diseases, such as syphilis, gonorrhea, and AIDS.

Vaginal barriers, such as the *diaphragm* and *cervical cap*, rely on similar principles. A diaphragm, the most popular form of vaginal barrier in use today, consists of a dome of latex rubber with a small metal hoop supporting the rim. Because vaginas vary in size, women choosing this method must be individually fitted. Before intercourse, the diaphragm is inserted so that it covers the cervical os. The diaphragm must be coated with a small amount of spermicidal (sperm-killing) jelly or cream to be an effective contraceptive. The failure rate of a properly fitted and used diaphragm is estimated at 5–6 percent. The cervical cap is smaller and lacks the metal rim. It, too, must be fitted carefully, but unlike the diaphragm, it can be left in place for several days.

An **intrauterine device (IUD)** consists of a small plastic loop or a T that is inserted into the uterine cavity. The mechanism of action remains unclear, but IUDs are known to stimulate prostaglandin production in the uterus, and some are effective for years after insertion. The resulting change in the chemical composition of uterine secretions lowers the likelihood of fertilization and subsequent *implantation* of the zygote into the uterine lining. (Implantation is discussed in Chapter 28.) In the United States, IUDs are in limited use, but they remain popular in many other countries.

"Natural family planning," also called "fertility awareness" and the **rhythm method**, involves abstaining from sexual activity on the days when ovulation might be occurring. The timing is estimated on the basis of previous patterns of menstruation and by monitoring changes in body symptoms of ovulation, including basal body temperature, cervical mucus texture, and, for some, urine tests for LH.

Sterilization is a surgical procedure that makes an individual unable to provide functional gametes for fertilization. Either sexual partner may be sterilized. In a **vasectomy** (vaz-EK-to-mē), a segment of the ductus deferens is removed, making it impossible for spermatozoa to pass from the epididymis to the distal portions of the reproductive tract (Figure 60b●). The surgery can be performed in a physician's office in a matter of minutes. The spermatic cords are located as they ascend

TABLE 35   Birth Control Guide

The Food and Drug Administration has approved a number of birth control methods. The choice of birth control depends on factors such as a person's health, frequency of sexual activity, number of sexual partners, and desire to have children in the future. Failure rates, based on statistical estimates, are another key factor. The most effective way to avoid both pregnancy and sexually transmitted disease is to practice total abstinence (refrain from sexual contact).

Failure rates in this chart are based on information from clinical trials submitted to the FDA during product reviews. This number represents the percentage of women who become pregnant during the first year of use of a birth control method. For methods that the FDA does not review, such as periodic abstinence, numbers are estimated from published literature. For comparison, about 85 out of 100 sexually active women who wish to become pregnant would be expected to become pregnant in a year.

Serious medical risks from contraceptives, such as stroke related to oral contraceptives, are relatively rare. This chart is a summary of important information, including risks, about drugs and devices approved by the FDA for contraception and sterilization. It is not intended to be used alone, and a health professional should be consulted regarding any contraceptive choice. Review product labeling carefully for more information on use of these products.

Type of Contraceptive	FDA Approval Date	Description	Failure Rate (number of pregnancies expected per 100 women per year)	Some Risks	Protection from Sexually Transmitted Diseases (STDs)	Convenience	Availability
Male condom (Latex/ polyurethane)	Latex: Use started before premarket approval was required Polyurethane: cleared in 1989: available starting 1995	A sheath placed over the erect penis blocking the passage of sperm.	11	Irritation and allergic reactions (less likely with polyurethane)	Except for abstinence, latex condoms are the best protection against STDs , including gonorrhea and AIDS.	Applied immediately before intercourse; used only once and discarded. Polyurethane condoms are available for those with latex sensitivity.	Nonprescription
Diaphragm with spermicide	Use started before premarket approval was required.	A dome-shaped rubber disk with a flexible rim that covers the cervix so that sperm cannot reach the uterus. A spermicide is applied to the diaphragm before insertion.	17	Irritation and allergic reactions, urinary tract infection. (c) Risk of toxic shock syndrome, a rare but serious infection, when kept in place longer than recommended.	None	Inserted before intercourse and left in place at least six hours after: can be left in place for 24 hours, with additional spermicide for repeated intercourse.	Prescription
Oral contraceptives —combined pill	First in 1960; most recent in 2003	A pill that suppresses ovulation by the combined actions of the hormones estrogen and progestin. A chewable form was approved in November 2003.	1–2	Dizziness; nausea; changes in menstruation, mood, and weight; rarely cardiovascular disease, including high blood pressure, blood clots, heart attack, and strokes	None	Must be taken on daily schedule, regardless of frequency of intercourse. Women using the chewable tablet must drink 8 oz. of liquid immediately after taking.	Prescription

27

TABLE 35   Birth Control Guide (*Continued*)

Type of Contraceptive	FDA Approval Date	Description	Failure Rate (number of pregnancies expected per 100 women per year)	Some Risks	Protection from Sexually Transmitted Diseases (STDs)	Convenience	Availability
Oral contraceptives —progestin-only minipill	1973	A pill containing only the hormone progestin that reduces and thickens cervical mucus to prevent the sperm from reaching the egg.	2	Irregular bleeding, weight gain, breast tenderness, less protection against ectopic pregnancy	None	Must be taken on daily schedule, regardless of frequency of intercourse.	Prescription
Postcoital contraceptives (Preven and Plan B)	1998–1999	Pills containing either progestin alone or progestin plus estrogen	Almost 80 percent reduction in risk of pregnancy for a single act of unprotected sex	Nausea, vomiting, abdominal pain, fatigue, headache	None	Must be taken within 72 hours of having unprotected intercourse.	Prescription
Injection (Depo-Provera)	1992	An injectable progestin that inhibits ovulation, prevents sperm from reaching the egg, and prevents the fertilized egg from implanting in the uterus.	less than 1	Irregular bleeding, weight gain, breast tenderness, headaches	None	One injection every three months.	Prescription
IUD (Intrauterine device)	1976	A T-shaped device inserted into the uterus by a health professional.	less than 1	Cramps, bleeding, pelvic inflammatory disease, infertility, perforation of uterus	None	After insertion by physician, can remain in place for up to 10 years, depending on type.	Prescription
Periodic abstinence	N/A	To deliberately refrain from having sexual intercourse during times when pregnancy is more likely.	20	None	None	Requires frequent monitoring of body functions (for example, body temperature for one method).	Instructions from health-care provider

27

TABLE 35 Birth Control Guide (*Continued*)

Type of Contraceptive	FDA Approval Date	Description	Failure Rate (number of pregnancies expected per 100 women per year)	Some Risks	Protection from Sexually Transmitted Diseases (STDs)	Convenience	Availability
Surgical sterilization—male	N/A	Sealing, tying, or cutting a man's vas deferens so that the sperm can't travel from the testicles to the penis. (h)	less than 1	Pain, bleeding, infection, other postsurgical complications	None	One-time surgical procedure.	Surgery
Trans-abdominal surgical sterilization—female Falope Ring, Hulka Clip, Filshie Clip	Before 1976	The woman's uterine tubes are blocked so the egg and sperm can't meet in the uterine tube, preventing conception.	less than 1	Pain, bleeding, infection, other postsurgical complications, ectopic (tubal) pregnancy.	None	One-time surgical procedure that requires an abdominal incision.	Surgery

from the scrotum on either side; after each cord is opened, the ductus deferens is severed. A 1-cm section is removed, and the cut ends are tied shut. The cut ends cannot reconnect, and in time, scar tissue forms a permanent seal. Alternatively, the cut ends of the ductus deferens are blocked with silicone plugs that can later be removed. This more recent vasectomy procedure may make it possible to restore fertility at a later date. After a vasectomy, a man experiences normal sexual function, because the secretions of the epididymis and testes normally account for only about 5 percent of the volume of semen. Spermatozoa continue to develop, but they remain within the epididymis until they degenerate.

The uterine tubes can be blocked by a surgical procedure known as a **tubal ligation** (Figure 60c●). The failure rate for this procedure is estimated at 0.45 percent. Because the surgery requires that the abdominopelvic cavity be opened, complications are more likely than with vasectomy. As in a vasectomy, attempts may be made to restore fertility after a tubal ligation.

Oral contraceptives, condoms, vaginal barriers, and sterilization are the primary contraception methods for all age groups. But the proportion of the population using a particular method varies by age group. Sterilization, for example, is most popular among older women, who may already have had children. Relative availability also plays a role. For example, a sexually active female under age 18 can buy a condom more easily than she can obtain a prescription for an oral contraceptive. But many of the differences are attributable to the relationship between risks and benefits for each age group. When considering the use and selection of contraceptives, many people simply examine the list of potential complications and make the "safest" choice. Many women, for example,

reconsidered their use of oral contraceptives because of media coverage of the risks. But complex decisions should not be made on such a simplistic basis, and the risks associated with the use of contraceptives must be considered in light of their relative efficiencies.

Although pregnancy is a natural phenomenon, it has risks, and the pregnancy-related mortality rate for women in the United States averages about 8 deaths per 100,000 pregnancies. That average encompasses a broad range: The rate is 7 per 100,000 women under age 20, but 40 per 100,000 among women over age 40. Although these risks are small, for pregnant women over age 35 the chances of dying from pregnancy-related complications are almost twice as great as the chances of being killed in an automobile accident and are many times greater than the risks associated with the use of oral contraceptives. For women in developing nations, the comparison is even more striking: The pregnancy-related mortality rate for women in parts of Africa is approximately 1 per 150 pregnancies. In addition to preventing pregnancy, combination birth control pills have been shown to reduce the risks of ovarian and endometrial cancers and fibrocystic breast disease.

Before age 35, the risks associated with oral contraceptive use are lower than the risks associated with pregnancy. The notable exception involves women who take the pill and also smoke cigarettes. Younger women are more fertile, so despite a lower mortality rate for each pregnancy, they are likely to have more pregnancies as the result of birth control failures.

After age 35, the risks of complications associated with oral contraceptive use increase, but the risks of using other methods remain relatively stable. Women over age 35 (smokers) or 40 (nonsmokers) are therefore often advised to use

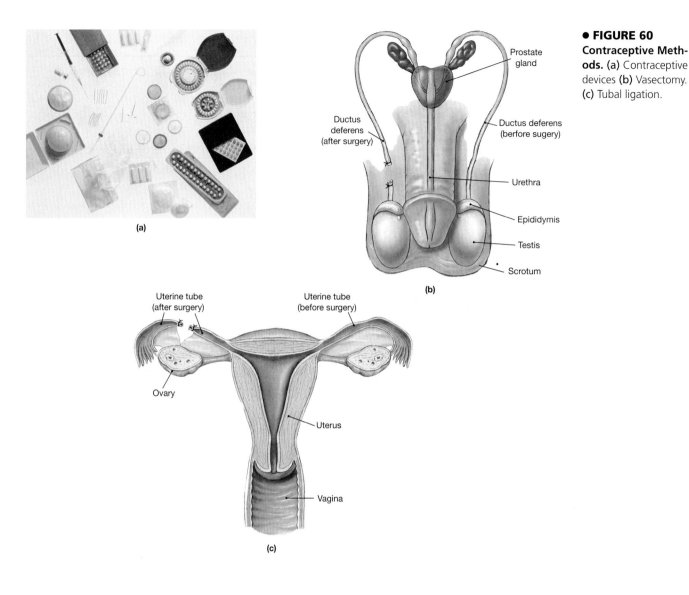

other forms of contraception. Because each contraceptive method has its own advantages and disadvantages, research on contraception continues.

■ EXPERIMENTAL CONTRACEPTIVE METHODS

A number of experimental contraceptive methods are being investigated. For example, one approach is to develop a method of blocking human chorionic gonadotropin (HCG) receptors at the corpus luteum. Produced by the placenta, HCG maintains the corpus luteum for the first three months of pregnancy. If the corpus luteum were unable to respond to HCG, normal menses would occur despite the implantation of a blastocyst. A vaccine against HCG that would require booster shots every three months is being tested, but questions regarding long-term effects and safety concerns remain to be answered. The progesterone antagonist *RU-486 (Mifepristone)* blocks the action of progesterone at the endometrial lining. Daily use of low doses inhibits ovulation and causes amenorrhea. A single higher dose is an effective emergency contraceptive. Still higher doses in the luteal phase can disrupt

early implantation, resulting in the degeneration of the endometrium and menstruation.

With HIV and other STDs a concern, improved barrier contraceptives are being developed. A chemical gel, *BufferGel*, that is a spermicide and microbicide (for protection from STDs) is in clinical testing.

Several male contraceptives are also under development:

- **Gossypol**, a yellow pigment extracted from cottonseed oil, produces a dramatic decline in sperm count and sperm motility after two months. Fertility returns within a year after treatment is discontinued, but permanent sterility occurs in up to 10 percent of users, making the drug unacceptable to the World Health Organization.

- Reversible suppression of gonadotrophins, spermatogenesis, and sperm count has been achieved with various strengths and forms of testosterone alone, as well as with testosterone (or other androgens) plus progesterones. Various changes in libido, shifts in levels of HDL cholesterol, and weight gain have been reported as side effects. Daily oral regimens or long-acting injections are approaching clinical trials and may eventually be available.

TECHNOLOGY AND THE TREATMENT OF INFERTILITY

HA p. 738

Infertility *(sterility)* is usually defined as an inability to achieve pregnancy after one year of appropriately timed intercourse. Problems with fertility are relatively common: An estimated 10–15 percent of U.S. married couples are infertile, and another 10 percent are unable to have as many children as they desire. It is thus not surprising that reproductive physiology has become a popular field, and that the treatment of infertility has become a major medical industry.

An infertile woman is unable to produce functional oocytes or support a developing embryo. An infertile man is incapable of providing a sufficient number of motile sperm capable of successful fertilization. Because the infertility of either sexual partner has the same result, the diagnosis and treatment of infertility must involve evaluations of both partners. Approximately 40 percent of infertility cases are attributed to the female partner, 40 percent to the male partner, and 20 percent to both partners.

Recent advances in our understanding of reproductive physiology are providing new solutions to fertility problems. The various problems, and the approaches to solving them—called **assisted reproductive technologies (ARTs)**—include the following (Figure 61●):

• **Low sperm count/abnormal spermatozoa.** In cases of male infertility due to low sperm counts, or unexplained infertility, semen from one or several ejaculates can be concentrated,

and the most motile sperm selected and introduced into the uterus. This technique, interuterine insemination or artificial insemination, is often combined with hormonal treatment of the woman to induce ovulation. Normal fertilization and pregnancy may then occur. In cases in which a male's spermatozoa are unable to penetrate the oocyte, single-sperm fertilization (intracytoplasmic sperm injection or ICSI) can be accomplished by microscopically manipulating the sperm, oocyte, and the corona radiata. The fertilized egg is then used in an IVF procedure (discussed in a later section). If the male cannot produce functional spermatozoa, donor sperm obtained from a *sperm bank* (which screens the donor and stores the sperm) can be introduced into the female reproductive tract (artificial insemination). In this case intrauterine insemination may not be necessary.

• **Hormonal problems.** If the problem involves the woman's inability to ovulate because her gonadotropin or estrogen levels are low, or if she is unable to maintain adequate progesterone levels after ovulation, these hormones can be provided to induce ovulation and/or maintain the luteal endometrium.

• **Problems with oocyte production.** *Fertility drugs,* such as clomiphene (*Clomid*), stimulate ovarian oocyte production. Clomiphene blocks the feedback inhibition of the hypothalamus and pituitary gland by estrogens. As a result, circulating FSH levels rise, so more follicles are stimulated to complete their development. Injected purified gonadotropins, such as *Pergonal* (FSH and LH) and *Metrodin* (FSH), are also used to accelerate ovum development. The chance that a single oocyte will be fertilized through well-

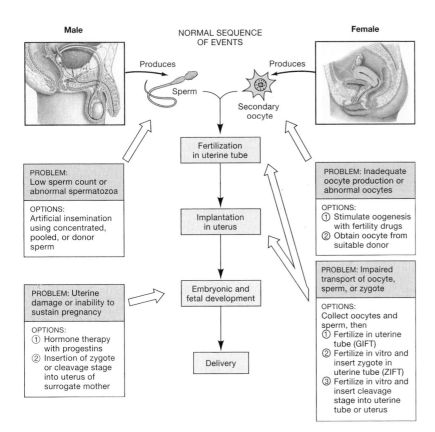

● FIGURE 61
Infertility Problems and Treatment Options

Male — NORMAL SEQUENCE OF EVENTS — Female

Produces → Sperm → Secondary oocyte ← Produces

PROBLEM: Low sperm count or abnormal spermatozoa

OPTIONS: Artificial insemination using concentrated, pooled, or donor sperm

PROBLEM: Uterine damage or inability to sustain pregnancy

OPTIONS:
① Hormone therapy with progestins
② Insertion of zygote or cleavage stage into uterus of surrogate mother

Fertilization in uterine tube

Implantation in uterus

Embryonic and fetal development

Delivery

PROBLEM: Inadequate oocyte production or abnormal oocytes

OPTIONS:
① Stimulate oogenesis with fertility drugs
② Obtain oocyte from suitable donor

PROBLEM: Impaired transport of oocyte, sperm, or zygote

OPTIONS:
Collect oocytes and sperm, then
① Fertilize in uterine tube (GIFT)
② Fertilize in vitro and insert zygote in uterine tube (ZIFT)
③ Fertilize in vitro and insert cleavage stage into uterine tube or uterus

timed sexual intercourse is about one in three. Increasing the number of oocytes released raises the odds of fertilization and therefore the odds of a pregnancy. It is not easy to determine just how much ovarian stimulation is needed, however, so treatment with fertility drugs commonly results in multiple mature eggs. Careful monitoring of follicle development and avoidance of fertilization if too many eggs are present reduces the chances of a multiple birth.

- **Blocked uterine tubes.** Blockage of or damage to the uterine tubes can interfere with oocyte, sperm, and zygote transport. Normally, fertilization occurs within the uterine tube. However, this site is not essential, and fertilization can also take place *in vitro*—in a test tube or petri dish (*vitro,* glass). The ovaries are stimulated with injected hormones, and a large "crop" of mature oocytes is "harvested" from tertiary follicles by inserting a long needle through the vaginal wall (with ultrasound guidance) into the follicles. The individual oocytes are examined for defects, sperm and a carefully controlled fluid environment are provided, and if fertilization occurs, early development will proceed normally. The zygote can be maintained in an artificial environment through the first two to five days of development. The embryo is then placed directly into the uterus through a tube introduced into the cervix. The process, called *in vitro fertilization,* or **IVF**, was involved in 98 percent of ART procedures in the United States in 2001. Almost 50 percent of these IVF procedures involved ICSI as well.

- **Abnormal oocytes.** Unlike sperm, which have a short existence and are continually being produced, a woman's oocytes age with her. Age-related changes in the characteristics and quality of the oocytes, rather than changes in hormone levels or uterine responsiveness, seem to be the primary cause of infertility in older women. Viable donor eggs can be obtained from younger women. Through treatment with fertility drugs, the donor's ovaries are stimulated to produce a large crop of oocytes, which are then collected and fertilized in vitro, generally by the spermatozoa of the recipient's male partner. After cleavage has begun, the pre-embryo is placed in the recipient's uterus, which has been synchronized with the donor's menstrual cycle by ovulatory hormones and "primed" by progesterone therapy. Although the mother has no genetic relationship to the embryo, pregnancy proceeds normally.

The live birth rate of this procedure is roughly 45 percent for women over age 40, while using their own eggs give a live birth rate of 10 percent.

- **An abnormal uterine environment.** If fertilization and transport occur normally, but the uterus cannot maintain a pregnancy, the problem may involve low levels of progesterone secretion by the corpus luteum. Hormone therapy may solve this problem. If a woman's uterus simply cannot support development, an IVF procedure can introduce a zygote or cleavage-stage embryo into the uterus of a *surrogate mother.* If the embryo survives and makes contact with the endometrium, development will proceed normally even though the surrogate mother may have no genetic relationship to the embryo. The egg may come from the woman who cannot carry the pregnancy, the surrogate, or another egg donor.

Surrogate motherhood, which sounds relatively simple and straightforward, has proven to be one of the most explosive solutions in terms of ethics and legality. Since 1990, several court cases have resulted from disputes over surrogate motherhood and who merits legal custody of the infant. Legal battles have also erupted over a variety of other complex infertility-related questions, some of which will take years to sort out. To understand the problem, consider the following questions:

- Do parents share property rights over frozen and stored zygotes? May a husband have any of the stored zygotes implanted into the uterus of his second wife without the consent of his first wife, who provided the oocytes? May a wife use her husband's stored spermatozoa to become pregnant after his death?

- If both donor oocyte and donor sperm are used, do adoption laws apply?

- If the husband provided the spermatozoa that fertilized the oocyte of a donor who is not his wife, for implantation into a surrogate mother, should the wife, the surrogate mother, or the oocyte donor have custody of the child if the father dies?

- If a hospital stores frozen pre-embryos or spermatozoa, or zygotes, but the freezer breaks down, what is the hospital's liability? What is the monetary value of a frozen pre-embryo?

If you use your imagination, you can probably think of even more complex problems, many of which will probably be debated in courtrooms over the next decade.

DISORDERS OF DEVELOPMENT

Fetal development is a complex process, and developmental disorders are extremely diverse. Figure 62● surveys representative disorders of development.

■ TERATOGENS AND ABNORMAL DEVELOPMENT *HA p. 745*

Teratogens (TER-a-tō-jenz) are stimuli that disrupt normal fetal development by damaging cells or altering their chromosomal structure or that interfere with normal induction. **Teratology** (ter-a-TOL-ō-jē)—literally, the "study of monsters"—deals with extensive departures from the pathways of normal development. Teratogens that affect the embryo in the first trimester can disrupt cleavage, gastrulation, or neurulation. The embryonic survival rate will be low, and most survivors will have severe anatomical and physiological defects that affect all the major organ systems. Errors introduced into the developmental process during the second or third trimester are likely to affect specific organs or organ systems, for the major organizational patterns are already established. Nevertheless, the alterations reduce the chances for long-term survival of the infant.

We encounter many powerful teratogens in everyday life. The location and severity of the resulting defects vary with the nature of the stimulus and the time of exposure. Radiation is a powerful teratogen that can affect all cells. Even the x-rays used in diagnostic procedures can break chromosomes and produce developmental errors; thus, nonionizing procedures such as ultrasound are used to track embryonic and fetal development. Fetal exposure to the microorganisms responsible for syphilis or rubella ("German measles") can also produce serious developmental abnormalities, including congenital heart defects, mental retardation, and deafness.

Some chemical agents are teratogenic only if they are present at a time when embryonic or fetal targets are vulnerable to their effects. Thousands of critical inductions are under way during the first trimester, initiating developmental sequences that will produce the major organs and organ systems of the body. In almost every case, the nature of the inducing agent remains unknown, and the effects of unusual compounds within the maternal circulation cannot be predicted. As a result, virtually any unusual chemical that reaches an embryo has the potential for producing developmental abnormalities. For example, during the 1960s, the European market was strong for **thalidomide**, a drug that is effective in promoting sleep and preventing nausea. Thalidomide was commonly prescribed for women in early pregnancy, with disastrous results. The drug crossed the placenta and entered the fetal circulation, where, for a few days, it interfered with the induction process responsible for limb development. Many infants exposed to the drug were born without limbs or with drastically deformed ones. Thalidomide was not approved by the U.S. Food and Drug Administration (FDA), and it could not be sold legally in the United States. Although the FDA is often criticized today for the slow pace of its approval process,

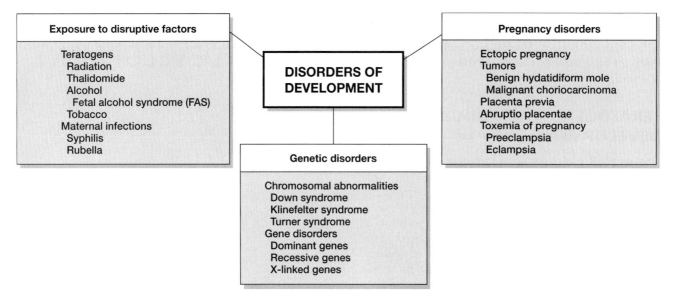

● **FIGURE 62**
Disorders of Development

in this case the combination of rigorous testing standards and complex bureaucratic procedures protected the public.

However, even when extensive testing is performed with laboratory animals, uncertainties remain because the chemical nature of the inducer responsible for a specific process may vary from one species to another. For example, thalidomide produces abnormalities in humans and monkeys, but developing mice, rats, and rabbits are unaffected by the drug.

More powerful teratogens have an effect regardless of the time of exposure. Pesticides, herbicides, and heavy metals are common around agricultural and industrial environments, and these substances can contaminate the drinking water in the area. A number of prescription drugs, including certain antibiotics, tranquilizers, sedatives, steroid hormones, diuretics, anesthetics, and analgesics, also have teratogenic effects. Pregnant women should read the "Caution" label before using any drug without the advice of a physician. Most "natural" herbs and substances have not been tested, and their chemical composition may vary from source to source, so their effects during pregnancy are unknown, and they should not be considered uniformly safe to use during pregnancy. (We know that some plants produce teratogens and store them in their leaves, presumably as a defense against herbivorous animals.)

Fetal alcohol syndrome (FAS) occurs when maternal alcohol consumption produces developmental defects in the fetus, such as skeletal deformation, cardiovascular defects, and neurological disorders. Fetal mortality rates can be as high as 17 percent, and the survivors are plagued by problems in later development. The most severe cases involve mothers who consume the alcohol content of at least 7 ounces of hard liquor, 10 beers, or several bottles of wine each day. However, because the effects produced are directly related to the degree of exposure, there is probably no level of alcohol consumption that can be considered completely safe. Fetal alcohol syndrome is the number one cause of mental retardation in the United States today, affecting roughly 7500 infants each year.

Smoking presents another major risk to the developing fetus. In addition to introducing potentially harmful chemicals, such as nicotine, smoking lowers the P_{O_2} of maternal blood and reduces the amount of oxygen that reaches the placenta. A fetus carried by a smoking mother will not grow as rapidly as one carried by a nonsmoking mother, and smoking increases the risks of spontaneous abortion, prematurity, and fetal death. The rate of infant mortality after delivery is also higher when the mother smokes, and postnatal development can be adversely affected as well.

■ INDUCTION AND SEXUAL DIFFERENTIATION

The physical (phenotypic) sex of a newborn infant depends on the hormonal cues it receives during development, not on the genetically determined sex of the individual. If something disrupts the normal inductive processes, the individual's genetic and anatomical sexes may be different. Such a person is called a **pseudohermaphrodite** (soo-do-her-MAF-rō-dīt). For example, if a female embryo becomes exposed to male hormones, it will develop the sexual characteristics of a male. Such situations are relatively rare. The most common cause is the hypertrophy of the fetal adrenal glands and their production of androgens in high concentrations; in some cases, this condition has been linked to genetic abnormalities. Maternal exposure to androgens, in the form of anabolic steroids or as a result of an endocrine tumor, can also produce a female pseudohermaphrodite.

Male pseudohermaphrodites may result from an inability to produce adequate concentrations of androgens due to some enzymatic defect. In **testicular feminization syndrome**, the infant appears to be a normal female at birth. Typical physical changes occur at puberty, and the individual develops the overt physical and behavioral characteristics of an adult woman. Menstrual cycles do not begin, however, because the vagina ends in a blind pocket, and there is no uterus. Biopsies performed on the gonads reveal a normal testicular structure, and the interstitial cells are busily secreting testosterone. The problem apparently involves a defect in the cellular receptors that are sensitive to circulating androgens. Neither the embryo nor the adult tissues can respond to the testosterone produced by the gonads, so the person develops as, and remains physically, a female.

If detected in infancy, many cases of pseudohermaphroditism can be treated with hormones and surgery to produce males or females of normal appearance. Depending on the arrangement of the internal organs and gonads, normal reproductive function may be more difficult to achieve. Sex hormones also affect the brain's development, and sex assignment is behavioral as well as anatomical. In one instance, a genetic male who had been raised as a female after undergoing surgery on indeterminate external genitalia later adopted a male name, attire, and behavior after hormonal changes occurred at puberty.

Pseudohermaphroditism is one example of a developmental problem caused by hormonal miscues or by an inability to respond appropriately to hormonal instructions. Another example is male infertility associated with maternal exposure to *diethylstilbestrol* (DES), a synthetic steroid prescribed in the 1950s to prevent miscarriages. An estimated 28 percent of male offspring produced abnormally small amounts of semen with marginal sperm counts at maturity. Daughters exposed to DES also have higher-than-normal infertility rates, due to uterine, vaginal, and uterine tube abnormalities, and they have an increased risk of developing vaginal cancer.

■ ECTOPIC PREGNANCIES *HA p. 747*

Implantation normally occurs at the endometrial surface that lines the uterine cavity. The precise location within the uterus varies, although in most cases implantation occurs in the uterine body. By contrast, in an **ectopic pregnancy**, implantation occurs outside the uterus.

The incidence of ectopic pregnancies is approximately 0.6 percent of all pregnancies. Women who douche regularly have a 4.4 times higher risk of experiencing an ectopic pregnancy, presumably because the flushing action pushes the zygote away from the uterus. If the uterine tube has been scarred by a previous episode of pelvic inflammatory disease, the risk of an ectopic pregnancy increases. Although implantation may occur within the peritoneal cavity, in the ovarian wall, or in the cervix, 95 percent of ectopic pregnancies involve implantation within a uterine tube. Because it cannot expand enough to accommodate the developing embryo, the tube normally ruptures during the first trimester. At that time, the bleeding that occurs in the peritoneal cavity can be severe enough to pose a threat to the woman's life.

In a few instances, the ruptured uterine tube releases the embryo with an intact umbilical cord, so further development can occur. About 5 percent of these abdominal pregnancies actually complete full-term development; normal birth cannot occur, but the infant can be surgically removed from the abdominopelvic cavity. Because abdominal pregnancies are possible, it has been suggested that men as well as women could act as surrogate mothers if a zygote were surgically implanted into the peritoneal wall. It is not clear, though, how the endocrine, cardiovascular, nervous, and other systems of a man would respond to the stresses of pregnancy.

■ PROBLEMS WITH PLACENTATION
HA p. 750

In a **placenta previa** (PRE-ve-uh; "in the way"), implantation occurs in or near the cervix. This condition causes problems as the growing placenta approaches the internal os (internal cervical orifice). In a **total placenta previa**, the placenta extends across the internal os, whereas a partial placenta previa only partially blocks the internal os. The placenta is characterized by a rich fetal blood supply and the erosion of maternal blood vessels within the endometrium. Where the placenta passes across the internal os, the delicate complex hangs like an unsupported water balloon. As the pregnancy advances, even minor mechanical stresses can be enough to tear the placental tissues, leading to massive fetal and maternal bleeding.

Most cases of placenta previa can be diagnosed by ultrasound in the second trimester. As the uterus enlarges, the placenta may retract from covering the cervical os. If it does not, then by the seventh month, the placenta reaches its full size, the cervical canal is dilated, and the uterine contents push against the placenta where it spans the internal cervical os without the support of the uterine wall. Minor, painless bleeding may occur. The treatment of total placenta previa involves bed rest for the mother until the fetus reaches a size at which cesarean delivery can be performed with a reasonable chance of survival of the neonate (newborn).

In an **abruptio placentae** (ab-RUP-shē-ō pla-SEN-tē), part or all of the placenta tears away from the uterine wall sometime after the fifth month of gestation. This is a serious and dangerous condition, and the bleeding and pain are usually sufficient to prompt an immediate visit to a physician. In severe cases, the bleeding leads to maternal anemia, shock, and kidney failure. Although maternal mortality is only 0.5 to 1 percent, the fetal mortality rate from this condition ranges from 20 to 35 percent, depending on the severity of the fetal blood loss.

■ PROBLEMS WITH THE MAINTENANCE OF A PREGNANCY
HA p. 755

The rate of maternal complications during pregnancy is relatively high. Pregnancy stresses maternal systems, and the stresses can overwhelm homeostatic mechanisms. The term

28

toxemia (tok-SĒ-mē-uh) **of pregnancy** refers to disorders that affect the maternal cardiovascular system. Chronic hypertension is the most characteristic symptom, but fluid imbalances, proteinuria, and central nervous system (CNS) disturbances, leading to coma or convulsions, can also occur. Some degree of toxemia occurs in 6–7 percent of third-trimester pregnancies. Severe cases account for 20 percent of maternal deaths and contribute to an estimated 25,000 neonatal deaths each year. Prenatal care involves monitoring the mother's vital signs and urine to detect early signs of toxemia so that treatment can prevent further progression of the condition.

Toxemia of pregnancy includes **preeclampsia** (prē-ē-KLAMP-sē-uh) and **eclampsia** (ē-KLAMP-sē-uh). Preeclampsia is most common during a woman's first pregnancy. The mother's systolic and diastolic pressures become elevated, reaching levels at or above 180/110. Other symptoms include fluid retention and edema, along with CNS disturbances and changes in kidney function. Roughly 4 percent of women with preeclampsia develop eclampsia.

Eclampsia is heralded by the onset of severe convulsions lasting one to two minutes, followed by a variable period of coma. Other symptoms resemble those of preeclampsia, with additional evidence of liver and kidney damage. The mortality rate from eclampsia is approximately 5 percent; the mother can be saved only if the fetus is delivered immediately. Once the fetus and placenta have been removed from the uterus, symptoms of eclampsia disappear over a period of hours to days.

COMPLEXITY AND PERFECTION

HA p. 758

The expectation of prospective parents that every pregnancy will be idyllic and every baby will be perfect reflects deep-seated misconceptions about the nature of the developmental process. These misconceptions lead to the belief that when serious developmental errors occur, someone or something is at fault and that blame might be assigned to maternal habits (such as smoking, alcohol consumption, or improper diet), maternal exposure to toxins or prescription drugs, or the presence of other disruptive stimuli in the environment. The prosecution of women who give birth to severely impaired infants for "fetal abuse" (exposing a fetus to known or suspected risk factors) is an extreme example of this philosophy.

Although environmental stimuli can indeed lead to developmental problems, such factors are only one component of a complex system that is normally subject to considerable variation. Even if every pregnant woman were packed in cotton and confined to bed from conception to delivery, developmental accidents and errors would continue to occur with regularity.

Spontaneous mutations are the result of random errors in replication; such incidents are relatively common. At least 10 percent of fertilizations produce zygotes with abnormal chromosomes. Because most spontaneous mutations fail to produce visible defects, the actual number of mutations must be far larger. Most of the affected zygotes die before completing development, and only about 0.5 percent of newborns show chromosomal abnormalities that result from spontaneous mutations.

Due to the nature of the regulatory mechanisms, prenatal development does not follow precise, predetermined pathways. For example, much variation exists in the pathways of blood vessels and nerves, because it does not matter how blood or neural impulses get to their destinations, as long as they do get there. If the variations fall outside acceptable limits, however, the embryo or fetus fails to complete development. Very minor changes in heart structure can result in the death of a fetus, whereas large variations in venous distribution are common and relatively harmless. Virtually everyone can be considered abnormal to some degree, because no one has characteristics that are statistically average in every respect. An estimated 20 percent of your genes are subtly different from those found in the majority of the population, and minor defects such as extra nipples or birthmarks are quite common.

Current evidence suggests that as many as half of all conceptions produce zygotes that do not survive the cleavage stage. These zygotes disintegrate within the uterine tubes or uterine cavity; because implantation never occurs, there are no obvious signs of pregnancy. Preimplantation mortality is commonly associated with chromosomal abnormalities. Of those embryos that implant, roughly 20 percent fail to complete five months of development, with an average survival time of eight weeks. In most cases, severe problems affecting early embryogenesis or placenta formation are responsible. Figure 63● graphically shows the relation of prenatal mortality to gestational age.

Prenatal mortality tends to eliminate the most severely affected fetuses. Those with less extensive defects may survive, completing full-term gestation or arriving via premature delivery. **Congenital malformations** are structural abnormalities that are present at birth and that affect major systems. Spina bifida, hydrocephalus, anencephaly, cleft lip, and Down syndrome are among the most common congenital malformations; we described those conditions in earlier chapters of the main text. The incidence of congenital malformations at birth averages about 6 percent, but only 2 percent are categorized as severe. Of these congenital problems, only 10 percent can be attributed to environmental factors in the absence of chromosomal abnormalities or genetic factors, including a family history of similar or related defects.

Medical technology continues to improve our abilities to understand and manipulate physiological processes. Genetic analysis of potential parents can now provide estimates of the likelihood of specific problems, although the problems themselves remain outside our control. But even with a better understanding of the genetic mechanisms involved, we will probably never be able to control every aspect of development and thereby prevent spontaneous abortions and congenital malformations. Too many complex, interdependent steps are involved in prenatal development, and malfunctions of some kind are statistically inevitable.

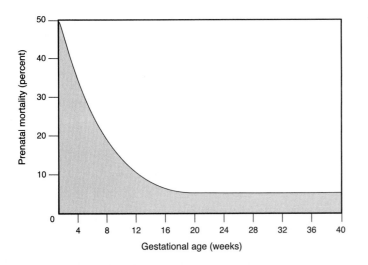

MONITORING POSTNATAL DEVELOPMENT   *HA p. 758*

Each newborn is closely scrutinized after delivery. The maturity of the newborn may be determined prior to delivery by means of ultrasound or amniocentesis (Table 36). Immediately on delivery, the newborn is checked and assigned an **Apgar rating**, which evaluates the heart rate, respiratory rate, muscle tone, response to stimulation, and color at one and five minutes after birth. In each category, the infant receives a score ranging from 0 (poor) to 2 (excellent), and the scores are then totaled. An infant's Apgar rating (0–10) has been shown to be an accurate predictor of newborn survival and of the presence of neurological damage. For example, newborn infants with cerebral palsy tend to have a low Apgar rating.

In the course of this examination, the newborn's breath sounds, depth and rate of respiration, and heart rate are noted. Both the respiratory rate and the pulse rate are considerably higher in infants than in adults. (See Table 1, p. 2.) Later, a more complete physical examination of the newborn focuses on the status of its vital systems. Inspection of the infant normally includes the following:

* The head of a newborn may be misshapen after vaginal delivery, but it generally assumes its normal shape within the next few days. However, the size of the head must be checked to detect hydrocephalus, and the cranial vault is checked to ensure that cerebral hemispheres are present and anencephaly does not exist.

* The eyes, nose, mouth, and ears are inspected for reflex responses and for obstruction.

* The abdomen is palpated to detect abnormalities of internal organs.

* The heart and lungs are auscultated to check for breath sounds and heart murmurs.

* The external genitalia are inspected. The scrotum of a male infant is checked to see if the testes have descended.

* Cyanosis of the hands and feet is normal in newborns, but the rest of the body should be pink. A generalized cyanosis may indicate congenital circulatory disorders, such as erythroblastosis fetalis, patent foramen ovale, ductus arteriosus, or tetralogy of Fallot.

Measurements of the neonate's body length, head circumference, and body weight are taken. A weight loss in the first 48 hours is normal, because fluid shifts occur as the infant adapts to the change from weightlessness (floating in amniotic fluid) to normal gravity. (Comparable fluid shifts occur in astronauts returning to Earth after extended periods in space.)

The nervous and muscular systems of newborns are assessed for normal reflexes and muscle tone. Reflexes commonly tested include the following:

* The **Moro reflex** is triggered when support for the head of a supine infant is suddenly removed. The reflex response consists of extension of the trunk and a rapid cycle of extension–abduction and flexion–adduction of the limbs. This reflex normally disappears at an age of about 3 months.

* The **stepping reflex** consists of walking movements triggered by holding the infant upright, with a forward slant, and placing the soles of the feet against the ground. This reflex normally disappears at an age of about 6 weeks.

* The **placing reflex** can be triggered by holding the infant upright and drawing the top of one foot across the bottom edge of a table. The reflex response is to flex and then extend that leg. This reflex also disappears at an age of about 6 weeks.

* The **sucking reflex** is triggered by stroking the lips. The associated *rooting reflex* is initiated by stroking the cheek, and the response is to turn the mouth toward the site of stimulation. These reflexes persist until age 4–7 months.

* The **Babinski reflex** is positive, with fanning of the toes in response to stroking the side of the sole of the foot. This reflex disappears at about age 3 years as descending motor pathways become established.

These procedures check for the presence of anatomical and physiological abnormalities. They also provide baseline information that is useful in assessing postnatal development. In addition, newborn infants are typically screened for genetic or metabolic disorders discussed previously, such as

TABLE 36 Tests Performed during Pregnancy and on the Neonate

Diagnostic Procedure	Method and Result	Representative Uses
Amniocentesis	A needle, inserted through abdominal wall into uterine cavity, collects amniotic fluid for analysis	Detects chromosomal abnormalities and level of hemolysis in erythroblastosis fetalis; determines fetal lung maturity; detects birth defects such as spina bifida; evaluates fetal distress
Pelvic ultrasonography	Standard ultrasound	Detects multiple fetuses, fetal abnormalities, and placenta previa; estimates fetal age, growth, and sex
External fetal monitoring	Monitoring devices on external abdominal surface measure fetal heart rate and force of uterine contraction	Detects irregular heart rate or fetal stress
Internal fetal monitoring	Electrode is attached to fetal scalp to monitor heart rate; catheter is placed in uterus to monitor uterine contractions	As above
Chorionic villi biopsy	Test performed during weeks 8–10 of gestation; small pieces of villi are suctioned into a syringe	Detects chromosomal abnormalities and biochemical disorders

Laboratory Test	Normal Values	Significance of Abnormal Values
Amniotic fluid analysis		
Karyotyping	Normal chromosomes	Detects chromosomal defects such as those in Down syndrome
Bilirubin	Traces only	Increased values may indicate amount of hemolysis of fetal RBCs by mother's Rh antibodies
Meconium	Not present	Present in fetal distress
Lecithin/sphingomyelin ratio (L/S ratio)	>2 : 1 ratio	Ratio below 2:1 indicates fetal immaturity
Creatinine	>2 mg/dl of amniotic fluid indicates week 36 of gestation.	Less than 2 mg/dl indicates fetal immaturity (not as accurate as L/S ratio)
Alpha-fetoprotein (AFP)	Week 16 of gestation: 5.7–31.5 ng/ml (lowers with increasing gestational age)	Increased values indicate possible neural tube defect, such as spina bifida
Blood tests		
Human chorionic gonadotropin (HCG)	Nonpregnant females (serum): <0.005 IU/ml	Determines pregnancy; used in home pregnancy tests
(maternal serum or urine)	Nonpregnant females (urine): negative Pregnant females (urine): >500,000 IU over 24 hours	
NEONATES		
Blood from umbilical cord	Blood typing titer	Detects maternal Rh antibody
Bilirubin	<12 mg/dl	Increased levels occur in jaundice due to immaturity of newborn's liver
Phenylalanine (serum)	1–3 mg/dl	>4 mg/dl occurs in phenylketonuria

phenylketonuria (PKU), congenital hypothyroidism, galactosemia, and sickle cell anemia.

The excretory systems of the newborn infant are assessed by examining the urine and feces. The first urination may be pink, owing to the presence of uric acid derivatives. The first bowel movement consists of a greenish-black mixture of epithelial cells and mucus called *meconium*.

Pediatrics is a medical specialty focusing on postnatal development from infancy through adolescence. Infants and young children cannot clearly describe the problems they are experiencing, so pediatricians and parents must be skilled observers. Standardized testing procedures are also used to assess developmental progress. In the **Denver Developmental Screening Test (DDST)**, infants and children are checked repeatedly during their first five years. The test checks gross motor skills, such as sitting up or rolling over, language skills, fine motor coordination, and social interactions. The results are compared with normal values determined for individuals of similar age. These screening procedures assist in identifying children who may need special teaching and attention.

Too often, parents tend to focus on a single ability or physical attribute, such as the age when the infant takes the first step or the growth rate compared against standardized growth charts. This kind of one-track analysis has little practical value, and parents can become overly concerned with how their infant compares with the norm. *Normal values are statistical averages,* not absolute realities. For example, most infants begin walking at 11 to 14 months of age. But about 25 percent start before then, and another 10 percent have not started walking by the 14th month. Walking early does not indicate true genius, and walking late does not mean that the infant will need physical therapy. The questions on screening tests such as the DDST are intended to identify *patterns* of developmental deficits. Such patterns appear only when a broad range of abilities and characteristics is considered.

LIVING AND DYING

Despite exaggerated claims, few cases of individuals who have reached an age of 120 years have been substantiated. Estimates for the life span of individuals born in the United States during 2000 are 74.1 years for males and 79.5 years for females. Interestingly enough, the causes of death vary with the age group. Consider the graphs shown in Figure 64●, which indicate the mortality statistics for various age groups. The major cause of death in young people is accidents; in adults over age 40–45, it is cardiovascular disease. More specific information about the major causes of death is given in Table 37. Many of the characteristic differences in mortality result from changes in the individual's functional capabilities which are linked to development or senescence. The picture differs significantly for those living in countries and cultures with different genetic and environmental pressures.

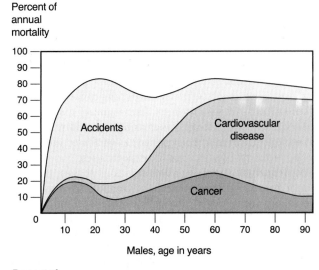

Males, age in years

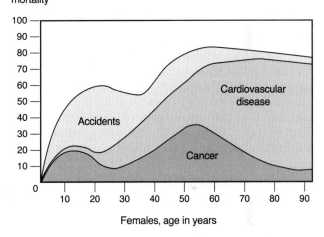

Females, age in years

● **FIGURE 64**
Major Causes of Postnatal Mortality

TABLE 37	**The Five Major Causes of Death in the U.S. Population (2001)**			
	Age 1–14	**Age 15–44**	**Age 45–64**	**Age 64**
Rank				
1	Accidents	Accidents	Cancer	Heart disease
2	Cancer	Cancer	Heart disease	Cancer
3	Congenital anomalies Cancer	Heart disease	Accidents	Cerebrovascular disease
4	Homicide	Suicide	Cerebrovascular disease	COPD*
5	Heart disease	Homicide	COPD	Pneumonia, influenza

*COPD = chronic obstructive pulmonary disease.

28

The differences in mortality values for males and females are related to differences in the accident rates among young people and in the rates of heart disease and cancer among older individuals. For instance, an upswing in cancer rates among females reflects a rising incidence of breast cancer for those over age 34, whereas lung cancer is the primary cancer killer of older men. Among women, the incidence of lung cancer and related killers, including pulmonary disease, heart disease, and pneumonia, has been steadily increasing as the number of female smokers has risen. This change has narrowed the difference between male and female life expectancies.

Experimental evidence and calculations suggest that the human life span has an upper limit of about 150 years. As medical advances continue, research must focus on two related issues: (1) extending the average life span toward that maximum and (2) improving the functional capabilities of long-lived individuals. The first objective may be the easiest from a technical standpoint. It is already possible to reduce the number of deaths attributed to specific causes. For example, new treatments promote remission in a variety of cancer cases, and anticoagulant therapies reduce the risks of death or permanent damage after a stroke or heart attack. Many defective organs can be replaced with functional transplants, and the use of controlled immunosuppressive drugs increases the success rates of these operations. Artificial hearts have been used, though with limited success thus far, and artificial kidneys and endocrine pancreases are under development.

The second objective poses more of a problem. Few people past their mid-90s lead active, stimulating lives, and most would find the prospect of living another 50 years rather horrifying unless the quality of their lives could be significantly improved. Our abilities to prolong life now involve making stopgap corrections in systems that are on the brink of complete failure. Reversing the process of senescence would entail manipulating the biochemical operations and genetic programming of virtually every organ system. Although investigations continue, breakthroughs cannot be expected in the immediate future.

In the interim, we are left with some serious ethical and moral questions. If we could postpone the moment of death almost indefinitely with some combination of resuscitators and pharmacological support, how would we decide when it is appropriate to do so? How can medical and financial resources be allocated fairly? Who gets the limited number of organs available for transplant? Who should be selected for experimental therapies of potential significance? Should we take into account that health care of an infant or child may add decades to a life span, whereas the costly insertion of an artificial heart in a 60-year-old may add only months to years? How shall we allocate the costs of sophisticated procedures that can reach hundreds of thousands of dollars per individual over the long run? Are these societal, individual, or family responsibilities? Will only the rich be able to survive into a second century of life? Should the government provide the funds? If so, what will happen to tax rates as the baby boomers become elderly citizens? And what about the role of the individual? If you decline treatment, are you mentally and legally competent? Could your survivors bring suit if you were forced to survive or if you were allowed to die? These and other difficult questions will not go away. In the years to come, we will have to find answers we are content to live and die with.

28

SCANS

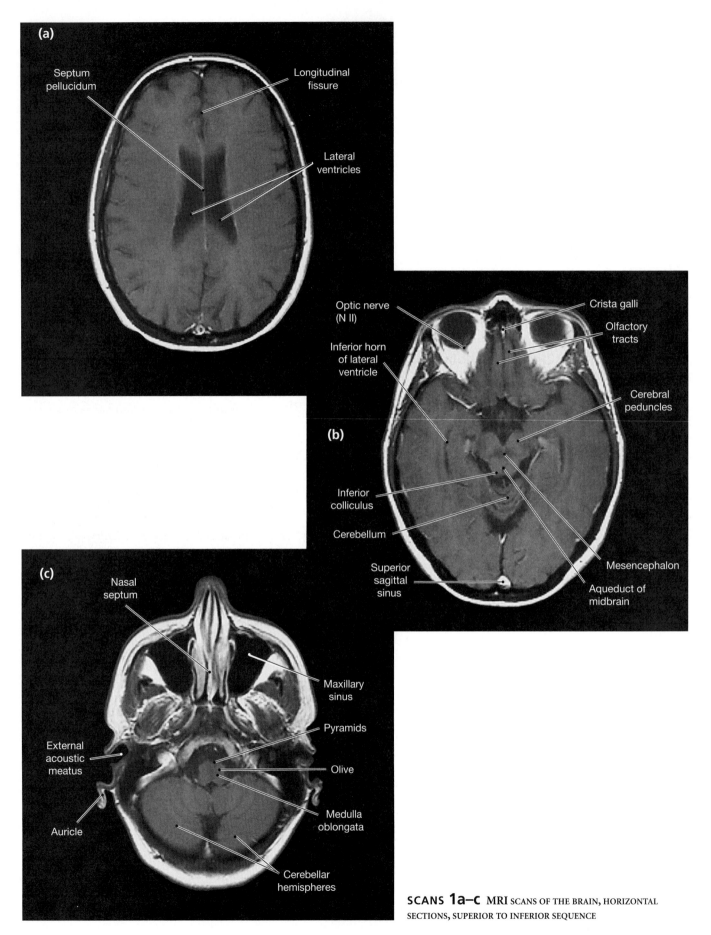

(a)

Septum
pellucidum

Longitudinal
fissure

Lateral
ventricles

(b)

Optic nerve
(N II)

Inferior horn
of lateral
ventricle

Crista galli

Olfactory
tracts

Cerebral
peduncles

Inferior
colliculus

Cerebellum

Superior
sagittal
sinus

Mesencephalon

Aqueduct of
midbrain

(c)

Nasal
septum

Maxillary
sinus

Pyramids

External
acoustic
meatus

Olive

Auricle

Medulla
oblongata

Cerebellar
hemispheres

SCANS 1a–c MRI SCANS OF THE BRAIN, HORIZONTAL
SECTIONS, SUPERIOR TO INFERIOR SEQUENCE

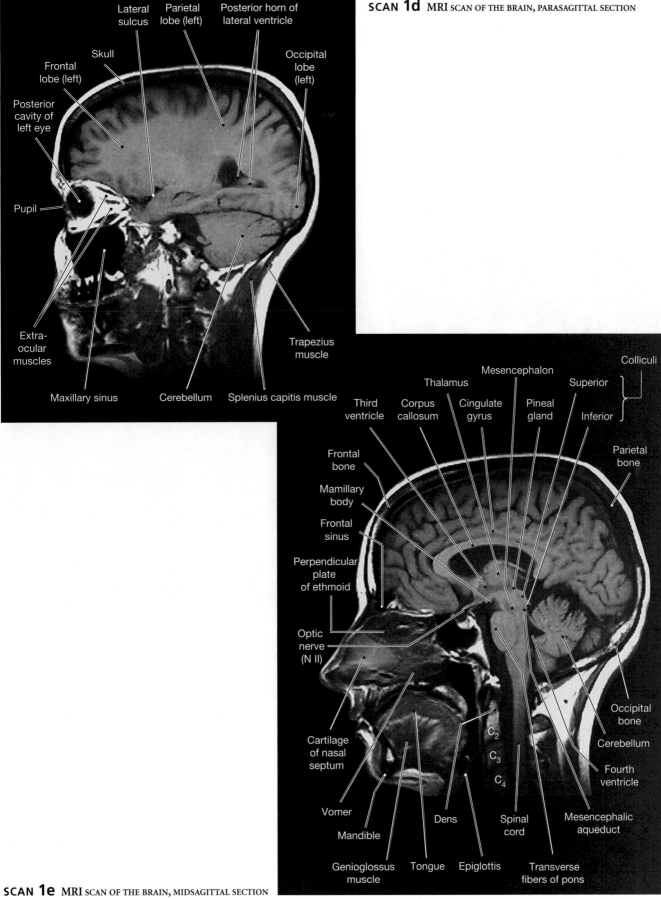

SCAN 1d MRI SCAN OF THE BRAIN, PARASAGITTAL SECTION

Lateral sulcus
Parietal lobe (left)
Posterior horn of lateral ventricle
Frontal lobe (left)
Skull
Occipital lobe (left)
Posterior cavity of left eye
Pupil
Extra-ocular muscles
Maxillary sinus
Cerebellum
Splenius capitis muscle
Trapezius muscle

SCAN 1e MRI SCAN OF THE BRAIN, MIDSAGITTAL SECTION

Third ventricle
Corpus callosum
Thalamus
Cingulate gyrus
Mesencephalon
Pineal gland
Superior
Inferior
Colliculi
Parietal bone
Frontal bone
Mamillary body
Frontal sinus
Perpendicular plate of ethmoid
Optic nerve (N II)
Cartilage of nasal septum
Vomer
Mandible
Genioglossus muscle
Tongue
Epiglottis
Dens
C_2
C_3
C_4
Spinal cord
Transverse fibers of pons
Occipital bone
Cerebellum
Fourth ventricle
Mesencephalic aqueduct

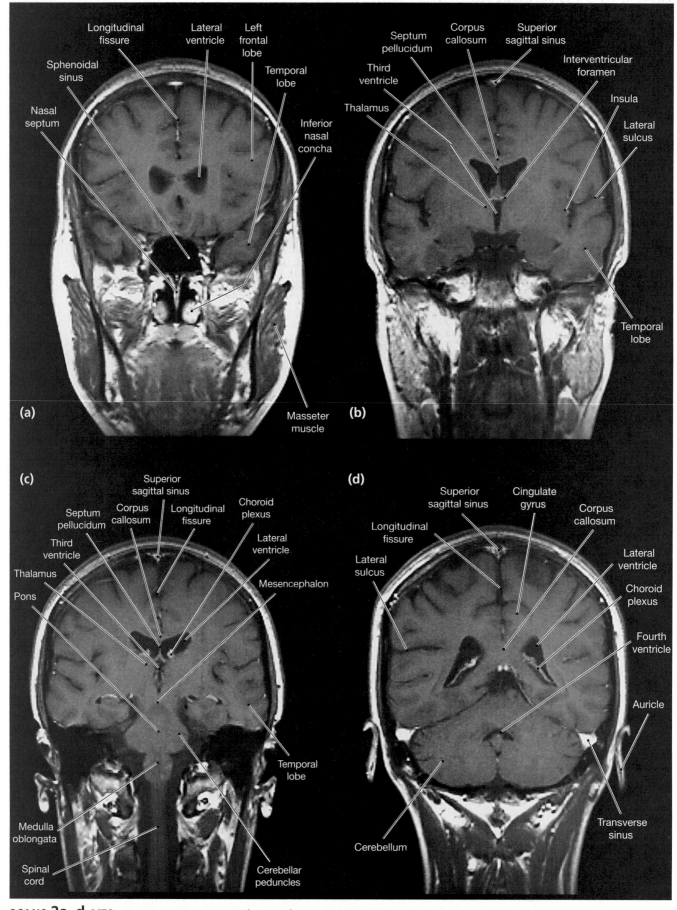

(a)

Sphenoidal sinus
Nasal septum
Longitudinal fissure
Lateral ventricle
Left frontal lobe
Temporal lobe
Inferior nasal concha
Masseter muscle

(b)

Septum pellucidum
Third ventricle
Thalamus
Corpus callosum
Superior sagittal sinus
Interventricular foramen
Insula
Lateral sulcus
Temporal lobe

(c)

Septum pellucidum
Third ventricle
Thalamus
Pons
Medulla oblongata
Spinal cord
Corpus callosum
Superior sagittal sinus
Longitudinal fissure
Choroid plexus
Lateral ventricle
Mesencephalon
Temporal lobe
Cerebellar peduncles

(d)

Longitudinal fissure
Lateral sulcus
Superior sagittal sinus
Cingulate gyrus
Corpus callosum
Lateral ventricle
Choroid plexus
Fourth ventricle
Auricle
Transverse sinus
Cerebellum

SCANS 2a–d MRI SCANS OF THE BRAIN, FRONTAL (CORONAL) SECTIONS, ANTERIOR TO POSTERIOR SEQUENCE

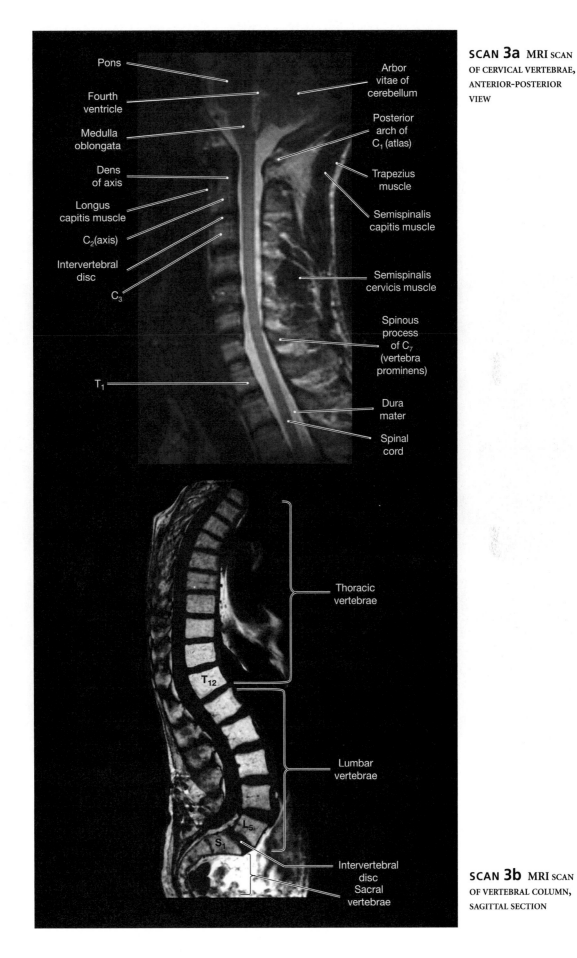

Pons

Fourth ventricle

Medulla oblongata

Dens of axis

Longus capitis muscle

C₂(axis)

Intervertebral disc

C₃

T₁

Arbor vitae of cerebellum

Posterior arch of C₁ (atlas)

Trapezius muscle

Semispinalis capitis muscle

Semispinalis cervicis muscle

Spinous process of C₇ (vertebra prominens)

Dura mater

Spinal cord

SCAN 3a MRI SCAN OF CERVICAL VERTEBRAE, ANTERIOR–POSTERIOR VIEW

T₁₂

L₅

S₁

Thoracic vertebrae

Lumbar vertebrae

Intervertebral disc
Sacral vertebrae

SCAN 3b MRI SCAN OF VERTEBRAL COLUMN, SAGITTAL SECTION

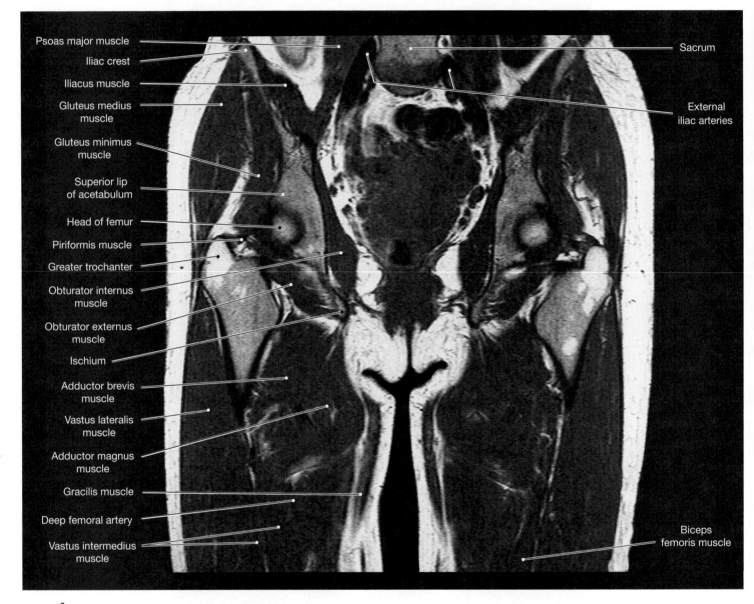

Psoas major muscle

Iliac crest

Iliacus muscle

Gluteus medius
muscle

Gluteus minimus
muscle

Superior lip
of acetabulum

Head of femur

Piriformis muscle

Greater trochanter

Obturator internus
muscle

Obturator externus
muscle

Ischium

Adductor brevis
muscle

Vastus lateralis
muscle

Adductor magnus
muscle

Gracilis muscle

Deep femoral artery

Vastus intermedius
muscle

Sacrum

External
iliac arteries

Biceps
femoris muscle

SCAN 4 MRI SCAN OF THE PELVIS AND HIP JOINT, FRONTAL SECTION

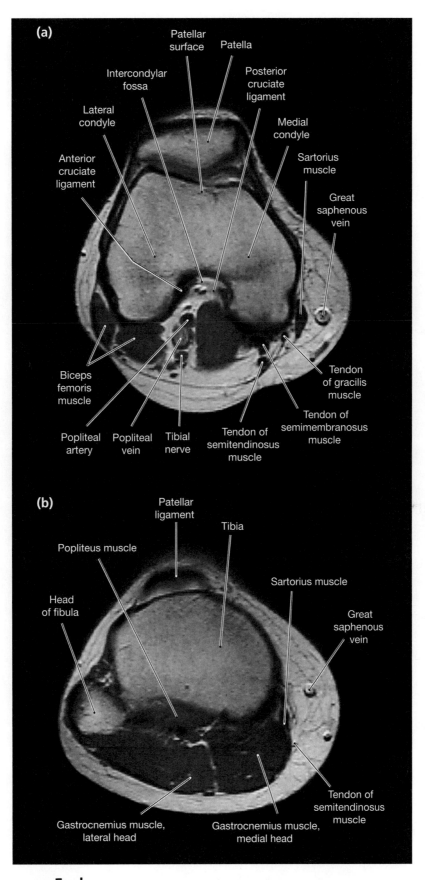

SCANS 5a–b MRI SCANS OF THE RIGHT KNEE JOINT, HORIZONTAL SECTIONS, SUPERIOR TO INFERIOR SEQUENCE

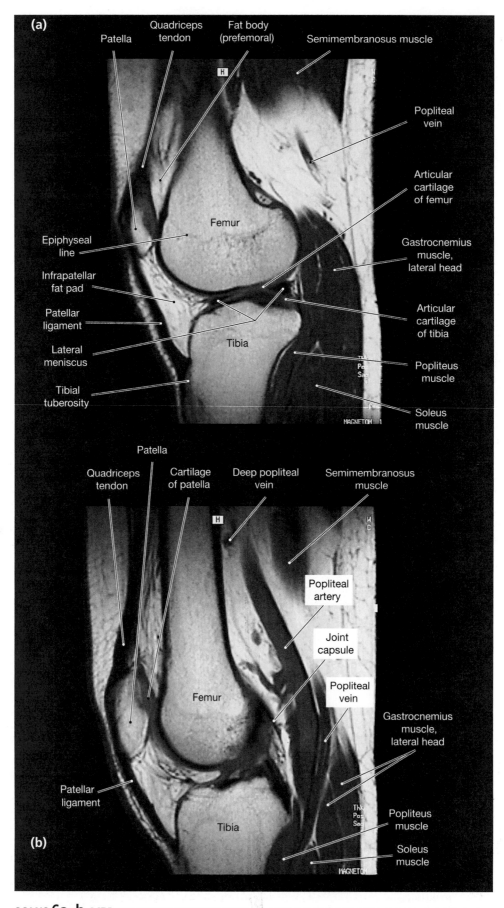

(a)

Patella

Quadriceps tendon

Fat body (prefemoral)

Semimembranosus muscle

Popliteal vein

Articular cartilage of femur

Femur

Epiphyseal line

Gastrocnemius muscle, lateral head

Infrapatellar fat pad

Articular cartilage of tibia

Patellar ligament

Tibia

Lateral meniscus

Popliteus muscle

Tibial tuberosity

Soleus muscle

(b)

Patella

Quadriceps tendon

Cartilage of patella

Deep popliteal vein

Semimembranosus muscle

Popliteal artery

Joint capsule

Popliteal vein

Femur

Gastrocnemius muscle, lateral head

Patellar ligament

Popliteus muscle

Tibia

Soleus muscle

SCANS 6a–b MRI SCANS OF THE RIGHT KNEE JOINT, PARASAGITTAL SECTIONS, LATERAL TO MEDIAL SEQUENCE

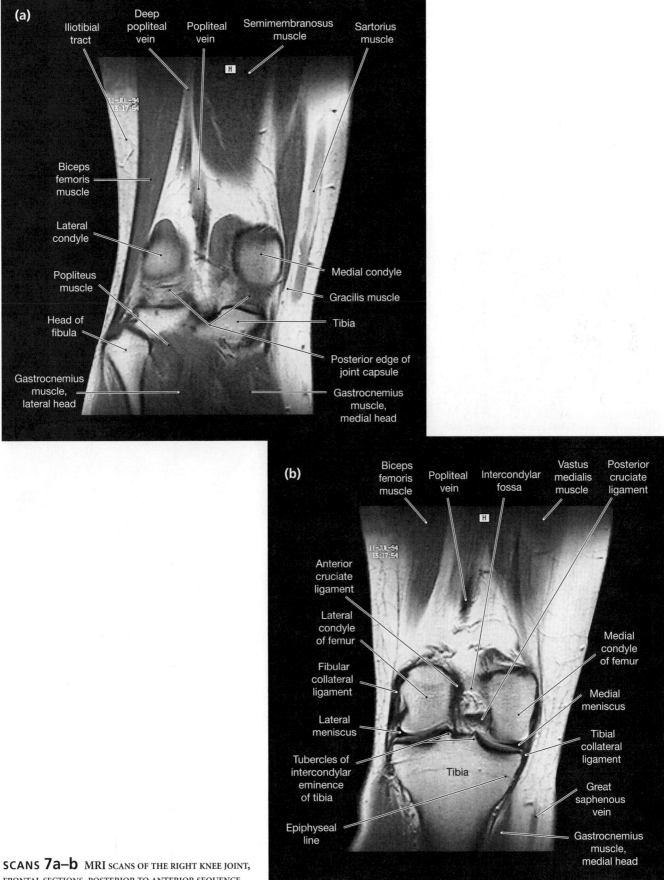

(a)

Iliotibial tract

Deep popliteal vein

Popliteal vein

Semimembranosus muscle

Sartorius muscle

Biceps femoris muscle

Lateral condyle

Popliteus muscle

Head of fibula

Gastrocnemius muscle, lateral head

Medial condyle

Gracilis muscle

Tibia

Posterior edge of joint capsule

Gastrocnemius muscle, medial head

(b)

Biceps femoris muscle

Popliteal vein

Intercondylar fossa

Vastus medialis muscle

Posterior cruciate ligament

Anterior cruciate ligament

Lateral condyle of femur

Fibular collateral ligament

Lateral meniscus

Tubercles of intercondylar eminence of tibia

Epiphyseal line

Tibia

Medial condyle of femur

Medial meniscus

Tibial collateral ligament

Great saphenous vein

Gastrocnemius muscle, medial head

SCANS 7a–b MRI SCANS OF THE RIGHT KNEE JOINT, FRONTAL SECTIONS, POSTERIOR TO ANTERIOR SEQUENCE

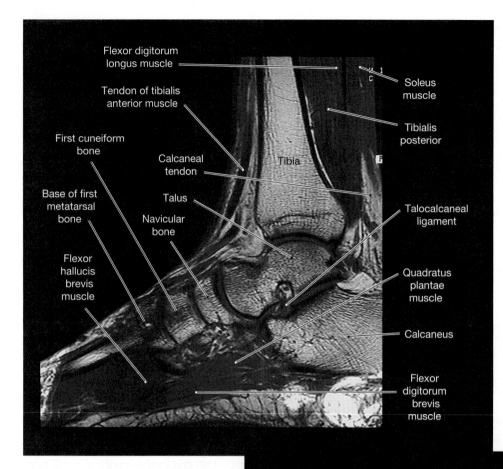

Flexor digitorum longus muscle

Tendon of tibialis anterior muscle

First cuneiform bone

Calcaneal tendon

Base of first metatarsal bone

Talus

Navicular bone

Flexor hallucis brevis muscle

Soleus muscle

Tibialis posterior

Tibia

Talocalcaneal ligament

Quadratus plantae muscle

Calcaneus

Flexor digitorum brevis muscle

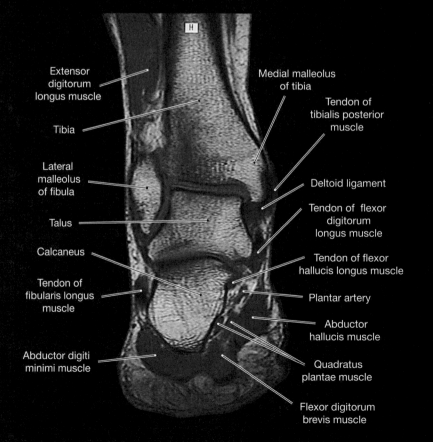

Extensor digitorum longus muscle

Tibia

Lateral malleolus of fibula

Talus

Calcaneus

Tendon of fibularis longus muscle

Abductor digiti minimi muscle

Medial malleolus of tibia

Tendon of tibialis posterior muscle

Deltoid ligament

Tendon of flexor digitorum longus muscle

Tendon of flexor hallucis longus muscle

Plantar artery

Abductor hallucis muscle

Quadratus plantae muscle

Flexor digitorum brevis muscle

SCAN 8b   MRI SCAN OF THE RIGHT ANKLE
JOINT, FRONTAL SECTION

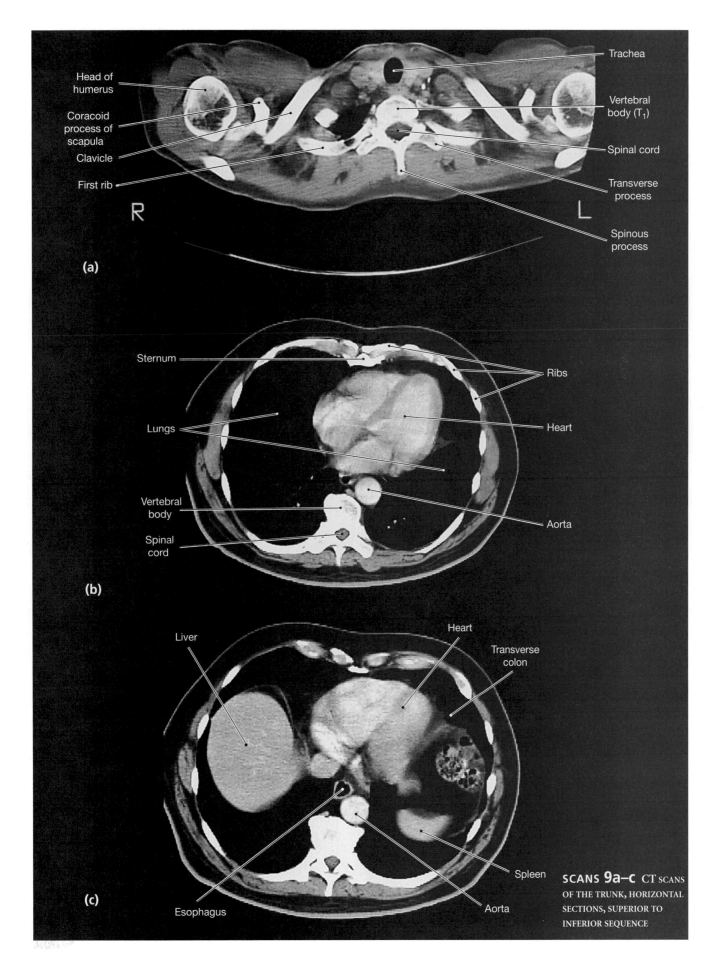

(a)

R L

Head of humerus

Coracoid process of scapula

Clavicle

First rib

Trachea

Vertebral body (T₁)

Spinal cord

Transverse process

Spinous process

(b)

Sternum

Lungs

Vertebral body

Spinal cord

Ribs

Heart

Aorta

(c)

Liver

Heart

Transverse colon

Spleen

Aorta

Esophagus

SCANS 9a–c CT SCANS OF THE TRUNK, HORIZONTAL SECTIONS, SUPERIOR TO INFERIOR SEQUENCE

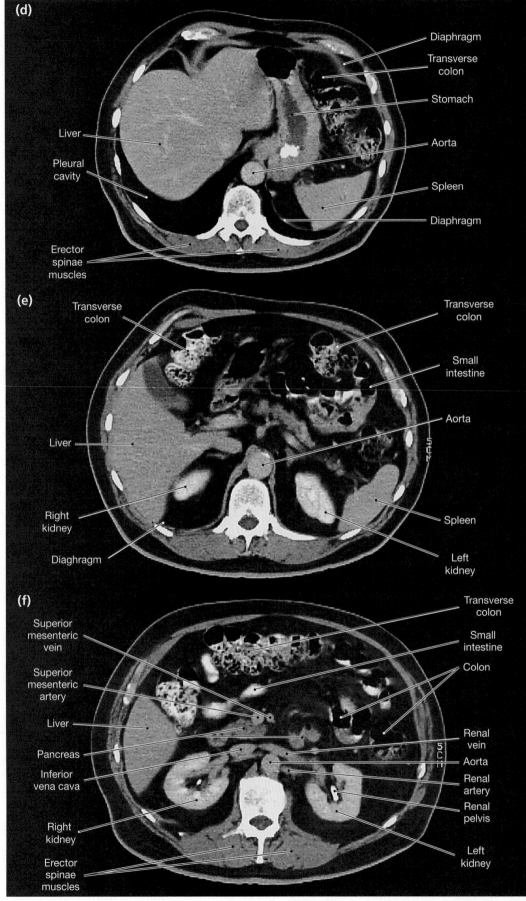

(d)

Diaphragm

Transverse colon

Stomach

Liver

Aorta

Pleural cavity

Spleen

Diaphragm

Erector spinae muscles

(e)

Transverse colon

Transverse colon

Small intestine

Liver

Aorta

Right kidney

Spleen

Diaghragm

Left kidney

(f)

Transverse colon

Superior mesenteric vein

Small intestine

Colon

Superior mesenteric artery

Liver

Renal vein

Pancreas

Aorta

Inferior vena cava

Renal artery

Renal pelvis

Right kidney

Left kidney

Erector spinae muscles

SCANS 9d–f
CT SCANS OF THE TRUNK, HORIZONTAL SECTIONS, SUPERIOR TO INFERIOR SEQUENCE (*CONTINUED*)

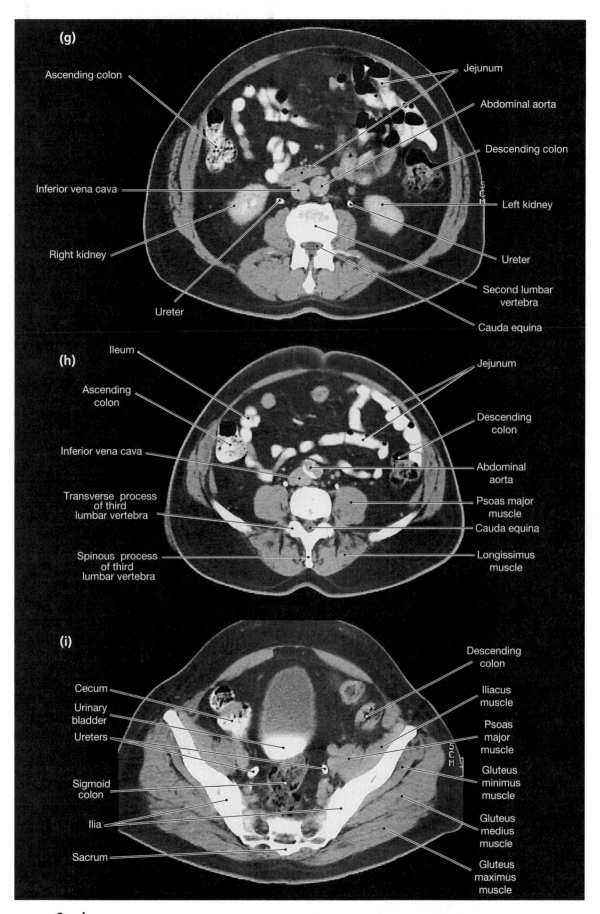

(g)

Ascending colon

Jejunum

Abdominal aorta

Descending colon

Inferior vena cava

Left kidney

Right kidney

Ureter

Ureter

Second lumbar vertebra

Cauda equina

(h)

Ileum

Jejunum

Ascending colon

Descending colon

Inferior vena cava

Abdominal aorta

Transverse process of third lumbar vertebra

Psoas major muscle

Cauda equina

Spinous process of third lumbar vertebra

Longissimus muscle

(i)

Descending colon

Cecum

Iliacus muscle

Urinary bladder

Psoas major muscle

Ureters

Gluteus minimus muscle

Sigmoid colon

Gluteus medius muscle

Ilia

Sacrum

Gluteus maximus muscle

SCANS 9g–i CT SCANS OF THE TRUNK, HORIZONTAL SECTIONS, SUPERIOR TO INFERIOR SEQUENCE

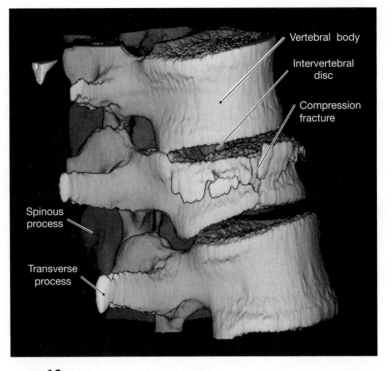

SCAN 10a 3-DIMENSIONAL CT SCAN SHOWING A FRACTURE OF THE BODY OF A LUMBAR VERTEBRA

Vertebral body

Intervertebral disc

Compression fracture

Spinous process

Transverse process

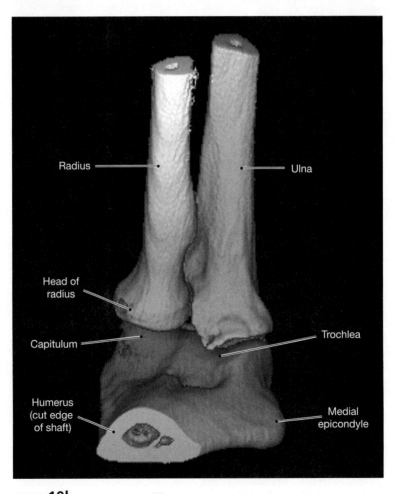

SCAN 10b 3-DIMENSIONAL CT SCAN OF THE ELBOW JOINT, SUPERIOR VIEW

Radius

Ulna

Head of radius

Capitulum

Trochlea

Humerus (cut edge of shaft)

Medial epicondyle

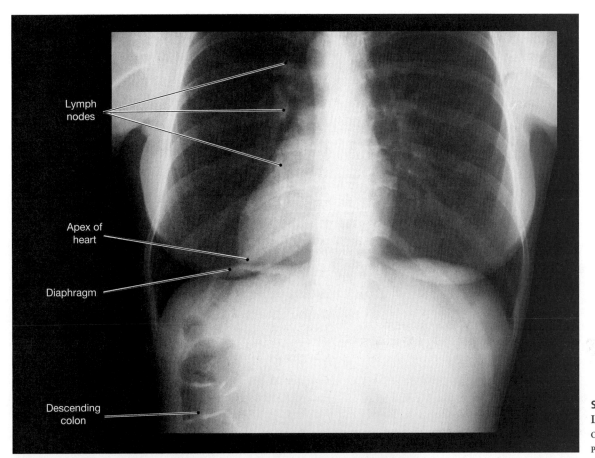

Lymph
nodes

Apex of
heart

Diaphragm

Descending
colon

SCAN 10c
LYMPHANGIOGRAM
OF THORAX,
POSTERIOR VIEW

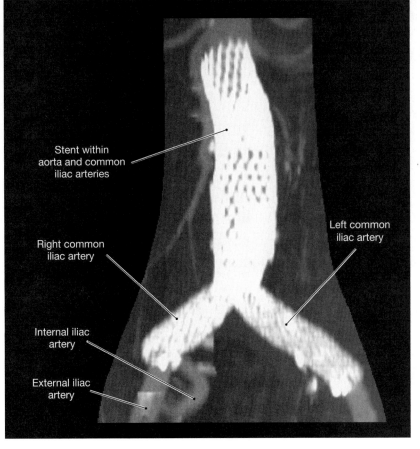

Stent within
aorta and common
iliac arteries

Right common
iliac artery

Left common
iliac artery

Internal iliac
artery

External iliac
artery

SCAN 10d 3-DIMENSIONAL CT SCAN SHOWING THE
USE OF A STENT WITHIN THE ABDOMINAL AORTA AND
COMMON ILIAC ARTERIES

PHOTO CREDITS

INDEX